CORE AND ATM NETWORKS

Proceedings of the European Conference on Networks and Optical Communications 1997 (NOC '97)

Published in three parts:

Broadband Access Networks

Core and ATM Networks

Photonic Networks, Optical Technology and Infrastructure

Editors: D.W. Faulkner and A.L. Harmer

Core and ATM Networks

NOC '97

Edited by

D.W. Faulkner

BT Laboratories, Ipswich, United Kingdom

and

A.L. Harmer

I.E.C., Geneva, Switzerland

IOS Press

Ohmsha

Amsterdam, Berlin, Oxford, Tokyo, Washington, DC

ISBN 90 5199 340 4 (IOS Press)
ISBN 90 5199 342 0 (IOS Press, three parts)
ISBN 4 274 90165 3 C3055 (Ohmsha)
ISBN 3 905084 21 X (AKM AG, Basel)
Library of Congress Catalog Card Number 97-72894

Publisher
IOS Press
Van Diemenstraat 94
1013 CN Amsterdam
Netherlands

Distributor in the UK and Ireland
IOS Press/Lavis Marketing
73 Lime Walk
Headington
Oxford OX3 7AD
England

Distributor in Germany
IOS Press
Spandauer Strasse 2
D-10178 Berlin
Germany

Distributor in the USA and Canada
IOS Press, Inc.
P.O. Box 10558
Burke, VA 22009-0558
USA

Distributor in Japan
Ohmsha, Ltd.
3-1 Kanda Nishiki-cho
Chiyoda-ku
Tokyo 101
Japan

PRINTED IN THE NETHERLANDS

Contents

Plenary Session

Restructuring Towards the Information Infrastructure; a Belgian Perspective
D. Frimout 1

From User Needs to Telecommunication Services
F.C. Vagliani 9

Optical Technologies for the Information Superhighway
P. Demeester 15

High Capacity Systems

Long Distance 10Gbit/s Soliton Transmission over Standard Fibre with Periodic Dispersion Compensation
P. Harper, F.M. Knox, D.S. Govan, P.N. Kean, I. Bennion and N.J. Doran 18

Performance Improvement of 10-Gbit/s Soliton Transmission over Dispersion Managed Standard Fiber by Using the Nonlinearity of Dispersion Compensating Fiber
X.Y. Tang 25

High Bit Rate Optical Transmission Systems Using a Midpoint OPC and DSCF's
M. Hanawa, B. Yao, M. Takahara and J. Yamawaku 29

Soliton WDM Transmission in Long-Haul Dispersion Managed Fiber Systems
X.Y. Tang, D. Breuer and K. Petermann 33

Investigation of RZ- and NRZ-Format 40Gb/s Transmission over Non Dispersion Shifted Fibre using Chirped Fibre Grating Dispersion Management
K. Ennser, R.I. Laming and M.N. Zervas 40

Field Trial of 210 Km in the German National Network Using a Lumped Amplifier Transmission System at 1310 nm
A. Mattheus, J.J.E. Reid, P.I. Kuindersma, B. Teichmann, C. Schulien, L.C. Garcia and R. Leppla 46

Polarisation Mode Dispersion and WDM

Polarisation Effects in Transmission Fibres
A.S. Siddiqui 53

Characterization of 11000 Km of Installed Standard Monomode Fibre and Statistical Analysis in View of Network Design
H. Gruhl, G. Herchenröder, A. Mattheus and J. Vobian 59

Equalization of Bit Distortion Induced by Polarization Mode Dispersion
H. Bülow 65

Crosstalk Evaluations of Optical Add/Drop Multiplexer in WDM-ring Networks
C. Caspar, A. Gladisch, H.-M. Foisel and B. Strebel 73

Design of Wavelength Branching Units in Long Haul WDM Networks
A. Lord 81

Planning of SDH Transport Networks

Numerical Analysis of Transparency Limitations in Optical Core Networks
R. Freund and B. Strebel 88

Design Techniques for SDH Mesh-Restorable Networks
F. Poppe, M. Pickavet, P. Arijs and P. Demeester 94

A Comparison of 2 and 4 Fibre MS SPRings
M. Robinson and I. Wright 102

A Set of Metrics for the Evaluation of SDH Transport Networks
A. Antonopoulos, J.J. O' Reilly and P. Lane 110

Distributed-Controlled Multiple-Ring Architecture with Classified Path Restoration for Nation-Wide Trunk-Transmission Networks
M. Tomizawa, S. Matsuoka, Y. Uematsu and K. Hagimoto 118

Optical Transmission in Utility Companies

High Capacity SDH Networks for New Operators – Using the National Electricity Distribution Network
A.M. Henderson 126

ATM Networks, Poster Session

Full-service Access Network Architectures: an Analysis of Different Options
M. D'Agrosa, D. Lanzoni and L. Zanetta 134

Cell Broadcasting in ATM Photonic Switches
M. Casoni and C. Raffaelli 139

An Accelerated Simulation Technique for Evaluating Communications Systems Utilising FEC
S. Fragiacomo, Y. Bian, A. Popplewell, C. Matrakidis and J. O' Reilly 145

Multimedia System Development Using a CORBA Based Platform
P. Ray 149

A Hybrid Rate/Reservation MAC Protocol for HFC Networks
M.S. Nunes and M.M. Macedo 153

Implementing Dynamic Unicast VCC Routing in ATM Networks
P. Aranzulla and J. Mellor 157

Core Networks, Poster Session

PMD Measurements Below 0.05 ps by the Fixed Analyser Technique
F. Ravet, Y. Choquet, B. Heens, X. Wang, L. Meuleman and M. Blondel 161

MO-16, a Modular Optical 16-Channel Transmission System – a First Step Towards a Photonic Network
U. Kohn, E. Kremers, G. Müller, E.-U. Scheuing and F.-C. Tischer 165

Simulation and Modeling of an Interconnected Network using Time and Wavelength Division Multiplexing Scheme
B. Boulet, F. Tillerot, J. Salaün and M. Le Ligné 169

Gaussian Pulse Transmission Considered with Higher Order Optical Effects
K. Kinjyou, M. Hanawa and M. Takahara 174

Nonlinear Jitter of Optical WDM-Systems Due to Cross-Phase-Modulation
N. Hanik 178

A Wavelength Division Multiaccess Protocol for Optical Fibre Ring Networks
A.E. Kamal and G.K. Janssens 182

IP Networking and ATM

Internet, a Future Information Superhighway
D. de Maeyer 186

Performance Comparison of Classical IP and LAN Emulation over the Mons-Valenciennes ATM Network
L. Meuleman, J. Hanoteau, N. Point, L. Wuilmart, M. Lamquin, G. Bisiaux and A. Mayeur 195

An SDL Model for Simulation and Evaluation of LAN Emulation over ATM
J.R. Vidal and J. Mataix 203

ATM Planning and Routing

Strategic Network Planning for ATM Deployment – A Bell Canada Perspective
S.G. Bartlett and A. Sultan 211

Efficiency and Usability of Equivalent Bandwidth Algorithms for ATM Network Dimensioning
M. Naldi and P. Pelusi 218

Designing ATM Networks Using LINDA
J. Frings 226

Hybrid STM/ATM Network Architectures – A First Assessment
A. Sultan 234

Comparison of IP and PNNI Routing for the Use in Intelligent Routing with Respect to Survavibility, Congestion and Mobility
I. Lievens and P. Demeester 242

Routing in Integrated SDH-ATM Networks
G. Knight and R. Shi 250

Multimedia Trials and Applications

The AMUSE Residential Multimedia Trials: Network Architecture and First Results
J. Charzinski, J. Färber, J. Enssle and M. Frank 258

Guidelines for the Evolution of Core Network Capabilities for the Support of Advanced Multimedia Services
C.D. Anagnostakis, Th.B. Zahariadis and I.S. Venieris 266

Broadband Video Conference in an Integrated IN/B-ISDN Architecture
H. Brandt, C. Tittel and P. Todorova 274

Experiments of Integrated Traffic Control for ATM in the ACTS Project EXPERT
M. Lorang, T. Renger, H. Pettersen, E. Aarstad, L. Jaussi and M. Potts 279

Survivability and Reliability

Integrated Dynamic Routing, Load Balancing and Survivability in ATM-based IBCN
Y. T'Joens, P. Georgatsos, S. Sartzetakis and D. Ranc 287

Experimental Results on Distributed Restoration in an ATM Laboratory Testbed (RACE•IMMUNE)
L. Nederlof, H. Vanderstraeten, D. Chantrain and L. Van Hauwermeiren 295

PANEL – Protection Across NEtwork Layers
P. Demeester, M. Gryseels, K. Struyve, K. Vandoorselaere, A. Autenrieth, C. Brianza, G. Signorelli, R. Clemente, M. Ravera, A. Jajszczyk, M. Roszkiewicz, G. Kalbe, Y. Harada, T. Yada and A.G. Rhissa 303

Optimal Space Capacity Design for Various ATM Protection Switching Methods
T. Bauschert and T. Frisanco 312

An Algorithmic Approach for Survivability in ATM Networks: Leap VP
I. Lievens and P. Demeester 320

An Approach to QoS Monitoring in ATM Networks
H.R. Wiltfang 328

Author Index 337

Core and ATM Networks
D.W. Faulkner and A.L. Harmer (Eds.)
IOS Press 1997 © AKM

Restructuring towards the information infrastructure; a Belgian perspective

by Dr Dirk FRIMOUT, Customer Program Director - Technology Scan
BELGACOM, Technology management
177 Boulevard Jacqmain, B-1030 Brussels, Belgium

1. Introduction

In the past public national operators (PNOs) have enjoyed telecommunication services monopoly. But the world is changing.

Facing competition will be the main challenge of the year 1998. To be successful a PNO has to focus on the customer and on services provision. Those issues are discussed here below.

2. Delighting the customer ...

For establishing an efficient and interesting communication with somebody it is required to understand his language, his interests, his needs, ... So is the human being.

The telecom operator (TO) has to adopt the same behaviour towards its clients than two persons communicating.

Key issues are:

- ✓ client segmentation,
- ✓ distribution channels,
- ✓ customer care.

2.1 Who are you?

Up to now the PNO had not really a strong knowledge of its market. The main parameter for the customer segmentation was the telephony bill. That latter is quite efficient as far as the voice and private lines business are concerned.

New developments have brought new clients requirement (e.g. ATM grade of service, mobility, ...) leading to differentiated customer approaches.

Knowing the clients' profile is much than a marketing act. It decides of the global customer care to offer him/her the services basket matching his/her needs.

Clients segmentation is the first step to follow to achieve a fruitful service offer. A real product strategy has to be defined to allow clients' needs satisfaction. This segmentation does no longer rely on the telephony bill. It depends to the education level, the yearly expenses for entertainment, the professional activity, ...

A real product strategy is required varying from services to services. By example Value Added Services (VAS) could use a given segmentation while Internet or entertainment would require another client segmentation.

	Package 1	***Package 2***	**....**	***Package M***
Segment 1				
Segment 2				
....				
Segment N				

Environmental conditions (like competition) will impact that strategy.

The client segment being identified it is required to determine the right channel to communicate with him/her.

2.2 How do you want to communicate?

The relation with the customer has to become more than a bi-monthly bill.

Customised points of sale (e.g. according to the business area, the consumption habits, ...) have to be set-up to establish a direct contact with the telecom operator and the clients.

Alliances with established commercial companies are now common to benefit of scale economy: supermarket, software re-sellers, ... This allows the TO to benefit of the experience and skills developed by those companies.

Up to now the services package definition was easy.

It was telephony, leased lines or other standard products. Because every client are different they required specific services bundles. The telecom operator (TO) targets now the “one stop shopping”.

Customised distribution channels have then to establish a direct contact between the services provider and the services consumer.

2.3 What may I propose you?

Once the first contact has been established the dialogue can start.

Because the telecommunication world is rapidly evolving the customer needs satisfaction requires the TO to maintain a close relation with him/her.

The TO commit himself to provide the client with a cost effective solution to his/her problems.

The telecom operator’s client is no longer the end-user only. The TO gets new customer: the Internet Services Provide (ISP), the On-line Services Provider (OSP), the content provider, ...

This required many new skills: content development, hosting, packaging, ...

Competition gives then a nice opportunity to the PNO to diversify its activities. Positioned more and more in the services packaging the operator can generate new revenues by working higher levels of the value chain.

The TO evolves then from a network provider to a services provider, occupying new positions in the value chain. This allows fruitful business transition to high value added areas.

The here below graph illustrates a possible business migration for the TO.

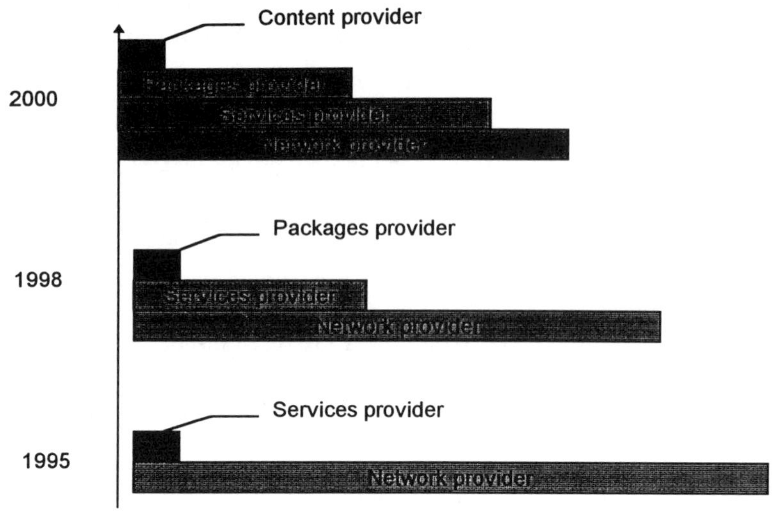

3. Better, faster and cheaper

To satisfy customer's needs efficient services platforms have to be defined. Those platforms rely on access and backbone networks performances.

3.1 Access network

The access network is the first section connecting the client to the TO's services platform.
Different technical solutions are possible depending to the segments/services target.

Technology	Suppliers	Typical users	Typical access speed
Optical fibre	PNO	Business & Residential	2Mbit/s and above
Wireless	PNO Mobile operator	Residential Mobile workforce	128 kbit/s
LMDS	PNO	Business & Residential	400 kbit/s and above
DTH	PNO	Business & Residential	30 Mbit/s
CATV modems	CATV operators	Residential	shared 2 to 27 Mbit/s
ADSL	PNO	Business	1 to 8 Mbit/s
ISDN-BA	PNO (CATV operators)	Business & Residential	128 kbit/s
Dial-up PSTN + modem	PNO (CATV operator)	Residential	9.6 to 33.6 kbit/s

All those technologies are or will be applied according to the services grade request and the price the client is willing to pay.

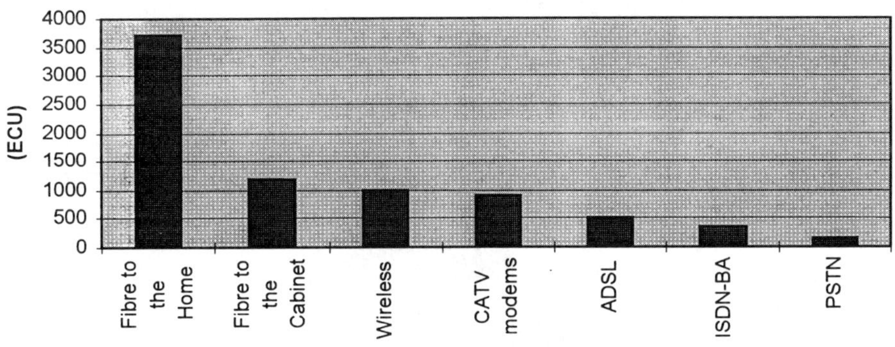

Annualised cost for Internet provision (OAM & capital costs) - source EITO

One of the most important challenge of the PNO is to realise the fibre migration as smoothly as possible taking into account the available revenues from the customer.

Services forecast (and the related revenues) is the main unknown deciding how far the fibre has to be deployed.

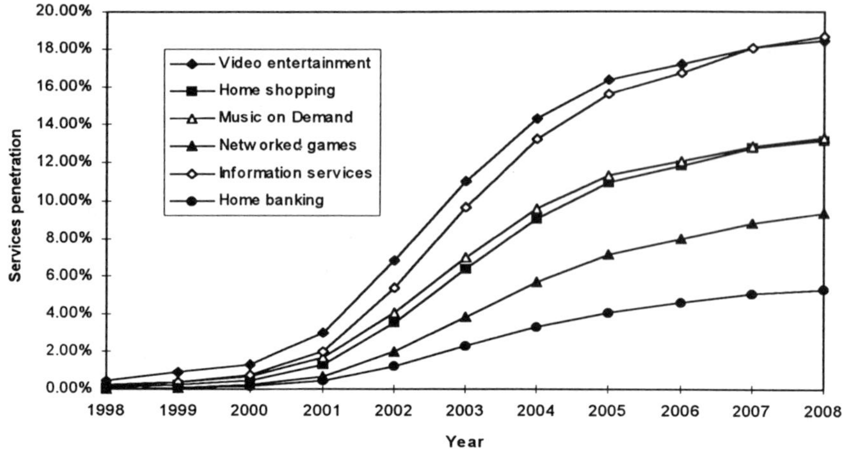

Services penetration forecasts for the residential market (source: Analysys)

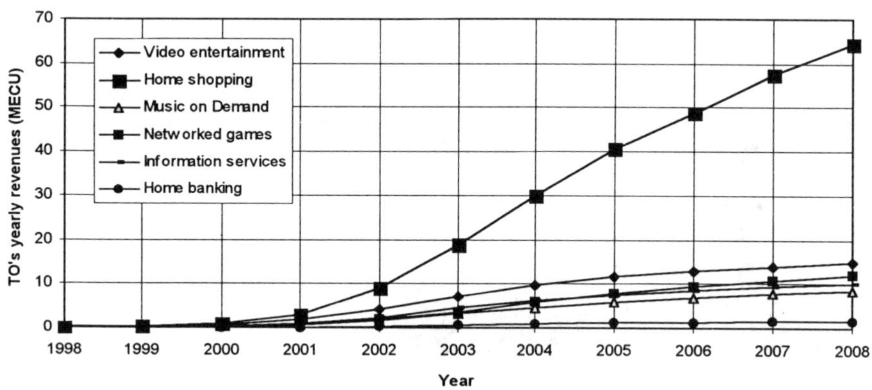

Revenues generated for the TO (source: Analysys)

Alternative technologies like ADSL, broadband wireless, DTH can also be used for transition or definitive access platforms.

All these technical solutions allow to build the high speed access to the information highway.

3.2 Backbone network

The optical fibre backbone is currently designed to support STM-16 and even 4xSTM-16 transmission.
The main problem with backbone resources is not their availability but their price.

By example, more and more TO bill the backbone resources used for Internet on a volume basis. This could lead to a slower development of Internet-based services.

Fortunately new technologies like ATM with its different possible grades of service as well as the Integrated Services Internet would lead to a general private lines tariffs decrease.

ATM and IP (and Frame relay for n x 64 kbit/s) integrated backbones are the key drivers for developing a flexible and cost effective core platform enabling existing and new (multimedia) services.

This also helps to merge the telecom and datacom communities requirements.

At the backbone level alliances are required between national TO in order to offer low cost gateways for competing in a de-regulated environment. More and more TOs are now joining their efforts: that is the "Super Carriers" concept.
Currently the Super Carriers are building the IP high speed backbone enabling Europe to develop new services using Internet.

Those alliances allow also the Super Carriers to export competition all over Europe.

3.3 Intelligent network

The intelligent network will allow the telecom operator to implement his marketing strategy as well as the customer follow-up.

An intelligent layer allows for re-configuring the network or activating some special features (e.g. for billing purposes) related to services provision.

Switches digitisation and integrated services management systems are the absolutely necessary tools for clients' satisfaction.

The required features could be, for instance, the "by second" billing, local fee tariff, flat fees, marketing acts for target segments, telephony based value added services, ...

4. People, team and skills

The clients and their needs being identified, the services platform being in place, the TO employees' skills may then operate.

The telecommunication world is a fast changing environment. So must be a telecom operator.

Evolving from a 100% government owned operator to a competitive company the TO has to define new processes using the manpower resources available.

This means people reconversion programmes have to be defined and launched for increasing customer satisfaction at every level of the company.

This is maybe the most difficult target ...

5. Conclusions

Telecom operator are now completing their conversion programmes at different levels.

Customer approach strategy and client care processes have been identified and already launched for traditional services.

The fibre migration programmes are defined and the mass deployment of the most efficient technologies will take place. Fibre based access as well as xDSL technologies will offer turnkey solutions for the new services platform.
That access network will be supported by an integrated ATM and IP backbone.

New skills are required for such a fast changing business. People management and experiences valorisation will also be one of the success keys.

Core and ATM Networks
D.W. Faulkner and A.L. Harmer (Eds.)
IOS Press 1997 © AKM

From User Needs to Telecommunication Services

Federico Carlo VAGLIANI
Italtel, Via Alessio di Tocqueville 13, 20100 Milano, Italy

Abstract. This paper deals with the process of the specification of Telecommunication Services that meet the users expectations of the future. It starts from a layered scheme of the need-application-service-network chain. Then it gives an ICT Service model and focusses on the service/network boundary, addressing the concept of the service components/attributes and their relations with the user needs, and skipping the layers of applications and services. Finally, the implications of the service components/attributes on the underlying network are briefly reviewed.

1. Introduction

The current evolution of Information and Communication Technology and the related Business are putting service provision entrepreneurs and the ICT Industry in general, in front of very complex planning and implementation issues. There are different reasons for this complexity, all related to current and future proliferation in terms of applications, players and their roles, architectures and technologies. In particular, the telecommunication services industry, one of the areas where investment life cycles are longer, and wide interoperability and back compatibility are a must, is challenged with problems of planning and implementation of unprecedented difficulty, also because of the accelerating speed of the innovation process. The telecom hardware and software supply industry is sharing, of course, these same problems with their clients, and even under more "time to market" pressure.

A necessary phase of the planning process in Telecommunications is that of devising the new services needed in future by the Users and to conceive the underlying new systems, platforms and standards, in as future-proof as possible a way.

In the following some considerations are made about the mutual implications of users' needs and Telecommunication Services, and about the related planning and implementation problems.

2. Applications and services

End user needs involving ICT are satisfied through ICT Applications and Services. ICT Applications may be "telecommunication only" applications (e.g. teleconference) or "other" applications (e.g. information retrieval.) involving anyhow telecommunication. In this latter case, the telecommunication service is provided both to the end user and to the other service provider (in some cases called Value Added Service Provider). The Applications are the objects most closely related to the end users needs. The Service defines the relationship between the Service Provider and his Customer. If the service is a VA Service the Service Provider is totally involved with the application. If the service is a

Telecommunication Service the application layer is out of the service provision domain and is of interest to the end users only. In any case the relationship between the end user needs and the Telecommunication Service is mediated by the Application, that acts as a (partial) shield. This enables to some extent to plan and implement Services that are open to satisfy "ranges" of user needs, through Applications that match needs to services (a typical example is that of the Fax and of the telephone set, that enable or, better, "are", two different applications, of the POTS service). A layered approach is understood above, where Needs are above Applications, that are placed above Services. Continuing this approach, below the Service layer, a Service Component/Attribute layer may be identified, that provides elements that may be aggregated in different combinations in order to provide different services (Service *component/attribute* is a somewhat wider concept than service *attribute*, as used e.g. in [1]). This layer is even less strictly related to the specific user needs than the one above. Below the Service Component layer the concept of service is lost and the area of the specific network functions enabling the service components is entered.

In the following the three layers of applications, services and service components are considered with the purpose of better understanding the links between user needs and service components and how critical needs are with respect to the planning and implementation process.

3. An ICT service model

A few service modelling considerations are needed at this point, general enough to be open to a long term evolution of the services. Figure 1 gives a picture of an ICT Service as a relationship between a Service Customer and a Service Provider, involving three kinds of interaction: commercial, functional and management. The **commercial interaction** refers to the transactions between Customer and Provider about the Service subscription contract, billing, documentation, payments, and the like. The **functional interaction** is related directly with the fulfilment of the needs of the Customer (through the Application) or, more specifically, of the Service User(s) enabled by the Service Customer to use the Service. The functional interaction is the one that characterises the service as an ICT one. For ICT services that are not just of the Communication type, that interaction is performed, *by definition* through a Telecommunication Network and thanks to a Telecommunication Service. A third type of interaction that may be considered in between is the **service management interaction** that includes any action that, without changing the commercial terms, changes the terms of the functional interaction. Two examples of it may be the management of a Centrex service operated by the user, and the downloading of a new software release, into a Set Top Box operated by a Multimedia Information Service Provider. The commercial interaction may be performed, totally or partially or not at all, through the same interface used for the functional interaction. Two very popular examples of the degree of integration of the commercial interaction into the functional interface are the residential POTS (no integration) and the credit card enabled telephones (total integration). The management interaction may be performed through the functional interface, too, or through a specialised one.

The three above types of interaction are a first level grouping of "service components". The definition of an ICT Service, to be complete, must include all pertinent kinds of interaction, and for each of them the specification of all the relevant components and their relationships. The above general model can be used, of course, for the analysis of the Telecom services.

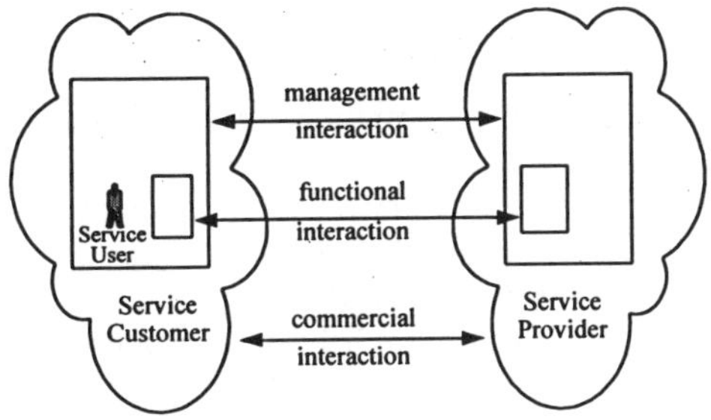

Figure 1 - ICT Service Provider/Service Customer-User interaction model

4. From user needs to telecom service components/attributes specification

Service components/attributes are reviewed below more in detail and checked against the trend of the user needs.

functional interaction

Key items in the functional interaction of telecom services are:

- the type of physical interface between the User and the Telecom Service Provider
- the message/signal structure, timing and delivery,
- the access to the service
- the addressing aspects
- the quality of service and its reporting

- type of physical interface

The physical interface may be of three kinds:

- Human machine (like in a telephone service where the set belongs to the Service provider)
- Wired
- Wireless

As far as the user needs are concerned, on the wired/wireless alternative the trend is clearly in favour of the wireless, because of the physical constraints reduction it affords. Some other aspects of users' needs are, instead, more or less directly, in favour of the wired solution, mainly: bandwidth and territorial density of active users (or, in other words, the territorial traffic density), electromagnetic pollution, and battery management. It goes without saying that, so far, the physical mobility need seems of outstanding importance. Human-machine interaction related needs at the physical level, instead, are less and less a matter of interest of the Telecom Service Providers, since liberalisation of the Terminals.

- signal/message structure, timing and delivery

There are, and there will be two basic classes of traffic generated by the users, mostly related to two classes of applications:

The class of "human real time communication applications" (audio and visual) with its needs of time continuity in input and output, tightly controlled delay of transmission and session-duration, puts a set of requirements on signals/messages structure and timing, leading essentially to continuous bit rate (variable or constant), limited transmission delay and jitter, generating "signal flow" type of traffic

The class of most of the other applications (like most machine to machine and human non real time), that have in general no specific needs in terms of structure and timing, apart from, in some cases, some not too stringent needs on delay of transmission, generating "message burst" type of traffic.

Whether these two classes should be unified at the application level in order to appear at the service level as a single class of traffic, or not, is an issue having great implications deep into the network. Also the delivery, e.g. type of connectivity, (topology, directionality, etc.) and timing of connectivity (real time, store and forward, etc.). is strictly related to the above aspects.

Irrespective of the above classes, the service "capacity", or the quantity of information per unit time associated to a service, be it bit-rate or throughput, depends strongly upon the specific application and will occupy a wider and wider range of values, from a few bits/s to tens of Mbit/s or even more. The user needs will ceaselessly produce both classes of message/signal structures and increase of the highest bit rates, and, in general, of the traffic. However, it is not to be overlooked that a large number of very important applications will not need necessarily very high bit rates.

- access to the service and addressing aspects

The evolution of the needs in this respect regards

- service mobility: terminal and personal (user/customer)
- geographical level, up to world-wide, of the service availability
- advanced addressing /numbering systems (e.g. multiservice, multioperator, universal addressing)
- extension of VPN-like concepts, to "lower" classes of users
- intelligent guidance to the choice of service options, where more are available

From the point of view of the user needs this area is critical and it may be the playground where different operators play the most of their competitive advantage, in particular, with highly demanding end users.

- quality of service and relevant reporting

The needs of the end users on this matter, are clearly diversified and will be even more so in the future. The lesson of Internet teaches both that: i) low confidence-low quality communication may be of interest if it results in very low cost, ii) some very popular applications like e-mail may need much more reliability and better reporting at least for some classes of users, or perhaps a range of options.

Commercial and management interactions

The need of the user to get any service at any moment, anywhere, can lead to the concept of pay per consumption, with any provider, case by case, as opposed to the concept of subscription to a service available from a specific provider. This is the introduction of electronic commerce (or Digital Money, or Transactional Services) into the Telecom Services domain, through the **implementation of commercial interactions through the user terminal.** When the services available and the service Providers are highly diversified, this becomes a very natural way of interaction with the service provider.

In a similar way, when services become highly sophisticated, in particular for small business and professional users, the need to perform **management operations directly from the terminal** becomes high.

5. From telecom services components specification to architectural issues

The Telecom Operators and Manufacturers are faced by the problem of defining standards, investing in research and products development, and investing in networks and services deployment, much in advance with respect to the actual use of standards, products, networks and services. The specification of the services components/attributes is the focal point where the envisaged user needs (and the relevant revenues) on one side, and the resources that will enable the needs fulfilment (and the relevant investments) on the other one, will meet. It is there that central issues of the planning and implementation process are concentrated.

It is not the purpose of this paper to develop the complex subject of the relationships between the service components/attributes and the network architectures, however, a brief glance is given on it. To this purpose the main types of Service components/attributes reviewed above, are summarised below:

1) physical interface (wired, wireless)
2) service capacity
3) "signal flow" transport
4) "message burst" transport
5) connectivity (topology, symmetry, directionality, timing, etc.)
6) terminal mobility
7) user/customer mobility
8) extension of service availability (geographical and Operators "spaces")
9) addressing and numbering (universal, VPN-like options, etc.)
10) service guidance
11) quality of service
12) QoS reporting
13) commercial interactions
14) management interactions

They may be classified, according to the type of architectural issues they imply, as follows:

1) and 2) have physical layer implications on the Access, that is the most ubiquitous part of the network, and the most costly.

3), 4), 5), 11), and 12) have implications in all the message/signal-handling elements of the network, just above the very basic layers . The very important problems of how to deal with "signal flow" traffic and "message burst" traffic, that is the connectionless/connection-oriented issue and the role of ATM, relate to this group of service components. The two basic architectures of traditional TLC networks and Internet are at stake with this issue, the relevant architectural implications are as fundamental as they can be for a network.

5), 6), 7), 9), 10), have implications on Control and Real Time Intelligence, whereas 13) and 14) have implications on Non Real Time Intelligence.

8) has implications on interworking between different Operators, therefore it is highly dependent on standards. The interworking between Operators is not strictly an issue of service components, however it is worth mentioning. In particular, the cost accounting paradigm based on "bandwidth, time, distance", traditional for TLC is now challenged by the new paradigms introduced by the Internet. This is not at all a Business management only issue, it is also strictly related to the architecture behind the service which, in turn, has basic business implications. The very profitability of some services for the Operators is at stake, so is the existence of the services themselves.

Physical layer and message/signal handling implications affect the hardware, that is, the network deployment investments, where planning errors are most costly and cumbersome to correct. Intelligence and Control implications affect the software, therefore they are less critical from the point of view of the deployment than the above, but perhaps more critical from the point of view of development investments.

Standardisation is important for practically all the above list for the usual three reasons: interoperability, economy of scale, competitive market.

6. Conclusion

The goal of this paper was to find a rationale on how to define services fulfilling the users' needs of the future, allowing a sound planning and implementation process of the necessary resources. The layered approach proposed aims at postponing the specification of the services to the last phases of the process of planning and implementation, by specifying, rather, service components/attributes and by defining functions of use as general as possible (i.e. little user need dependence) for the "heavy-investment, long life cycle" types of resources, and leaving the most user-needs dependent ones to the most "agile" and flexible types of resources, as far as possible. This method is intended to minimise the dependence on projections, visions, scenarios, etc. However it is recognised that a minimum of guess and "visionary" work is inevitable.

Acknowledgements

This paper was written within the CONVAIR project activities. It is so far a contribution of the Author to the project, and does not give necessarily views shared by the other project partners. CONVAIR is a project on Strategies in Telecommunications belonging to the ACTS Program, subsidised by the European Commission. The author is grateful to Mario Bonatti and Michele Morganti for the helpful discussions on this paper's subject.

References

[1] CCITT Blue Book, volume 3, Rec.I 121 - Broadband aspects of ISDN

Core and ATM Networks
D.W. Faulkner and A.L. Harmer (Eds.)
IOS Press 1997 © AKM

Optical Technologies for the Information Superhighway

P. Demeester

IMEC – University of Gent, Department of Information Technology
Sint-Pietersnieuwstraat 41, B-9000, Gent, Belgium
tel.: +32-9-264 3334, fax: +32-9-264 3593
e-mail: demeester@intec.rug.ac.be

Abstract

The Information Superhighway is a vague concept which covers a wider range of topics. These include software and hardware related issues, such as : object orientation, service creation, network management and design, internetworking, optical technology, Si microelectronics, etc. In this paper we will address the optical technologies which play a major role in the development of the Information Superhighway. Although issues related to optical recording and displays will be briefly discussed, major emphasis will be put on optical communication technologies.

1. Why do we need an Information Superhighway ?

In order to answer that question, one should know the meaning of Information Superhighway[1]. Is it the infrastructure which will provide every single user with total mobility and an unlimited bandwidth, or is it the infrastructure which gives every user an Internet connection of 2 Mbit/s, or is it … ? These are all valid visions but they require a different infrastructure.

The answer to the question why we need an Information Superhighway is not clear today although a number of indications can be given (which change however over time). A few years ago it seemed clear that video-on-demand was the killer application to introduce a broadband infrastructure, both in the backbone network and the access network. Now it seems that the Internet is becoming the application for developing the Information Superhighway. The access platform will be mainly the PC with about 300 million PC's

[1] In the Newton's Telecom Dictionary (1995 version), the Information Superhighway is defined as follows : " A very vague concept which Senator Al Gore created in the early 1990s and which gained great popularity when he became vice president and the Clinton/Gore administration started pushing the concept. The Information Superhighway is a term sufficiently vague that it can mean anything to anyone. It can mean a gigantic Internet reaching everybody. It could just as easily mean a 500-channel interactive cable TV system to every household. Somewhere in all this is the idea that easy access to large amounts of information will enrich our lives immeasurably. "

worldwide and a penetration of about 1 PC in 3 households in the USA. This should be compared to about 600 million telephone lines worldwide. If we consider that the duration of an Internet session is about 5 to 10 times longer then the average duration of a telephone call (typical 4 to 5 minutes) and if we consider that 2 Mbit/s is an acceptable access rate to the Internet (compared to 64 kbit/s for digital telephony), then it is clear that (for a similar penetration of the two services) a large increase of the bandwidth of the access and backbone network is required.

2. Optical technologies for communications

The optical fibre has a very large impact on the development of communication networks. Optical fibre has penetrated both in the backbone, access and local area network where it replaces classical coaxial cable or twisted pair networks. Due to its unlimited bandwidth (many Tbit/s), its immunity against electromagnetic interference, its very low losses (below 0.5 dB/km) it became the preferred transmission medium for long distance or high bitrate connections. Cost for transmitting information over optical fiber has dropped by about 4 orders of magnitude over the last two decades, and at the same time the total bitrate increased by about 3 orders of magnitude. Some major developments have been the move from multimode to monomode optical fiber, the introduction of optical amplification (Erbium Doped Fibre Amplifier or EDFA), the use of wavelength division multiplexing or WDM and the development of soliton transmission. At this moment record values in laboratory environments have been reached which surpass 2 Tbit/s transmission capacity on a single fibre using WDM. In the field, telecommunication operators install commercial WDM systems with capacities in the order of 20 Gbit/s.

In order to use the capacity of the fibre, opto-electronic technologies had to be developed in order to generate, detect and manipulate the optical signals that are transmitted over the optical fibre. On the source and detector side, most opto-electronic components are using InP/InGaAsP III-V semiconductors because they can emit and detect light in the wavelength range around 1.3 and 1.55 μm, where the dispersion and transmission in the fibre is minimal. For medium and long distances, laser diodes are the preferred sources, whereas for shorter distances (e.g. LAN) LED's are also possible. Other functionalities are however also required such as : power splitting, wavelength (de)multiplexing and filtering, wavelength conversion, space switching, on-off switching, etc. These can be realised in the InP/InGaAsP material system but other possibilities such as Silica on Silicon, polymers, $LiNbO_3$, glass offer also attractive features for some of these components. The use of these additional functionalities make it possible to treat the optical signals in different ways resulting in the possibilities to realise optical cross-connects (similar to the electronic cross-connects used in SDH, however replacing the flexibility in the time domain by flexibility in the wavelength domain) and switches (similar to electronic ATM switches).

Optical technologies are also becoming more and more important in the field of very short distance communications, mostly called optical interconnection. Here we talk about optical communication between computer or communication equipment racks, between electronic printed circuit boards in a rack and even between chips. Because of the short distances to be bridged, other wavelengths can also be used, typically in the range of 850 nm (where GaAs is the preferred semiconductor material). The use of optical fibres to guide the light is also not necessary any more, resulting in the use of alternatives such as free space transmission, polymeric waveguides, etc. Note that in the case of optical interconnection, surface emitting laser diodes and LED's are playing an important role.

3. Optical technologies for displays and optical recording

The development of the Information Superhighway requires also the development of advanced terminals. Two important features are the possibility to store large amounts of information locally and to present the data to the users using full colour displays.

To store the data locally, hard discs using magnetic recording techniques and CD-ROM's are the preferred solutions. They have storage capacities in the order of a few Gbyte. Here a similar trend can be observed as with optical transmission : prices have dropped drastically and capacities have gone up very fast. One of the important features which limits the storage capacity on optical discs is the wavelength used to read (or eventually write) information. Currently GaAs laser diodes are used emitting in the 780 to 820 nm wavelength range. A new standard is now moving to the 630 nm range where InAlGaP/GaAs laser diodes are used. In future one hopes to move to blue (or UV) emitting semiconductor laser diodes because this could even more increase the storage density. In recent years there has been a breakthrough in this domain by the development of II-VI laser diodes and more recently GaN III-V laser diodes. In future capacities above 50 Gbyte may be expected.

A second important terminal feature is the possibility to represent the data to the user using a full colour display. Large flat screens with high resolution and based on liquid crystals, electroluminesence and other technologies are becoming available and are already widely used for portable applications. The advent of the very high efficiency blue LED based on GaN has made it possible for the first time to make large full colour screens based on LED's.

4. Conclusions

It is clear that optical technologies are one of the corner stones to the successful deployment of the Information Superhighway. They are used in the terminals and for the transfer of information at different distance and bitrate levels.

Core and ATM Networks
D.W. Faulkner and A.L. Harmer (Eds.)
IOS Press 1997 © AKM

Long Distance 10Gbit/s Soliton Transmission over Standard Fibre with Periodic Dispersion Compensation

P. Harper, F.M. Knox, D.S. Govan, P.N. Kean, I. Bennion and N.J.Doran

Photonics Research Group, Dept. EEAP, Aston University, Birmingham, B4 7ET, England

Abstract

We present results of an improved standard fibre transmission system. Our previous experiments have shown that in dispersion compensated standard fibre systems Q value can oscillate with propagation distance. By pre-chirping the data stream using a combination of standard and dispersion compensating fibre we have suppressed these oscillations and increased the propagation distance.

1 Introduction

There is currently a great deal of research being done into utilising standard fibre for high speed telecommunications links[1, 2, 3]. Previously we have considered upgrades to existing standard fibre links but there is also the possibility of using standard fibre for trans-oceanic links. The main problem that must be overcome is that standard fibre is optimised for use in the 1.3μm window where it has low dispersion. This is troublesome because modern systems operate at 1.5μm as this is the wavelength of operation of the erbium doped amplifier, which can be used in place of traditional regenerators to provide potentially bit rate transparent amplification.

A second problem for upgrades to existing links is that the older standard fibre is likely to suffer from high polarisation mode dispersion (PMD). Whilst this does not pose a problem at the current sub Gbit/s data rates it may become significant when upgrading to 10s of Gbit/s. By using solitons, which are more resilient to PMD than NRZ pulse formats this problem can be minimised [4]. This problem would be less pronounced in trans-oceanic systems where new fibre could be laid, but may still be significant.

One method of reducing the high dispersion of standard fibre links at 1.5μm is to include a suitable length of dispersion compensating fibre in the transmission path. Using such a scheme we have shown that it is possible to propagate 20ps 10Gbit/s soliton data over 5200km [2]. A second method of dispersion compensation is to use chirped fibre Bragg gratings. There has been some success using this technique [5] but the problem with these devices is that current fabrication techniques lead to prohibitively high PMD in the device.

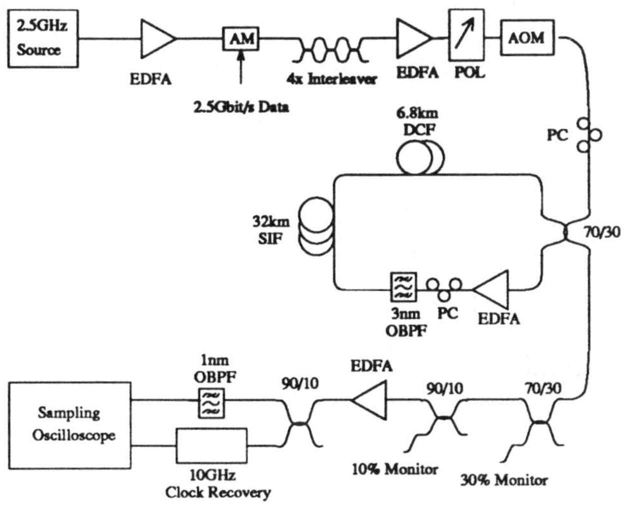

Figure 1: Schematic diagram of the experimental set-up. EDFA: Erbium Doped Fibre Amplifier, AM: 10GHz lithium niobate modulator, POL: Fibre Polariser, AOM: Aco:-sto Optic Modulator, PC Polarisation Controller, OBPF: Optical Band Pass Filter, SIF: Step-index Fibre, DCF: Dispersion Compensating Fibre. Numbers beside couplers denote coupling ratios.

Here we investigate the dynamics of soliton propagation in standard fibre using dispersion compensating fibre, and study the effects of matching the dispersion map to the source. For a given dispersion map the ideal source must have the correct sign and magnitude of chirp for its launch position in the dispersion map [6]. Here we compare results for three different launch positions using the same source.

2 Experimental Set-Up

Our initial experimental set-up is shown in Figure 1. The source was a 2.5GHz actively mode-locked, frequency stabilised erbium doped fibre ring laser operating at 1557nm. The time bandwidth product of this laser was 0.346 and the pulse width was 20ps. A 2^{31}-1 PRBS data pattern was imposed on the soliton pulse stream using a 2.5GHz pattern generator and a 10GHz lithium niobate amplitude modulator. This data stream was then passed through a passive, double Mach-Zender type, fibre interleaver to achieve the desired 10Gbit/s data rate.

The transmission system was a single span recirculating loop. The transmission fibre in this loop consisted of 31.6km of standard fibre (SIF) (with a dispersion of +16.75ps/nm/km at the operating wavelength) and 6.8km of dispersion compensating fibre (DCF) with a total dispersion of -522ps/nm. This combination of fibres gave an average dispersion of +0.2ps/nm/km over the 38.4km amplifier span. The signal was propagated in this loop until the required distance was reached before triggering the sampling oscilloscope and taking Q values. The trigger used was a 10GHz clock recovered signal. Using 10GHz ensured that patterning effects associated with interleaving were not masked out.

As expected for a dispersion managed system there was an enhancement of the soliton power [7]. In this case our launch power was 2.3mW, giving an enhancement factor was 5.4 over the average soliton power of the equivalent uniform dispersion system. There is also an associated decrease in Gordon-Haus jitter[8].

A schematic diagram of the transmission path is shown in Fig. 2 (a)

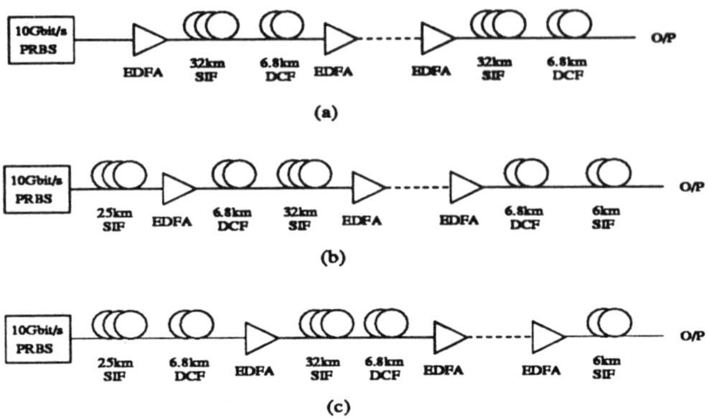

Figure 2: Transmission paths for the 3 experiments. In set-up (b) an initial length of 25km SIF is positioned before the first amplifier. In set-up (c) the initial length is 25km of SIF and 6.8km of DCF.

3 Experimental Results

Experimental results for set-up (a) are shown in Figure 3. These show that there was an rapid initial decrease in the Q value. It fell from the input value of 11 to ~8 within the first 1000km of propagation. The Q value then recovered and oscillated with a well defined period of ~1400km. The amplitude of these oscillations decreased with propagation distance with the Q finally falling below the minimum acceptable Q value of 6 (corresponding to bit error ratio of 10^{-9}) after 5275km.

The initial decrease in Q value is due to the near time bandwidth limited $sech^2(t)$ pulses injected into the loop adapting to the pulse shape and width most suitable for transmission in this dispersion map. The oscillations in Q value are due to further evolution of the pulses as they propagate in the system. Fig. 4 shows that these oscillations were also observed in the 10GHz component of the electrical spectrum of

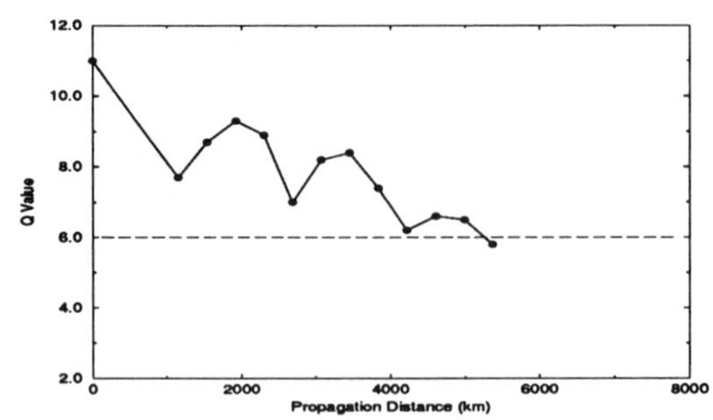

Figure 3: Experimental results for set-up (a) of Q value Vs propagation distance. The dotted line indicates the minimum acceptable Q value of 6.0. The maximum propagation distance was 5275km.

the signal. Although autocorrelations of the output could not be taken, pulse width variations could be monitored on the sampling oscilloscope. Again the variations in pulse width were similar to the Q oscillations, narrow pulses corresponding to high Qs and broad pulses corresponding to low Qs. As Figure 5 shows the pulses initially broaden before recovering to their initial width after 1750km (coincident with the first peak in Q value). The pulse width then continued to broaden and narrow as propagation distance increased.

These oscillations in Q are an indication that the launch position of the source was incorrect (an alternative way of looking at the situation is that the chirp of the source was incorrect for this dispersion map). In an attempt to reduce the oscillations the standard fibre in the loop was split into two sections, one of length 25.3km and the other of length 6.3km. The 25km section was positioned directly before the loop amplifier, as shown in Fig 2 (b), so that the data stream being injected into the loop passed through this section of fibre before being amplified in the loop.

The propagation experiment was repeated. As the results in Fig. 6, show the oscillations remained. Again there was a rapid initial degradation in the Q value, dropping from the back-to-back value of 12 to 8.4 within the first 1100km of propagation. In this experiment the oscillations were greater in amplitude and had a larger period of ~2000km. The oscillations were once more apparent in the 10GHz trace (see Fig.

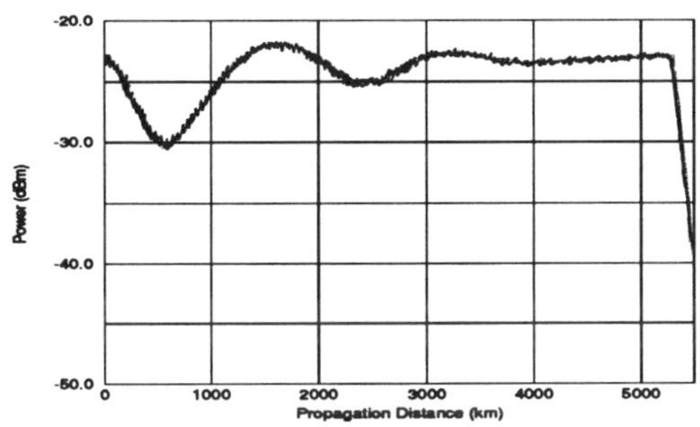

Figure 4: 10GHz electrical spectrum trace for set-up (a). Oscillations in power correspond to those in Q value.

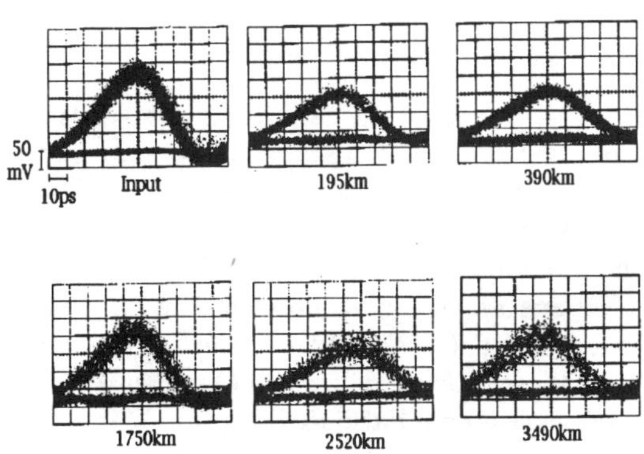

Figure 5: Eye diagrams taken after various propagation distances. The pulses initially broadened rapidly but had recovered to approximately the input width after ~1750km. Further broadening and contracting then took place.

7) and pulse width. The maximum propagation distance achieved before reaching a Q value of 6 was 6530km. The Q did however recover and at 7680km the Q value was 7.2 indicating that with this dispersion map propagation over 8000km should be possible.

A third dispersion map was tried. This time the DCF was positioned after the 25km SIF as shown in Fig. 2 (c). As the total dispersion of the DCF module was greater than that of 25km of SI fibre, there was over compensation for the standard fibre chirp leading to a net chirp of the opposite sign to that in the previous experiment on arrival at the amplifier.

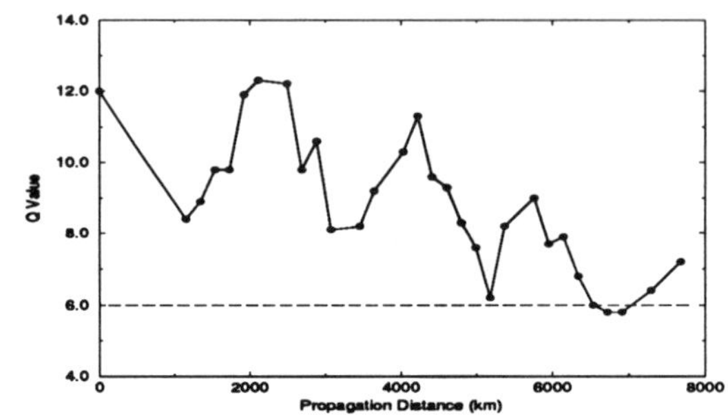

Figure 6: Experimental results for set-up (b) of Q value Vs propagation distance. The dotted line indicates the minimum acceptable Q value of 6.0. Q value is 7.2 at 7680km.

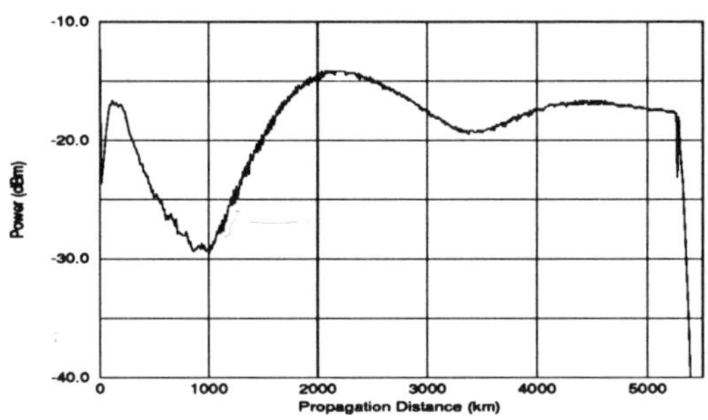

Figure 7: 10GHz electrical spectrum trace for set-up (b). Again, power oscillations are coincident with those of the Q value.

Q values are plotted in Fig. 8. This diagram shows that we have managed to suppress the oscillations in the Q value. There was also no rapid initial decrease in Q which is a good indication that the pulses were well matched to the dispersion map. The 10GHz electrical spectrum trace shows no oscillations (see Fig 9) and the pulse width was stable. The maximum propagation distance in this case was 5900km.

Numerical simulations are in good agreement with these experimental results. In simulations, set-ups (a) and (b) showed a rapid degradation of the Q in the first 1000km of propagation followed by oscilations in Q value. For set-up (c) there was only a gradual decrease in Q with no oscillations.

We have managed to increase our initial propagation distance of 5275km. Of the three experimental set-ups tried the second (Fig. 2 (b)) achieved the greatest propagation distance, 7680km. This did however have oscillations in the Q value and the Q dipped below 6 at ~6500km. By improving our set-up it was possible to suppress these oscillations but the maximum propagation distance in this case was 5900km. Although this maximum transmission distance was slightly less, the back-to-back Q value in this case was only 9.2 compared to the previous value of 12. By improving our back-to-back Q we should be able to increase our propagation distance for this set-up significantly.

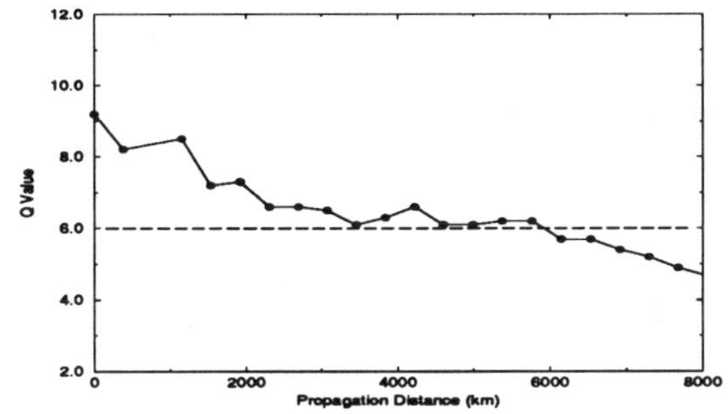

Figure 8: Experimental results for set-up (c) of Q value Vs propagation distance. The Q fell to the minimum acceptable value of 6.0 at 5900km. No oscillations were seen.

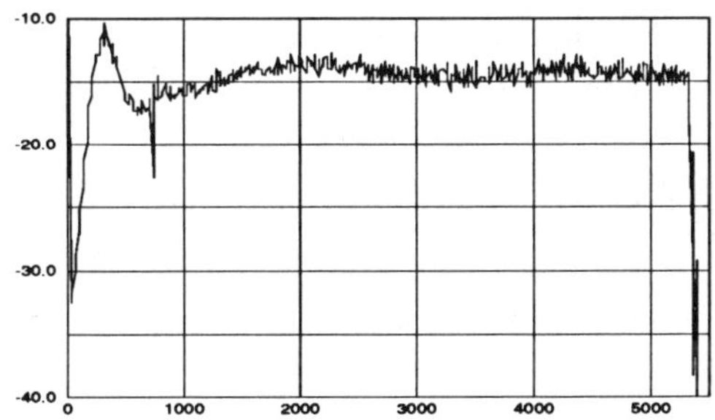

Figure 9: 10GHz electrical spectrum trace for set-up (c). This trace shows no oscillations

4 Conclusion

We have shown that it is possible, by finding the optimum launch position for a given source, to suppress oscillations of Q value and pulse width with propagation distance. This led to an increase in total propagation distance, which though modest at present has the potential to be increased to trans-oceanic distances. Our maximum propagation distance was 7680km. Further increases in distance should also be possible with in-line experiments: by using a single span recirculating loop we suffered an an additional ~4dB loss per round trip. This is also an indication that a larger amplifier span could be tolerated. It should also be pointed out that the net dispersion of the system has not yet been optimised and further increases in propagation distance could be made by tuning the wavelength of the source to find the most suitable average dispersion.

For solitons, dispersion managed systems have advantages over uniform dispersion systems. For a given pulse width the energy of the stable pulses in dispersion managed systems is higher giving a better signal to noise ratio, there is also a decrease in the Gordon-Haus jitter. Using standard fibre with periodic dispersion compensation also has the advantage that whilst the average dispersion is low the local dispersion is high. This reduces the effects of four wave mixing in WDM system.

Acknowledgements

The authors gratefully acknowledge financial support from EPSRC and studentship support from Alcatel Submarine Networks.

References

[1] F.M. Knox, W. Forysiak, and N.J. Doran. 10Gbit/s soliton communication system over standard fibre at 1.55μm and the use of dispersion compensation. *IEEE J. Lightwave Technol.*, 13(11):1955–1692, 1995.

[2] P. Harper, F. M. Knox, P. N. Kean, I. Bennion, and N.J. Doran. 10Gbit/s soliton propagation over 5250km in standard fibre with dispersion compensation. In *Proc. OFC '97, paper ThN1*, 1997.

[3] T. Georges. Pre-chirping and dispersion compensation for long-haul 20 Gbit/s soliton transmission at 1.55μm on non-dispersion-shifted fibres. In *Proc. OFC '97, paper WH2*, 1997.

[4] L. F. Mollenauer, K. Smith, J. P. Gordon, and C. R. Menyuk. Resistance of solitons to the effects of polarisation dispersion in optical fibres. *Opt. Lett.*, 14(21):1219–1221, 1989.

[5] R.I. Laming, W.H. Loh, M.J. Cole, M.N. Zervas, K.E. Ennser, and V. Gusmeroli. Fibre gratings for dispersion compensation. In *Proc. OFC '97, paper ThA3*, 1997.

[6] N.J. Smith, N.J. Doran, F.M. Knox, and W. Forysiak. Energy-scaling characteristics of solitons in strongly dispersion managed fibres. *Opt. Lett.*, 21(24):1981–1983, 1996.

[7] N.J. Smith, F.M. Knox, N.J. Doran, K.J. Blow, and I. Bennion. Enhanced power solitons in optical fibres with periodic dispersion management. *Electron. Lett.*, 32(1):54–55, 1996.

[8] N.J. Smith, W. Forysiak, and N.J. Doran. Reduced Gordon-Haus jitter due to enhanced power solitons in strongly dispersion managed system. *Electron. Lett.*, 32(22):2085–2086, 1996.

Core and ATM Networks
D.W. Faulkner and A.L. Harmer (Eds.)
IOS Press 1997 © AKM

Performance improvement of 10-Gbit/s soliton transmission over dispersion managed standard fiber by using the nonlinearity of dispersion compensating fiber

X. Y. Tang

Technische Universität Berlin, Fachgebiet Hochfrequenztechnik,
Einsteinufer 25, D-10587 Berlin, Germany

Tel: 49-30-31426816 , Fax: 49-30-31424626 , Email: tang@sun6hft.ee.tu-Berlin.DE

Abstract

10-Gbit/s soliton transmission over dispersion managed standard fiber is numerically investigated. It's shown that the performance of the system may be improved by introducing a section of dispersion compensation fiber after each amplifier.

Soliton transmission is currently competing with conventional systems for upgrading the capacity of future all-optical fiber communications. Dispersion compensation has been intensively investigated in conventional transmission systems as a promising technique to overcome the dispersion problem. In recent years, this approach has also been introduced into soliton communication [1-4]. Although most of the research work in soliton area were concentrated on the use of dispersion shifted fiber, the exciting experimental work has recently been reported on upgrading the embedded standard fiber (SMF) systems by the use of soliton [5]. So far, it has been believed that the nonlinearity of the dispersion compensation fibers (DCF's) would degrade soliton transmission in dispersion managed systems. For this reason, the previous work emphasized that DCF should be placed before each amplifier where the DCF could act as a purely linear compensator due to lower power [2] [6]. In this paper, we demonstrate through numerical simulation that the performance of soliton transmission over standard fiber may be greatly improved by introducing a section of DCF immediately after each amplifier.

In our numerical simulation, the amplifier spacing is set to 50 km. Every amplification period includes 42 km of SMF with dispersion of 16 ps/nm/km and 8 km of DCF with dispersion of -82.5 ps/nm/km. So the average dispersion of the transmission line is 0.24 ps/nm/km. The effective core areas of the SMF and DCF are assumed to be 80 μm^2 and 20 μm^2, respectively. The loss coefficients of the SMF and DCF are 0.21dB/km and 0.5dB/km, respectively. Each amplifier applies gain to compensate for the fiber loss of the preceding section and adds ASE noise with noise figure of 6 dB. The bit rate is 10 Gbit/s and the initial pulse width is τ=20 ps. The receiver is modelled as a fast photodiode followed by an electronic filter with a bandwidth of 8 GHz. The eye-closure penalty with respect to the back-to-back sensitivity is used to measure the system performance.

We discuss three types of system configuration. **Type 1**: 8 km DCF is placed before the amplifier. **Type 2**: at each amplifier site, the DCF is divided into 2 sections. 6 km of DCF is placed before the amplifier and the remained 2 km of DCF is after the amplifier. **Type 3**: 8 km DCF is placed after the amplifier.

Fig.1 (a) and (b) show the eye-closure penalty versus the peak input power after 1000 km and 3000 km, respectively. At low powers, the system performance is degraded by the accumulated ASE noise and the accumulated dispersion. With the increase of the input power, the performance is improved as the fiber dispersion is compensated for by the fiber nonlinearity through the soliton effect and also a higher signal-to-noise ratio (SNR) is obtained. But at very high power, the nonlinearity of the fiber degrades the performance severely. At the distance of 1000 km, in type 1, only in two narrow power windows, which are around 6 dBm and 16 dBm, respectively, the penalty less than 1 dB is possible. Whereas in type 2, we can have a very wide power window ranging from 4 dBm to 15 dBm for 1-dB penalty and nearly zero penalty is achievable if the optimum power of 10dBm is used. At the distance of 3000 km, in type 1, the power window for 1-dB penalty is around 10 dBm, still very narrow. In type 2, the power window is extended to the range from 7 dBm to 13 dBm. As long as the peak input power is less than 14 dBm, type 2 can provide better performance than type 1. Placing a section of DCF after the amplifier means that the nonlinearity of the DCF is introduced. So we conclude that the system performance can be improved by using some nonlinearity of DCF. We need to point out that at very high power, type 1 is superior to type 2 because the nonlinearity provided by the DCF in type 2 will further degrade the system where the strong nonlinearity of the SMF has already produced bad effect. In this range, the performance is very poor for both types, so we're not interested in the behavior in this power range.

Fig.2 show the evolution of the eye-closure penalty versus the transmission distance for three types of configuration at the peak input power of 10 dBm. We can see that the penalty doesn't increase monotonously with the distance, alternatively, oscillates with the distance. This behavior is due to the oscillation of the pulse width in this strongly dispersion perturbation system [5]. In type 1, the penalty increases with distance in the first 1800 km and reaches 1.8dB at 1800 km. But in type 2, attributed to the nonlinearity of the DCF, very stable transmission is obtained within 5000km, which is the distance we're interested in for upgrading the embedded SMF system. This result indicates that type 2 is preferable compared to type 1 for upgrading the embedded SMF system with maximum distance of 4000km. Of course, if the whole DCF is placed after the EDFA (type 3), the strong nonlinearity of the DCF will degrade the performance severely after 2000 km.

In conclusion, 10-Gbit/s soliton transmission over dispersion managed standard fiber is numerically investigated. We show that by introducing a section of DCF after each amplifier, the soliton systems over standard fiber is improved compared to the case where all the DCF's are placed before each amplifier. For the bit rate of 10-Gbit/s and the amplifier spacing of 50 km, the propagation up to 5,000 km at the eye-closure penalty of less than 1 dB is feasible.

Acknowledgment I wish to acknowledge the award of a research fellowship from the Alexander von Humboldt Foundation. I also would like to thank the useful discussion with D. Breuer and Prof. K. Petermann.

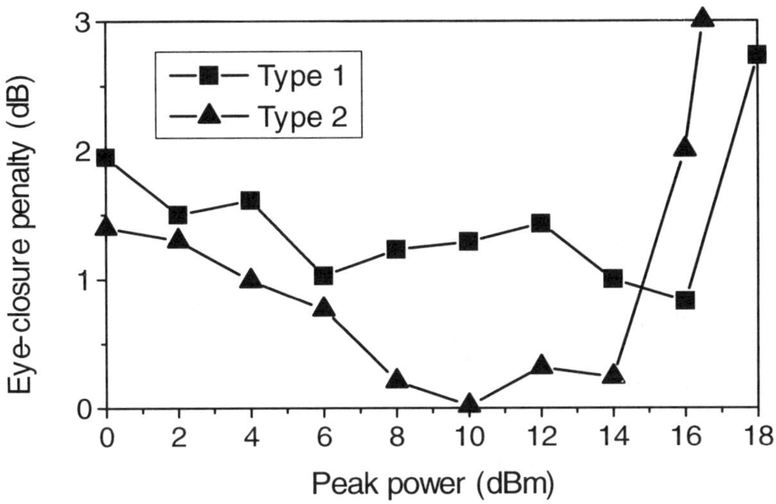

(a)

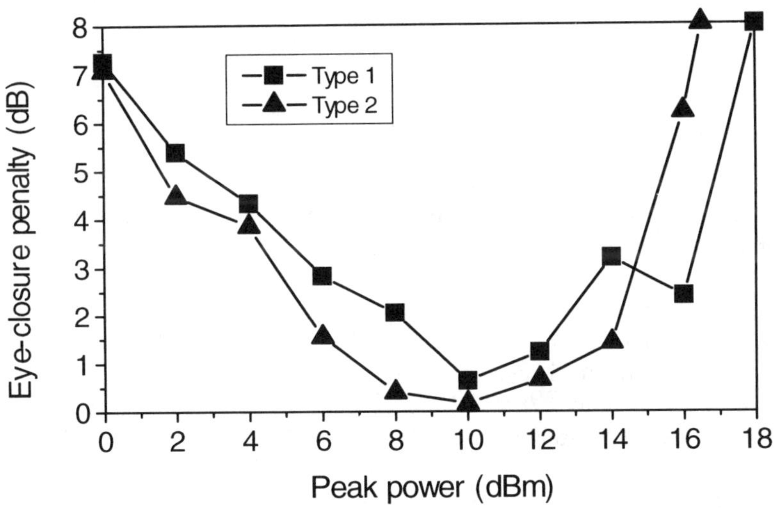

(b)

Fig.1 Eye-closure penalty versus peak input power after (a) 1000 km and (b) 3000 km for two configurations. (type 1: 8 km of DCF is placed before the amplifier. type 2: 6 km of DCF is placed before the amplifier and 2 km of DCF is after the amplifier.)

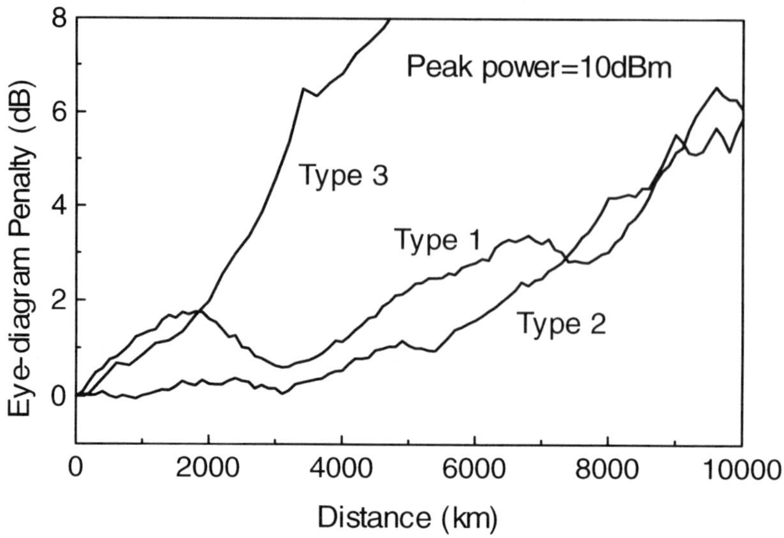

Fig. 2 Eye-closure penalty versus transmission distance for three configurations (type 1: 8 km of DCF is placed before the amplifier. type 2: 6 km of DCF is placed before the amplifier and 2 km of DCF is after the amplifier, type 3: 8 km of DCF is placed after the amplifier.)

References

[1] Nakazawa and H. Kubota, Optical soliton communication in a positively and negatively dispersion-allocated optical fiber transmission, *Electron. Lett.*, **31** (1995) 216-217.

[2] F. M. Knox, *et al.*, 10-Gbit/s soliton communication systems over standard fiber at 1.55µm and the use of dispersion compensation, *J. Lightwave Technol.*, 13 (1995) 1955-1962.

[3] J. Smith, *et al.*, Enhanced power solitons in optical fibers with periodic dispersion management, *Electron. Lett.*, **32** (1996) 54-56.

[4] M. Jacob, *et al.*, Experimental demonstration of soliton transmission over 28Mm using mostly normal dispersion fiber, *IEEE Photonics Technol. Lett.*, **9** (1997) 130-132.

[5] P. Harper, *et al.*, 10-Gbit/s soliton propagation over 5250km in standard fiber with dispersion management, *Optical Fiber Communication Conference*, 1997, ThN1.

[6] S. Wabnitz, *et al.*, Soliton transmission with periodic dispersion compensation: effect of radiation.

Core and ATM Networks
D.W. Faulkner and A.L. Harmer (Eds.)
IOS Press 1997 © AKM

High bit rate optical transmission systems using a midpoint OPC and DSCF's

Masanori HANAWA, Bing YAO, Mikio TAKAHARA and Jun YAMAWAKU
Yamanashi University, 4-3-11 Takeda, Kofu-shi, Yamanashi-ken 400, Japan

Abstract. A novel high bit rate optical transmission system which utilizes a midpoint optical phase conjugator and dispersion slope compensation fibers is proposed. Performance of the proposed system is compared with the dynamic soliton transmission system by amount of the timing jitter at the received point after 1,200km transmission.

1 Introduction

In these days, the most promising high capacity optical transmission system is the wavelength division multiplexing (WDM) system. Although one of the trend for the WDM system design is multiplexing many low capacity channels to realize very high capacity, so many channels may not be able to be allocated due to the bandwidth limit of the EDFA's when one intends to adopt the WDM system to long haul transmission system. So investigations of high capacity transmission systems with a single wavelength are still important.

There are two ways to improve channel capacity in a single wavelength channel. The one way is to raise the duty factor and the another is to utilize very narrow pulses (i.e. single pico-second pulses). However, the former causes inter symbol interference (ISI) and the latter induces higher order optical effects such as the third order dispersion. These cause serious pulse distortion and limit transmission distance. The dominant effects which limit transmission distance are pulse broadening due to chromatic dispersion and generation of new oscillations due to the third order dispersion effect which depends on the dispersion slope. So we propose utlization of a midpoint OPC and dispersion slope compensation fibers (DSCF's) to design very high bit rate (over several ten Gbit/s) and long haul (over one thousand km) optical pulse transmission systems.

2 System Configuration

Fig.1 shows the proposed system model. An OPC is inserted at the midpoint of the system so that chromatic dispersion and the self phase modulation (SPM) are almost compensated. Thus, although pulses are once seriously degraded by ISI when pulse duty ratio is high, it almost recovers the original waveform at the receiver side. In addition, a 25km DSF and a 5km DSCF are allocated in every amplifier spacing of 30km. So the third order dispersion is also compensated by DSCF's. From these features, the proposed system can achieve very high bit rate by using single pico-second pulses and a high duty factor.

Fig.2 shows a model of chromatic dispersion vs. wavelength characteristics of the DSF and the DSCF. The solid line is for the DSF and the broken line is for the DSCF. As the length of DSCF is one fifth of the system fiber, the absolute value of chromatic

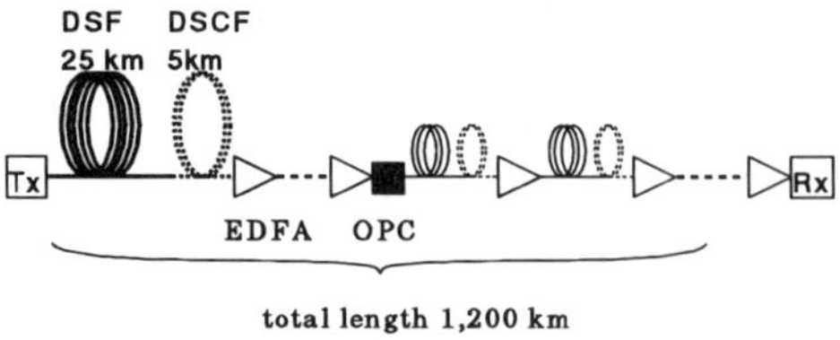

Fig.1 System model: an OPC is allocated at 600km point.

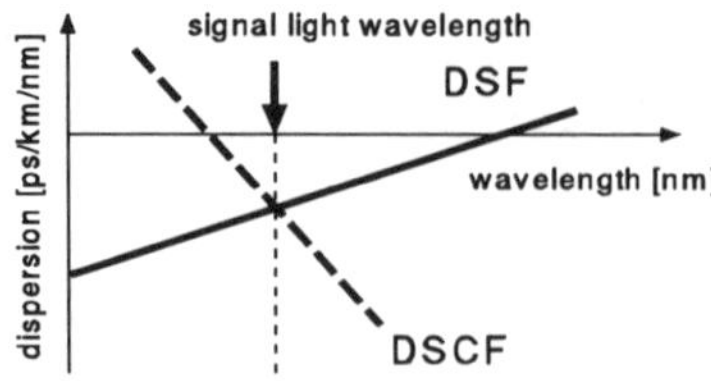

Fig.2 Concept of the DSCF.

dispersion and the dispersion slope of the DSCF is five times as large as one of the DSF. Moreover, it has an opposite sign of the one of the DSF.

3 Evaluation method of timing jitter

To investigate amounts of timing jitter of the proposed system, we applied a jitter evaluation method described in references [1] and [2]. Though the methods were proposed to compute the Gordon-Haus jitter of soliton systems, they can be easily extended to gaussian pulse transmission system.

Here, the number of EDFA's before and after the midpoint OPC are described as n_1 and n_2, amplifier spacing is Z_{amp}, the second order dispersion coefficient is β_2. Then the timing jitter $\langle \Delta t \rangle$ is computed from ensemble average of squared angular frequency $\langle \delta\omega^2 \rangle$ caused by amplifier spontaneous emission (ASE) noise.

$$\langle \Delta t \rangle = \sqrt{\left\{ \sum_{j=1}^{n_1} (-j + n_2)^2 + \sum_{k=1}^{n_2} k^2 \right\} \beta_2 Z_{amp} \langle \delta\omega^2 \rangle} \tag{1}$$

When we describe the pulse spectrum as $|\Phi|^2$, the pulse energy as W and the power spectrum density of ASE noise as S_η, $\langle \delta\omega^2 \rangle$ is given by

$$\langle \delta\omega^2 \rangle = \frac{2S_\eta}{W^2} \int_{-\infty}^{\infty} \omega^2 |\Phi|^2 d\omega. \tag{2}$$

where,

$$W = \int_{-\infty}^{\infty} |\Phi|^2 d\omega \tag{3}$$

As reference [1] treated soliton pulses, they assumed that $\langle \delta\omega^2 \rangle$ and W are constants. However, because pulses in the proposed system suffer various effects which change pulse spectrum, $\langle \delta\omega^2 \rangle$ and W is not constant in propagation. So we computed them from pulse spectrum obtained by numerical evaluation of the nonlinear schrödinger equation (NLSE). The NLSE used is,

$$\begin{aligned} \frac{\partial \phi}{\partial z} &= -\frac{\alpha}{2}\phi - \frac{i}{2}\beta_2 \frac{\partial^2 \phi}{\partial \tau^2} + i\gamma \mid \phi \mid^2 \phi \\ &+ \frac{1}{6}\beta_3 \frac{\partial^3 \phi}{\partial \tau^3} - \frac{2\gamma}{\omega_0}\frac{\partial}{\partial \tau}(\mid \phi \mid^2 \phi) - i\gamma T_R \phi \frac{\partial \mid \phi \mid^2}{\partial \tau} \end{aligned} \tag{4}$$

Table 1. Fiber constants

	DSF	DSCF
λ_0	1550.0nm	1521.4nm
loss	0.2dB/km	0.6dB/km
D	+1.0ps/km/nm	+1.0ps/km/nm
	±0.1ps/km/nm	±0.1ps/km/nm
$dD/d\lambda$	0.07ps/km/nm^2	-0.35ps/km/nm^2
fiber length	25.0km	5.0km
core diameter	10.0μm	5.0μm
Kerr coefficient	3.17×10^{-20}m^2/W	6.24×10^{-20}m^2/W

where the second row in Eq.(4) shows higher order effects and the most dominant term which causes serious pulse distortion is the first one (the third order dispersion term). In the term, β_3 is the third order dispersion coefficient given by

$$\beta_3 = \left(\lambda \frac{dD}{d\lambda} + 2D\right) \lambda^2 \tag{5}$$

where D is the chromatic dispersion at signal wavelength. We solved Eq.(4) using the beam propagation method and computed the timing jitter.

4 Simulation

Table 1 shows fiber constants used in our simulations. We assumed that the total length of the transmission line is 1,200km and DSCF's of 5km are allocated in every amplifier spacing of 30km. Fig.3 shows an example of pulse train at launching point. In this case, lauching peak power P_0 of pulses is 5dBm, bit rate R is 100Gbit/s (FWHM τ_w and duty factor r are 5ps and 0.5) and information bit pattern is "001001110". On the hand, Fig.4 shows pulses after 1,200km propagation. For comparison, a result obtained from a system which does not consist of DSCF's is also shown. The solid line shows a received pulse train of the proposed system and the broken line shows one of the OPC only system without DSCF's. In both results, although the effects of the second order dispersion and SPM is almost compensated by the midpoint OPC, pulses from the OPC only system get serious distortion caused by the third order dispersion. In contrast with this, the received pulses of the proposed system are almost the same with the launching ones.

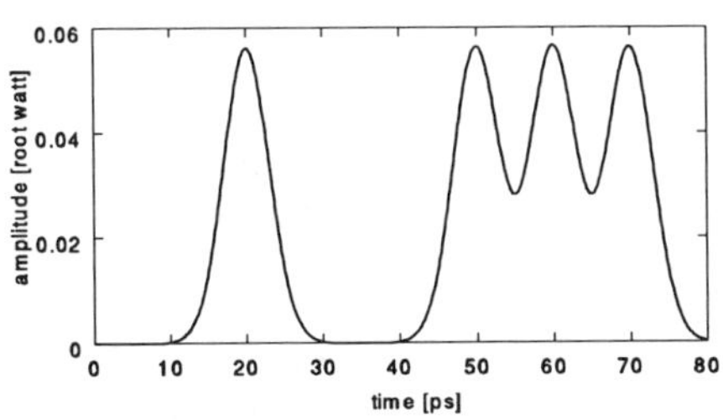

Fig. 3 Pulse train at 0km. (P_0=5dBm, τ_w=5ps, R=100Gbit/s)

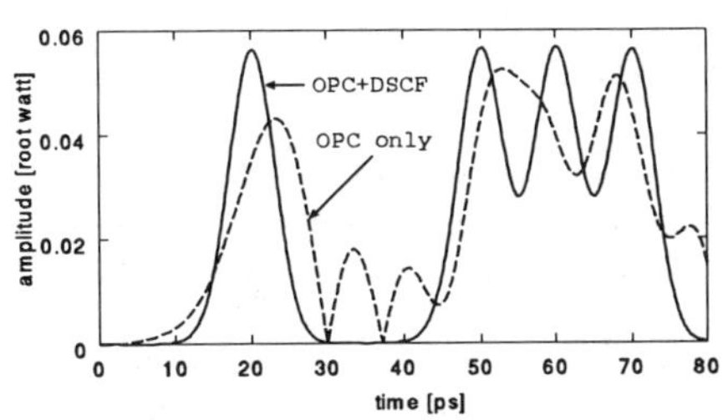

Fig. 4 Pulse trains at 1,200km. (P_0=5dBm, τ_w=5ps, R=100Gbit/s)

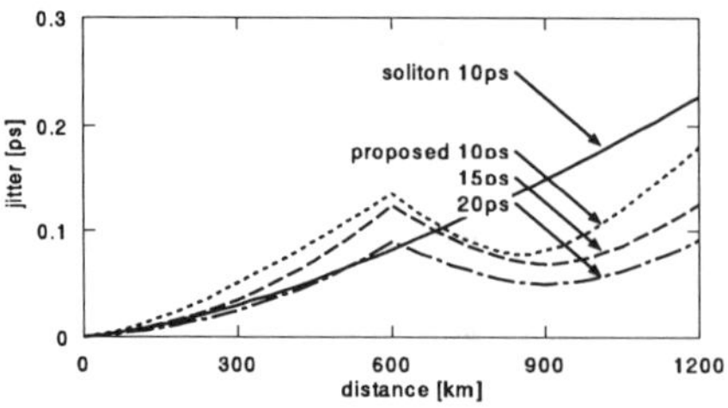

Fig. 5 Evolution of the timing jitter with propagation distance. (R=20Gbit/s)

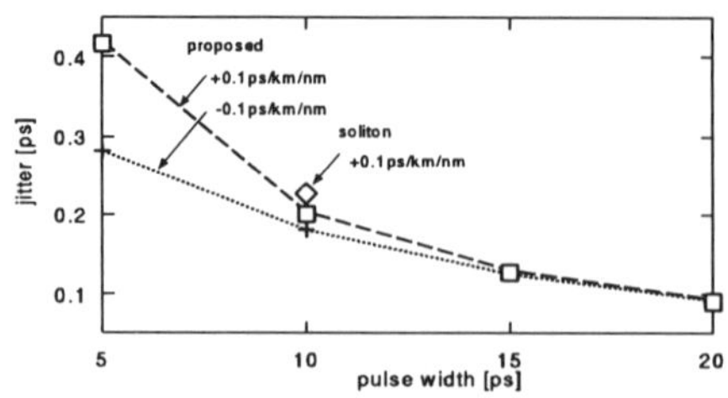

Fig. 6 Timing jitter vs. pulse width characteristics (R=20Gbit/s)

Next, we investigated timing jitter characteristics of the proposed system and compared them with the one of the dynamic soliton system of 20Gbit/s. In this evaluation, τ_w of the soliton system was fixed to 10ps and one of the proposed system was changed from 10ps to 20ps. Thus r of the proposed system was also changed from 0.2 to 0.4 to fix R to 20Gbit/s. Fig.5 shows timing jitter evolution vs. distance as a function of pulse width computed in the anormalous dispersion region (+0.1ps/km/nm). W of the proposed system at the launching point is set to equal with one of the soliton system, namely launching power $\bar{P}$ (not P_0) of all is equal. $\langle \Delta t \rangle$ of the proposed system is smaller than one of the soliton system and it becomes small as τ_w becomes big.

The timing jitter at 1,200km are completed in Fig.6. In this figure, results in the normalous dispersion region (–0.1ps/km/nm) are also plotted. The timing jitter in the normalous dispersion region is smaller than one in the anormalous region. From a comparison of propagated waveforms, we found that pulse compression occurs and bandwidth becomes wide in anormalous dispersion region. It makes $\langle \delta\omega^2 \rangle$ be big and $\langle \Delta t \rangle$ becomes big consequently. On the other hand, pulses were simply broaden and bandwidth did not become wide. So we should use the proposed system in the normalous dispersion region.

5 Conclusions

In this paper, we have proposed a novel high bit rate optical transmission system which utilizes a midpoint OPC and DSCF's. From simulated results, we have shown that the proposed system is effective to realize high bit rate and long transmission distance system. More precise evaluations are further works.

References

[1] C. Lorattanasane and K. Kikuchi, "Design of long distance optical transmission systems using midway optical phase conjugation", IEEE Photonics technorogy letters, Vol.7, No. 11, Nov. 1995.

[2] D. Marcuse, "An alternative deviation of the Gordon-Haus effect", Journal of lighwave technology, Vol.10, No.2, Feb. 1992. "Design of long distance optical transmission systems using midway optical phase conjugation", IEEE Photonics technorogy letters, Vol.7, No. 11, Nov. 1995.

[3] G.P.Agrawal, "Nonlinear Fiber Optics", Academic Press Inc. 1989.

Core and ATM Networks
D.W. Faulkner and A.L. Harmer (Eds.)
IOS Press 1997 © AKM

Soliton WDM Transmission in Long-Haul Dispersion Managed Fiber Systems

X.Y. Tang, D. Breuer, and K. Petermann

Technische Universität Berlin, Fachgebiet Hochfrequenztechnik,
Einsteinufer 25, D-10587 Berlin, Germany

Tel: 49-30-31426816 , Fax: 49-30-31424626 , Email: tang@sun6hft.ee.tu-Berlin.DE

Abstract

Based on the perturbation theory, we investigate the collision-induced frequency shift of WDM solitons in a dispersion managed system with alternating fiber segments of anomalous and normal dispersion. It's shown that by optimizing the dispersion management, the frequency shift may be greatly reduced. As for reducing the frequency shift, this dispersion management can have better result than the 4-step dispersion-decreasing fiber link. Through numerical simulation, we show that the system performance is significantly improved in comparison to the uniform fiber system with the same average dispersion.

1. Introduction

Soliton transmission using lumped erbium-doped fiber amplifiers (EDFA) is a promising candidate for realizing high-speed and long-haul fiber communication. Attributed to the large bandwidth of EDFA, the transmission capacity of soliton systems can be further enhanced through wavelength division multiplexing (WDM). A particular limitation to soliton-based WDM systems originates from the collisions of solitons in different channels [1]. In a fiber system with lumped amplifiers, the periodic power variation or dispersion variation can lead to asymmetric collision and thereby cause a net frequency shift of each soliton after each collision. For ultra-long distance systems, a soliton in one channel suffers from a large number of collisions with solitons in other channels at random positions. The accumulation of these frequency shifts will cause a jitter in the arrival time of solitons. Recently, Hasegawa *et al.* proposed that such frequency shift can be reduced by the use of a dispersion management (DM) scheme for which a stepwise dispersion-decreasing (DD) fiber link is used [2]. But the construction of such stepwise DD fiber link is troublesome as we have to control the length and dispersion of every fiber segments carefully. Moreover, this DM scheme can apply only to unidirectional transmission. In this paper we propose that the collision-induced frequency shift may be reduced by adopting another simple DM scheme with alternating fiber segments of opposite-sign dispersion. The study of soliton transmission in such DM fiber with both anomalous and normal dispersion fibers has already drawn great attention in recent years as it's a easy way to construct a transmission line with average dispersion of desired low anomalous value [3-5]. So far, however, little work has been done to understand the propagation characteristics of WDM solitons in such DM fiber link. Although the effect of dispersion variation on the collision-induced frequency shift was studied in the early work of Mollenauer *et al.* [1], in that study and in the recent work of Hasegawa *et al.* [2], the dispersion variations are within the anomalous region. To our knowledge, only Wabnitz has recently proposed through numerical simulations that a proper DM with compensating fibers of normal dispersion may suppress the instabilities of the sliding-filtered soliton WDM transmission [6]. But the effect of such DM scheme on the collision-induced frequency shift is not studied quantitatively. In this paper, based on the soliton perturbation approach, we investigate the frequency shift of WDM solitons in a unfiltered DM fiber link consisting of two fibers with opposite-sign dispersion. The method of optimizing such DM is presented. It's shown that the frequency shift can be effectively reduced through proper DM and therefore the system performance is improved though a dispersion perturbation is introduced into the system. Such DM scheme and the DM scheme using stepwise DD fiber link are also compared.

2. Theoretical model

We consider a system consisting of two fibers within one amplification period, with dispersion D_1, D_2 and lengths L_1, L_2, respectively. An analytical expression for the collision-induced frequency shift in the presence of periodic amplification and dispersion variation within anomalous region has been derived by Mollenauer *et al.* using the perturbation approach [1]. For a DM fiber link with both anomalous and normal dispersion fibers, if the length of the fiber segment L_i is smaller than its local dispersion length L_D, the change in pulse width within this fiber segment will be smaller than half the initial pulse width. Under this condition, Nakazawa *et al.* have shown that the soliton can propagate very stably over ultra-long distance [3]. As an example, if the pulse width τ_0=20ps, this condition requires D_iL_i<100ps/nm. For such weak DM system, the perturbation theory should also be applicable. So we extend the theory of Mollenauer to the case with normal dispersion fiber and get the net frequency shift as

$$\delta f = C\sum_{n=1}^{\infty} \mathrm{Im}\left[f_n \exp(jn\phi_c)\right]\frac{n^3x^4}{\sinh^2(nx)} \tag{1}$$

where $C = 8L_a / \tau_0 z_0 \pi^2$, $x = 2.8L_c / L_a$, z_0 is the average soliton period, L_a is the amplifier spacing, the collision length $L_c = 2\tau_0 / D_{ave}\Delta\lambda$, D_{ave} is the average dispersion, $\Delta\lambda$ is the channel spacing, $\phi_c = 2\pi\left(\int_0^{z_{col}} d(z)dz\right)/ D_{ave}L_a$, z_{col} is the distance from the collision center to the nearest amplifier, $d(z)$ is the normalized fiber dispersion, and

$$f_n = \frac{\alpha L_a}{1-\exp(-\alpha L_a)}\left[\frac{1-\exp(-\alpha L_a + in\phi_1)}{\alpha L_a + i2\pi n d_1} + \frac{\exp(-\alpha L_a + in\phi_1) - \exp(-\alpha L_a)}{\alpha L_a + i2\pi n d_2}\right] \tag{2}$$

where $\phi_1 = 2\pi d_1 L_1 / L_a$, $d_{1(2)}=D_{1(2)}/D_{ave}$, and α is the fiber loss.

3. Numerical results

In our numerical examples discussed below, the following parameters are used: the pulse width τ_0=20ps, the average dispersion D_{ave} is set at 0.5ps/nm/km, the fiber loss α=0.21dB/km, and the amplifier spacing L_a=60km.

Fig.1 shows the temporary frequency shift of the faster soliton during a collision in a uniform fiber and a DM fiber, respectively, where the collision center (z=0) is set at an amplifier. In the DM fiber the two solitons have actually collided several times before the two soliton are separated at last. But in the uniform fiber, the two soliton can meet each other only once. From Fig.1, we can see that before the collision center, the accumulated up-frequency-shifts are nearly the same for the two cases, whereas after the collision center, a smaller down-frequency-shift is produced in the DM fiber as the two solitons are separated more rapidly due to a larger local dispersion. Thus the net frequency shift after the collision is much smaller in the DM fiber than in the uniform fiber.

In Fig.2, we plot the frequency shift versus collision length for the uniform fiber, a DM fiber with D_1=4.5ps/nm/km, L_1=20; D_2=-1.5ps/nm/km, L_2=40km and an optimum 4-step DD fiber link used in Ref. [2]. The frequency shift depends on the relative position of the collision center and the amplifier, i.e., z_{col}, in Fig.1 each collision center is adjusted to yield the maximum frequency shift. Fig.1 shows that the frequency shift is greatly suppressed after adopting DM in comparison to the uniform fiber system. For L_c/L_a<1.5, the frequency shift is reduced by a factor of 20 after adopting DM. For larger (L_c/L_a>2.5), it doesn't matter whether such DM scheme can reduce the frequency shift or not as long as the introduction of DM into the system doesn't increase the frequency shift because the frequency shift is already negligible for uniform fiber at this range. The result also shows that this DM fiber is even better than a 4-step DD fiber link as for reducing the frequency shift.

The DM fiber used in Fig.2 is just a special example. To understand the dependence of the frequency shift on the dispersion profile, at first we define the dispersion strength of the normal (negative) dispersion fiber as $K_{-}=|D_{(-)}|L_{(-)}/D_{ave}L_a$, where $D_{(-)}$, $L_{(-)}$ is the dispersion and length of the normal dispersion fiber. Fig. 3 shows the net frequency shift versus the dispersion strength K_{-} for different lengths of the front fiber segment at L_c/L_a=0.7 in the two cases that the front segment is anomalous fiber or normal fiber. With the introduction of normal dispersion fiber into the system, as long as K_{-} is larger than 1.0, the frequency shift is reduced compared to the uniform fiber system regardless of the position and the lengths of the normal and anomalous fiber. This indicates that in a very large range, such DM scheme can reduce the frequency shift. Note that oscillations occur in the curves and those optimum points coincide with the integer K_{-}. In optimizing such DM fiber system, we suggest to choose K_{-}=2 because that at this point, the frequency shift is smaller and the weak DM condition is also satisfied so that the soliton can have ultra-long distance stable transmission.

Fig.4 shows the net frequency shift versus the length of the front fiber segment when K_{-}=2. There exists an optimum length, which is about 20km whenever the front segment is anomalous or normal fiber. If we choose L_1=20km, L_2=40km, and put the normal fiber before the anomalous fiber, the frequency shift is only 46MHz. For the back-direction transmission, we have L_1=40, L_2=20km, and the anomalous fiber becomes the front segment, the frequency shift is 240MHz, still reduced by a factor of 5 compared to the uniform fiber case. So this DM scheme can apply to bi-directional transmission.

Until now we just discussed the situation of a single collision, in ultra-long distance transmission, a soliton experiences many collisions, which cause a jitter in the soliton arrival time. To understand the effect of DM on the overall system performance, we carry out the numerical simulations of a two-channel WDM system where the bit rate of single channel is 10Gbit/s, the channel spacing is set at $\Delta\lambda$=1.9nm which gives L_c/L_a=0.7, and a pseudo-random sequence of 128 bits containing 64 solitons is used as the input. The receiver after demultiplexing is modelled as a fast photodiode followed by an electronic filter with a bandwidth of half the bit rate. In the simulations the amplified spontaneous emission (ASE) noise of the amplifiers is excluded in order to clarify the effect of pure collision process. Fig.5 (a) and (b) show the eye diagrams of one channel for a uniform fiber system and a DM

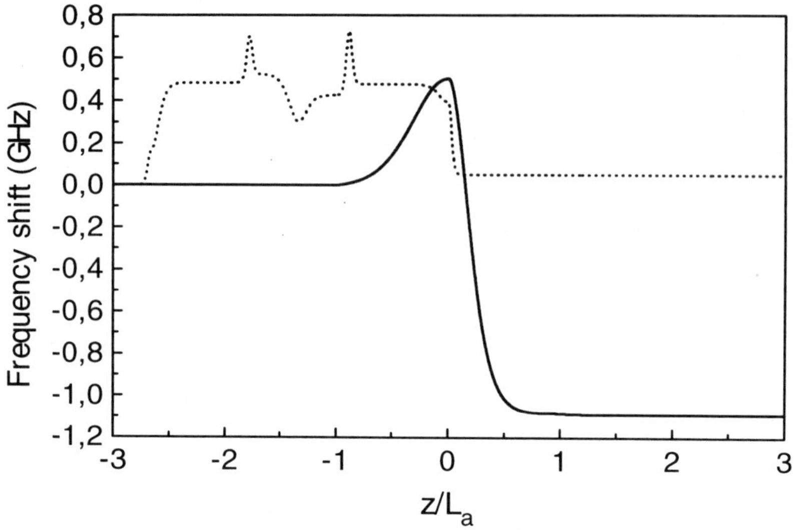

Fig.1 Temporary frequency shift during a collision (solid curve: DM fiber with D_1=4.5ps/nm/km, L_1=20km; D_2=-1.5ps/nm/km, L_2=40km. dashed curve: uniform fiber with D_{ave}=0.5ps/nm/km.)

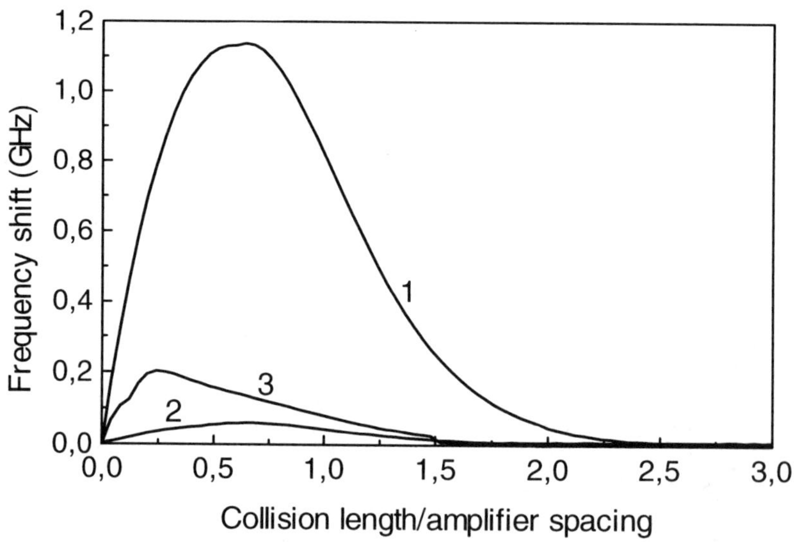

Fig.2 Frequency shift versus L_c/L_a. (Curve 1: uniform fiber with D_{ave}=0.5ps/nm/km. Curve 2: DM fiber with D_1=4.5ps/nm/km, L_1=20km and D_2=-1.5ps/nm/km, L_2=40km. Curve 3: 4-step DD fiber used in Ref. [2].)

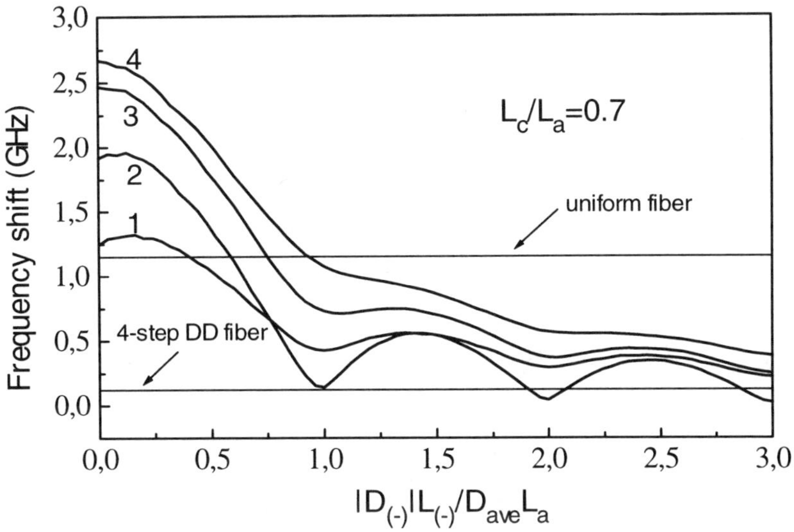

(a)

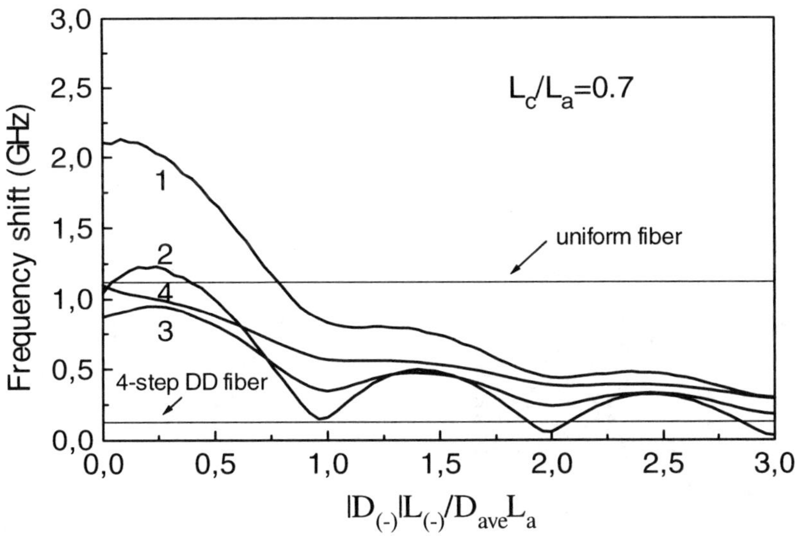

(b)

Fig.3 Frequency shift versus dispersion strength of the normal dispersion fiber for different lengths of the front fiber segment at L_c/L_a=0.7. (a) normal fiber is in the front. (b) anomalous fiber is in the front. (curve 1: L_1=5km, curve 2: L_1=20km, curve 3: L_1=40km; curve 4: L_1=55km.)

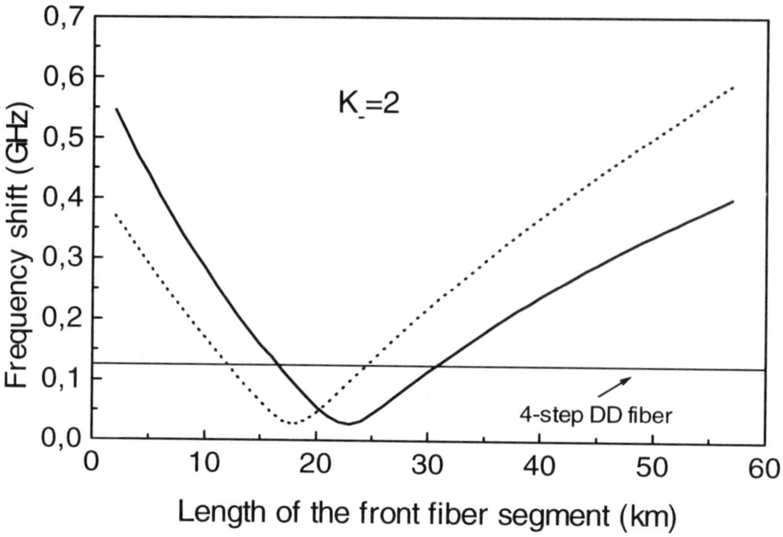

Fig.4 Frequency shift versus the length of the front fiber segment for K_-=2. (solid curve: anomalous fiber is in the front; dashed curve: normal fiber is in the front.)

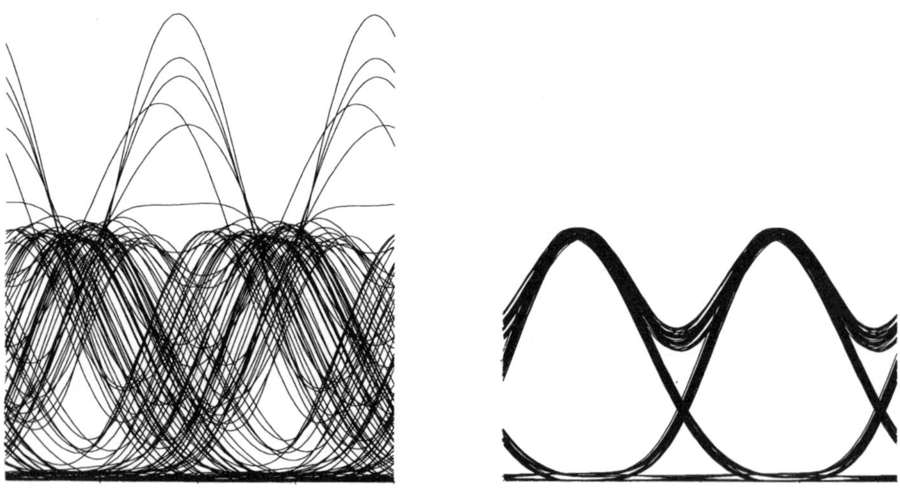

(a) (b)

Fig.5 Eye-diagrams at 3,000km for (a) uniform fiber with D_{ave}=0.5ps/nm/km and (b) DM fiber with D_1=4.5ps/nm/km, L_1=20km; D_2=-1.5ps/nm/km, L_2=40km. (L_c/L_a=0.7).

fiber system, respectively. For the uniform fiber system, a significant timing jitter is present, so it is impossible to transmit the signal over 3,000km at this channel spacing due to the larger collision-induced frequency shift. In order to get error-free transmission over this distance in uniform fiber, we have to reduce the channel spacing to decrease the collision-induced timing jitter. This means that a maximum channel spacing is limited by the collision-induced frequency shift. However, after adopting DM, clear eye diagram is obtained even at this larger channel spacing, eliminating the limitation on maximum channel spacing.

4. Conclusion

In conclusion, through the use of proper dispersion management with alternating fiber segments of anomalous and normal dispersion, the frequency shift resulting from soliton collisions in WDM systems is reduced and the system performance may be improved in comparison to the uniform fiber system with the same average dispersion. Especially in the case that the collision length is close to the amplifier spacing, the improvement of system performance by adopting DM is very significant. As for reducing the frequency shift, an optimum DM fiber with both anomalous and normal dispersion fiber can have better result than a 4-step dispersion-decreasing fiber link. Due to the reduction of frequency shift, the maximum channel spacing limited by the collision-induced frequency shift is also extended. This DM scheme is much easier to realize than the stepwise dispersion-decreasing fiber link and can apply to bi-directional transmission.

Acknowledgment *X. Y. Tang* acknowledges the award of a research fellowship from the Alexander von Humboldt Foundation.

References

[1] L. F. Mollenauer, S. G. Evangelides, and J. P. Gordon, Wavelength division multiplexing with solitons in ultra-long distance transmission using lumped amplifiers, *J. Lightwave Technol.*, **9** (1991) 362-367.
[2] A. Hasegawa, S. Kumar, and Y. Kodama, Reduction of collision-induced time jitters in dispersion-managed soliton transmission systems, *Opt. Lett.*, **21** (1996) 39-41.
[3] M. Nakazawa and H. Kubota, Optical soliton communication in a positively and negatively dispersion-allocated optical fiber transmission, *Electron. Lett.*, **31** (1995) 216-217.
[4] N. J. Smith, *et al.*, Enhanced power solitons in optical fibers with periodic dispersion management, *Electron. Lett.*, **32** (1996) 54-56.
[5] J. M. Jacob, *et al.*, Experimental demonstration of soliton transmission over 28Mm using mostly normal dispersion fiber, *IEEE Photonics Technol. Lett.*, **9** (1997) 130-132.
[6] S. Wabnitz, Stabilization of sliding-filtered soliton wavelength division multiplexing transmission by dispersion-compensating fiber, *Opt. Lett.*, **21** (1996) 638-640.

Core and ATM Networks
D.W. Faulkner and A.L. Harmer (Eds.)
IOS Press 1997 © AKM

Investigation of RZ- and NRZ-format 40Gb/s Transmission over Non Dispersion Shifted Fibre using Chirped Fibre Grating Dispersion Management

K Ennser[1], R I Laming and M N Zervas
Optoelectronics Research Centre, University of Southampton
Southampton SO17 1BJ, UK, Tel.: +44 1703-593139, Fax: +44 1703 593142,
E-mail: ke@orc.soton.ac.uk

Abstract: The RZ-format is shown to give a 2-fold increase in transmission distance compared with the conventional NRZ-format. In addition, a larger power margin is obtained at the expense of a reduced dispersion tolerance. RZ-transmission for super-Gaussian and hyperbolic-secant pulse shapes is studied and it is found that soliton-like transmission is less affected by amplifier noise for lower duty cycles.

1. Introduction

The feasibility of single-carrier full 40Gb/s system depends mainly on the speed of electrical and photonic circuits and the suppression of the fibre dispersion limitation. Most of the embedded fibre cable is standard single-mode fibre. Although the standard cable has zero-dispersion at 1.3μm, no all-optical amplifier is readily available. At the third window (1.55μm) where erbium-doped fibre amplifier (EDFA) operates, the high chromatic dispersion causes strong Inter Symbol Interference (ISI) limiting the transmission at 40Gb/s to less than 4km. Dispersion equalization techniques are the key to transmit beyond the dispersion-limited transmission length. In-line optical dispersion compensation methods include mid-span spectral inversion (MSSI), dispersion-compensating fibre (DCF) and linearly chirped fibre Bragg gratings.

In separate reports, 40Gb/s TDM dispersion compensated transmission over non-dispersion shifted fibre at 1.55μm has been demonstrated using either MSSI, DCF or gratings. Transmission over 202km was achieved using MSSI for 1.5dB penalty [1]. Unrepeatered transmissions using DCF have been shown over 80km [2] and 150km [3]. Long continuously chirped fibre gratings have been employed to achieve 109km transmission in a single repeater system [4].

Besides investigation of the dispersion equalization techniques, it is also important to determine the optimum modulation format and pulse duty cycle, because they affect the dispersion-induced broadening and residual dispersion tolerance. RZ-transmission has been shown to improve the performance of high bit-rate systems compared with the NRZ data format when using DCF fibres [5-6]. However, to date RZ-transmission using gratings has not been thoroughly investigated.

In this paper, we theoretically compare RZ- and NRZ-format data transmission over

---

[1] *On leave from the High-Frequency Institute, Berlin Technical University, Germany.*

standard fibre links employing chirped fibre grating dispersion compensators. It is shown, that the RZ-format can result in a 2-fold increase in maximum transmission distance compared with the NRZ-format. In addition, the RZ-format provides an increased power margin in the system and the dispersion tolerance decreases with the duty ratio. The impact of SPM, higher-order dispersion and amplifier noise for different pulse durations are evaluated for RZ- and soliton-based RZ-transmission. System guidelines are proposed relating the pulse width, equalizer spacing, input power and maximum transmission distance.

2. System Model

The system layout is illustrated in Fig 1. An IM/DD 40Gb/s single-channel system is investigated. The optical pulse propagation is solved by the Split-Step Fourier Method. A 2^7-1 pseudo-random test pattern is employed and the data is generated using a balanced Mach Zehnder modulator. The duty ratio is $\tau = T_{FWHM}/T_B$, where T_{FWHM} is the full-width of the initial pulses at half maximum and T_B the bit period. In soliton-based RZ-data, a hyperbolic-secant generator is considered and since the pulses are multiplexed, a random relative phase is applied. The standard fibre is characterised by a dispersion of 16ps/(nm.km) with a dispersion slope of 0.08ps/(nm^2.km), attenuation of 0.2dB/km and nonlinear coefficient of 1.31(W.km $^{-1}$). Loss of the fibre, grating and circulator(~2dB) are compensated by 6dB noise figure in-line amplifiers. The linearly chirped fibre Bragg grating transfer function is calculated by solving the coupled-mode equations. The grating is designed with a hyperbolic tangent apodisation profile and to give 95% peak reflectivity and 4.2nm bandwidth. At the receiver, the signal is optically filtered by a 160GHz bandwidth Bessel filter, before being detected and electrically filtered by a 28GHz bandwidth Bessel filter. The eye-opening penalty is evaluated with respect to the back-to-back sensitivity.

3. Comparison of NRZ- and RZ-Transmission

Figure 2 compares the performance of the NRZ- and RZ-data format for 50km and 100km equalizer spacings versus the average fibre-input power. The performance of low power (< -8dBm) is close to a linear transmission and a very small system penalty would be expected for a noiseless transmission, but as Fig. 2 shows, it is very sensitive to amplifier noise. For power higher than +2dBm, the fibre nonlinearities are the main limitation. The RZ-format is less affected by the accumulated noise

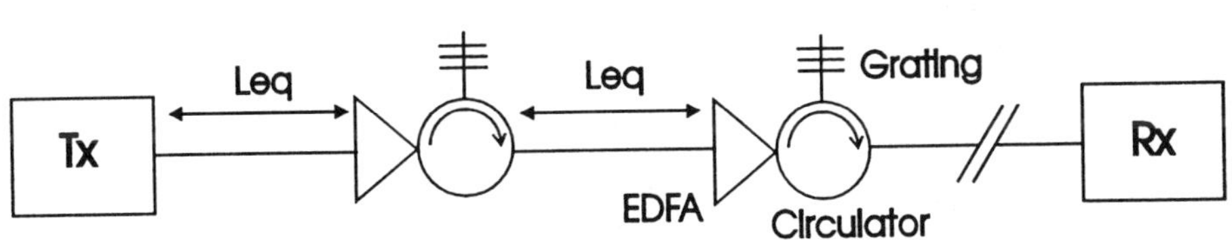

Fig 1: Block diagram of single-channel transmission system. EDFA: erbium-doped fibre amplifier with 6dB noise figure and L_{eq}: equalizer spacing.

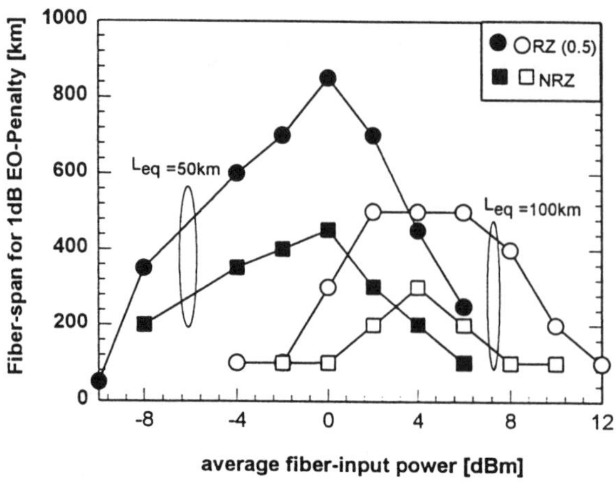

Fig 2: Maximum fibre length for 1dB eye-opening penalty vs. average fibre-input power for NRZ- and RZ-format transmission ($\tau = 0.5$) with amplifier noise figure of 6dB. Equalizer spacing (Leq) of 50km and 100km are considered.

in the propagation and has a larger margin of input power. The RZ-format can provide better performance than the NRZ-format whilst requiring only half the number of amplifiers, gratings and circulators.

The dispersion tolerance of the modulation formats is compared. Amplifier noise is neglected in this analysis since average input power is +2dBm, where the performance of noisy and noiseless transmission is near. The duty ratio of the RZ-signal is 0.5 in Fig. 3, which illustrates the EO-penalty against the residual fibre dispersion after ~200km. The fibre dispersion is varied, but not the total dispersion compensated by the gratings, so that the system has an accumulated residual dispersion at the end of the link. The tolerance of NRZ- and RZ-signal are ±35ps/nm (4.25km standard fibre span) and ±24ps/nm (3km standard fibre span), respectively. Although the NRZ-transmission has a larger tolerance, the system penalty close to zero residual dispersion is significantly better for the 0.5 duty ratio RZ data format.

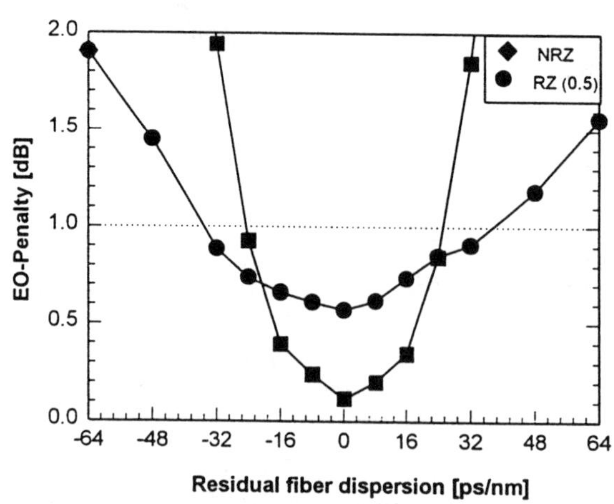

Fig 3: Eye-closure penalty against accumulated residual fibre dispersion over ~200km. RZ-signal has $\tau = 0.5$.

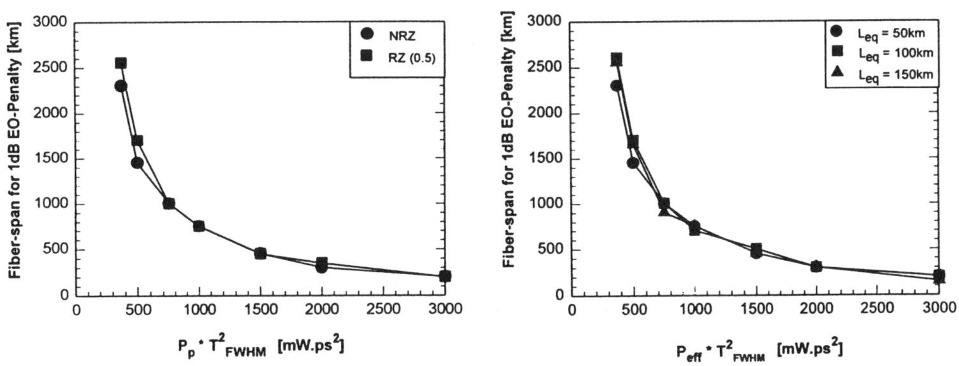

Fig 4 (a) ***Left:***Maximum fibre-span for 1dB eye-opening penalty vs. product of input peak-power (P_p) and square of input pulse full-width half-maximum (T^2_{FWHM}) for RZ- and NRZ-signal. Equalizer spacing of 50km. **(b)** ***Right:*** Maximum fibre-span for 1dB eye-opening penalty vs. product of effetive input peak-power (P_{eff}) and square of input pulse full-width half-maximum (T^2_{FWHM}) for NRZ-signal and different equalizer spacings (L_{eq}).

System guidelines have been determined by considering the effects of SPM and dispersion, but ignoring noise and higher-order dispersion. It might be expected that a relationship between nonlinear length (L_{NL}) and dispersion length (L_D) as in soliton transmission, whereby a balance of GVD, pulse width, SPM and fibre link-span could be obtained. It is found that the maximum transmission distance, L_{max}, is inversely proportional to the product of the transmission peak power, P_p, and the square of the pulse full-width half-maximum, T_{FWHM} [7].

$$L_{\max} \propto P_p \cdot T^2_{FWHM}$$

Fig. 4a confirms that the relation is valid for dispersion-compensating grating systems. While the grating is a quasi-linear device, it can be possible to apply the concept of the guiding-center soliton [8]. In other words, the product of effective input power (P_{eff}), effective length (L_{eff}) and inverse of amplifier/equalizer spacing (L_{eq}) should be constant.

$$P_{eff1} \cdot \frac{L_{eff1}}{L_{eq1}} = P_{eff2} \cdot \frac{L_{eff2}}{L_{eq2}}$$

Therefore, if input power is adjusted in proportion to the equalizer distance a similar performance should be found. Fig. 4b illustrates NRZ-transmission performance for an effective input peak-power according to the above equation. The numerical results show a good agreement. Thus we can predict the effects of increasing the amplifier/equalizer spacing or reducing the duty cycle on maximum transmission distance.

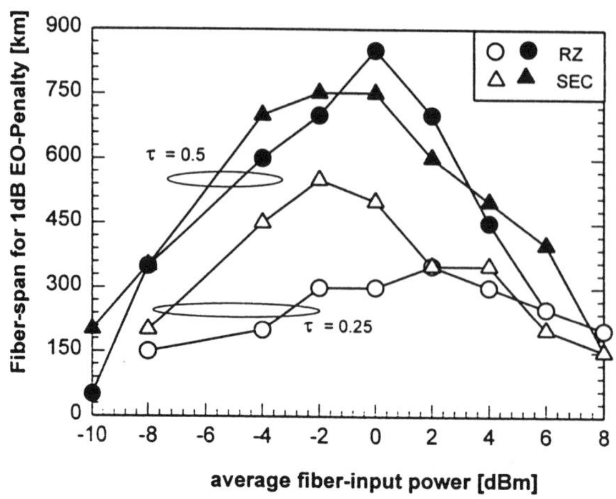

Fig 5: Maximum fibre-span for 1dB eye-opening penalty vs. average fibre-input power for RZ- and soliton-based RZ-signal. Equalizer spacing of 50km.

4. Soliton-based RZ-signal

The NRZ- and RZ-pulses modulated by the Mach-Zehnder modulator have a shape close to a super-Gaussian function. Because their formats are related, the comparison and guidelines were possible. However, the present technology of high-speed pulse generation (>10Gb/s) is able to produce a nearly transform-limited hyperbolic-secant pulse. For the same τ, the RZ-format has sharper tails avoiding ISI. In other side, the random phase variation of time-division multiplexing soliton-based RZ-signal reduces the attraction of neighbouring bits.

The maximum standard fibre length for 1dB EO-penalty versus average fibre-input power is plotted for RZ- and soliton-based RZ-transmission in Fig. 5. The equalizer spacing is 50km. The system guidelines predict an increased distance for a small duty ratio, but the effect of amplifier noise and higher-order dispersion degradates strongly the system performance. For $\tau = 0.5$, RZ- and soliton-based transmission show approximately the same results. For duty cycle $\tau = 0.25$, RZ-transmission is strongly affected by amplifier noise for average launched power below +5dBm. Therefore, the soliton-like format is more robust for lower duty ratios because the energy of the pulses is better distributed in the time slot and relative low ISI.

5. Conclusion

We have compared 40Gb/s NRZ- and RZ-format transmission over non-dispersion shifted fibre employing chirped fibre grating dispersion compensators. The RZ-format has been shown to result in a 2-fold increase in the maximum transmission length as well as allowing a larger range of average transmission powers. For equalizer spacings of 50km and 100km, transmission distances up to 850km and 500km, respectively may be obtained using the RZ-format. However, a drawback is that the RZ-signal is less tolerant to residual dispersion in a fibre link.

The system performance of the hyperbolic-secant pulse shape for a lower duty ratio ($\tau < 0.5$) is less affected by amplifier noise, but it has nearly the same results for $\tau = 0.5$ as super-Gaussian like RZ-signal.

Acknowledgments: K. Ennser acknowledges the Brazilian Council (CNPq) for financial support. This work was partially supported by Pirelli Cavi SpA. The ORC is an EPSRC-funded Interdisciplinary Research Centre.

References

[1] A.D. Ellis, M.C. Tatham, D.A.O.Davies, D.Nessett, D.G. Moodie and G. Sherlock, "40Gb/s transmission over 202km of standard fibre using midspan spectral inversion", *Electron. Lett.*, vol. 31, no. 4, pp. 299-301 (1995).

[2] A.D. Ellis and D.M. Spirit, "Unrepeatered transmission over 80km standard fibre at 40Gb/s", *Electron. Lett.*, vol. 30, no. 1, pp. 72-74 (1994).

[3] R. Ludwig et al., "Unrepeatered 40Gbit/s RZ single channel transmission over 150km of standard singlemode fibre at 1.55µm", *Electron. Lett.*, vol. 33, no. 1, pp. 76-77 (1997).

[4] L. Dong et al., "40Gbit/s 1.55µm Transmission over 109km of non-dispersion shifted fibre with long continuously chirped fibre gratings", in *Proc. OFC'97*, postdeadline paper PD6.

[5] D. Breuer, K.Ennser and K. Petermann, "Comparison of NRZ- and RZ-modulation format for 40Gb/s TDM standard-fibre systems", in *Proc. ECOC'96*, pp. 2.199-202, paper TuD3.3.

[6] N. Edagawa, I. Morita, M. Suzuki, S. Yamamoto, H. Taga and S. Akiba, "20Gbit/s 8100km Straight-Line Single-Channel Soliton-Based RZ Transmission Experiment Using Periodic Dispersion Compensation", in *Proc. ECOC'95*, Brussels, Belgium, pp. 983-986, postdeadline paper Th. A.3.5.

[7] K. Ennser and K.Petermann, "Performance of RZ- vs. NRZ-Transmission on Standard Single-Mode Fibers", *IEEE Photonics Technol. Lett.*, vol. 8, no. 3, pp.443-445 (1996).

[8] A. Hasegawa and Y. Kodama, "Guiding-center Soliton in Optical Fibers", *Opt. Lett.*, vol.15, no. 24, pp. 1443-1445 (1990).

Core and ATM Networks
D.W. Faulkner and A.L. Harmer (Eds.)
IOS Press 1997 © AKM

Field trial of 210 km in the German national network using a lumped amplifier transmission system at 1310 nm

A. Mattheus[1], J.J.E. Reid[2], P.I. Kuindersma[2], B. Teichmann[3], C. Schulien[3], L.C. Garcia[4], R. Leppla[5]

[1]*Deutsche Telekom AG, TZ, PO Box 64276, D-64295 Darmstadt, Germany,*
[2]*Philips Optoelectronics Research, Prof. Holstlaan 4, 5656 AA Eindhoven, The Netherlands,*
[3]*Lucent Technologies Network Systems GmbH, Thurn-und-Taxis Straße 10, D-90411 Nürnberg, Germany,*
[4]*Telefónica Investigación y Desarrollo, C/Emilio Vargas, 6. Madrid 28043, Spain,*
[5]*University of Kaiserlautern, Erwin-Schrödinger-Straße, D-67633 Kaiserslautern, Germany.*

Abstract. The succesful operation of a 210 km, 10 Gbit/s, 1314 nm field transmission test in RZ format between the German cities of Kassel and Hannover is reported. The link included a number of newly developed devices such as a 1314 nm DFB RZ laser with integrated electro-absorption modulator, flexible STM-64 terminal equipment suitable for NRZ and RZ transmission, polarisation insensitive in-line semiconductor amplifers, and optical repeater units which were managed by a service channel at 1510 nm. The robustness of the system was demonstrated in particular with respect to the insertion of additional fiber connectors with plane facets and the presence of a relatively large PMD.

1. Introduction. In the last few years several high capacity transmission techniques have been investigated in detail for upgrading the existing standard monomode fiber (SMF) based national core networks. Due to the excellent features of Erbium-doped fiber amplifiers (EDFA's) most efforts have been done in the 1550-nm window including NRZ and RZ transmission, the use of dispersion compensating fibers (DCF) and WDM techniques. Another alternative is the dispersion supported transmission (DST). All these techiques have to cope with the realively large chromatic dispersion (CD) of the SMF which is in the order of $17\ \mathrm{ps}/(\mathrm{nm}\cdot\mathrm{km})$ around 1550 nm wavelength. Much less efforts have been done in the minimum dispersion regime of the SMF around 1300 nm. No dispersion compensating techniques (such as DCF) are required there.

In the framework of the ACTS project UPGRADE the feasibility of optical transparent high capacity data transmission in the 1.3-μm window is investigated using newly developed polarisation insensitive semiconductor optical amplifiers (SOA) as in-line repeaters. A major milestone of this project was a 10 Gbit/s field trial along a 210 km link between the German cities Darmstadt and Hannover. The feasability of the optical transparent RZ transmission was shown as well as the successful monitoring and control of the five in-line semiconductor optical amplifiers (SOA) [1] via a 1510-nm Service Channel. The results of this experiment which was demonstrated at the CeBIT'97 is summarized in this article.

2. System Architecture. The architecture of the 210 km field trial between Kassel and Hannover is shown in Fig. 1. A unidirectional 10 Gbit/s data stream loaded with a narrow-band video channel and a PRBS sequence was transmitted from Kassel to the CeBIT in Hannover. The transmission equipment was installed at the Kassel central station and comprised the 1314 nm soliton source, 10 Gbit/s SDH devices - denoted as system interface unit 64 (SIU-64), and video facilities for real data traffic. At Trendelburg (the switching station between Kassel and Uslar) the incoming and outgoing fibers were bridged by DIN connectors with plane facets. Optical repeater units (ORU) were implemented at the next five switching stations between Uslar and Hildesheim. These devices included polarisation insensitive SOA's for the 1314 nm data channel and facilities for the optical Service Channel at 1510 nm. Some of the ORUs were additionally provided with 10 nm optical bandpass filters (BPF) centered around 1314 nm. During the installation phase before the CeBIT the input and output fibers at the Hannover switching station were bridged by DIN connectors. However, another ORU was inserted at the Hannover station (see Fig. 1) during the CeBIT in order to allow a larger system margin.

In Table 1 the parameters of upstream fibers between Kassel and Hannover are listed up. The individual link lengths ranged from 10 km to 37 km. Moreover, the first fiber span from Kassel to Uslar was as long as 53,11 km (for the fiber was just bridged in Trendelburg) with an attenuation of more than 19 dB. The individual zero dispersion wavelengths λ_0, the dispersion slopes and the chromatic dispersion at 1310 nm and 1314 nm are presented as well, respectively, with the exception of the Hannover-CeBIT link where a different fiber than the originally envisaged one had to be used. We had an average fiber attenuation of $\alpha = 0{,}38\ \mathrm{dB/km}$, and the path averaged chromatic dispersion between Kassel and

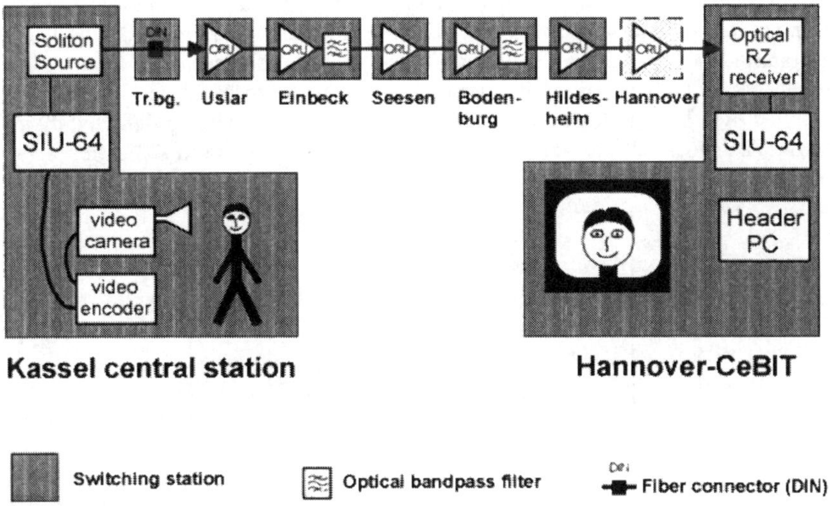

Figure 1. Set-up of the 210 km field trial Kassel-Hannover including video equipment, 10 Gbit/s SDH system interface units (SIU-64), optical repeater units (ORU) and the RZ source plus optical RZ receiver. The 1510 nm Service Channel is controlled by the Header PC.

Hannover was as large as $<D>= 0{,}28\ ps/(nm \cdot km)$ at 1314 nm wavelength. Polarisation mode dispersion (PMD) was less than 3 ps in all sections except for the Bodenburg-Hildesheim upstream fiber (13 ps). In Fig. 2 the variation of the chromatic dispersion at the laser wavelength of 1314 nm is additionally illustrated taking the figures from Table 1.

	dist.	Att. (1310 nm)	Att. (1310 nm)	λ_0	slope	Disp. (1310 nm)	Disp. (1314 nm)
	km	dB	dB/km	nm	$\frac{ps}{km \cdot nm^2}$	$\frac{ps}{nm \cdot km}$	$\frac{ps}{nm \cdot km}$
Kassel-Trendelburg	36,71	13,1	0,36	1312,9	0,0883	-0,256	0,097
Trendelburg-Uslar	16,4	6,0	0,37	1312,9	0,0883	-0,256	0,097
Uslar-Einbeck	33,24	11,34	0,34	1307,3	0,0909	+0,245	0,609
Einbeck-Seesen	31,42	11,78	0,38	1314	0,0876	-0,350	0,000
Seesen-Bodenburg	24,5	9,33	0,38	1311,6	0,0878	-0,140	0,211
Bodenburg-Hildesheim	21,76	8,52	0,39	1312,5	0,0878	-0,220	0,132
Hildesheim-Hannover	35,66	13,48	0,38	1306,9	0,0891	+0,276	0,633
Hannover-CeBIT	10,34	4,79	0,46	-	-	-	-

Table 1. Parameters of the upstream fiber between Kassel and Hannover-CeBIT. The total link length was 199,69 km from Kassel to Hannover and 210 km up to the CeBIT, respectively.

3. Data Transmission Equipment. In this section the equipment needed for data transmission is described in more detail. One narrow-band channel was used for video transmission and encoded into an STM-1 frame before being multiplexed into the STM-64 data stream. The STM-64 terminal multiplexers (SIU-64 in Fig. 1) at both ends of the link have been designed for the NRZ data format. However, relatively simple NRZ/RZ-converters were inserted in our experiment. Thus the SIU-64 is flexible in terms of using the NRZ or RZ format for the optical signal. The soliton source included a 1314 nm DFB laser diode with integrated electro-absorption modulator (EAM) followed by a SOA booster [2]. This source emitted the optical RZ data stream with 42 ps pulse width (FWHM), 6-10 dBm average output power and an extinction ratio of 8,2 dB. At the receiver side the input signal was preamplified by a SOA followed by 0,7 nm optical filter before being detected by a PIN photodetector. The specified receiver sensitivity ranges from -15 dBm to -23 dBm.

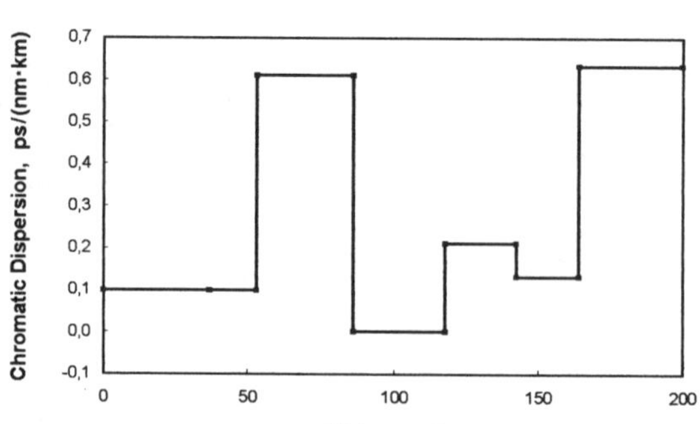

Figure 2. Chromatic dispersion of the upsstream fiber at 1314 nm along the Kassel-Hannover-link.

4. Optical Repeater Unit (ORU). In Fig. 3 the internal structure of the ORU (which was indicated by a triangle in Fig. 1) is presented. The ORU is a bidirectional terminal with two input and output fibers, respectively. The 1,3 μm data stream and the 1510 nm Service Channel for controlling the SOAs are carried by the same fiber in the same direction, respectively.

Let us now follow the signal transmission from East to West, i.e. from point A to point B in Fig. 3. The incoming 1,3 μm and 1510 nm signals are separated by a WDM coupler. Next the 1,3 μm data signal is amplified by a SOA and, in parallel, the monitored power levels at the input and output of the SOA - using 90/10 and 95/5 couplers - are launched into the semiconductor optical amplifier control unit (SOAC). Finally the outgoing data signal and service channel are combined by another WDM coupler.

The incoming 1510 nm Service Channel data stream - carrying the status information of the previous ORU - is detected by the link layer control (LLC). In parallel the status of SOA is fed electronically from the SOAC to the LLC. The bit stream of the new status information is processed by the LLC, emitted by a 1510 nm laser transmitter (Tx) and launched into WDM ouput couplers in both directions, EAST and WEST.

It is worthwhile to have a closer look at the internal losses of the ORU. In Fig. 4 the individual contributions to the losses are presented following the EAST-WEST signal path between the reference points A and B according to the notation in Fig. 3. The ORU is equipped with SC/APC connnectors (in view of a future field trial Madrid-Lisbon in 1998). However, the installed fibers at each switching station between Kassel and Hannover are terminated by DIN connectors. Therefore DIN-SC/APC fiber patch cords had to be inserted contributing up to 1,8 dB additional loss (2 connectors, 1 splice per connector). Fig. 4 figure reveals significant differences between this practical ORU suitable for field application and a

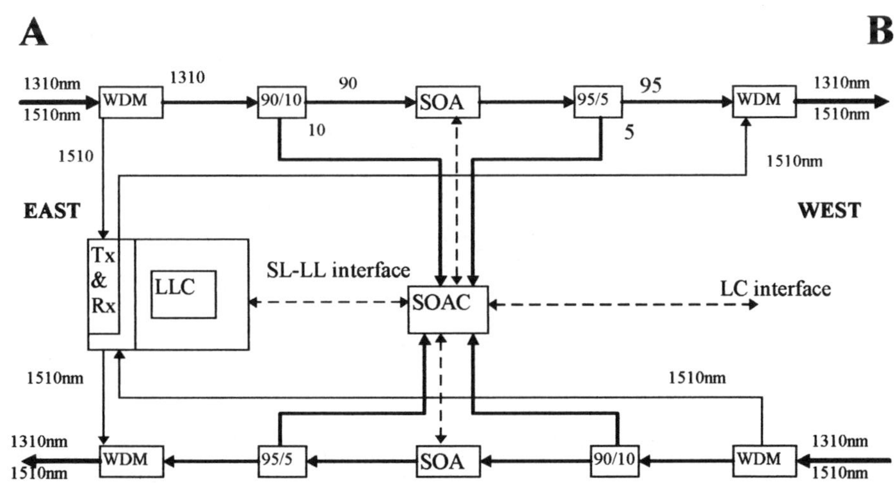

Figure 3. Internal structure of the optical repeater unit (ORU). Fibers (solid lines) and electrical (dashed lines) connections are indicated. The optical signal flow at 1310 nm and 1510 nm is distinguished by thick and thin solid lines. LLC = link layer control, SOAC = semiconductor laser amplifier control, SL-LL = SLAC interface with LLC, LC = link control.

common laboratory version of an optical amplifier. A laboratory model does not need both, fiber patch cords between the ORU and the transmission fiber and 1510 nm Service Channel facilities. Furthermore devices with the smallest coupling losses can be selected in laboratory experiments. For practical application minimum, typical and maximum losses of all components have to be taken into account. Table 2 gives the specified internal losses of the present ORU's.

Internal losses of the 1310 nm signal channel	min	typ	max
Coupling loss of SLA chip to fiber	-	2-3	-
Optical band pass filter (BPF) including connectors	-	3	-
Optical isolators (1 splice per isolator)	0.6	1.2	1.6
Service channel facilities	2.23	2.57	3.62
Terminal connectors (1 splice per connector)	1.2	1.2	1.2

Optical losses of the 1510 nm Service Channel	
from input (A) to receiver (RX)	from transmitter (TX) tou output (B)
• min 1.6 dB	• min 0.8 dB
• max 2 dB	• typ 0.9 dB
	• max 1.1 dB

Table 2. Specification of internal losses of the ORU for both, the signal and the Service Channel. In case of splices and connectors only the maximum specifications were used.

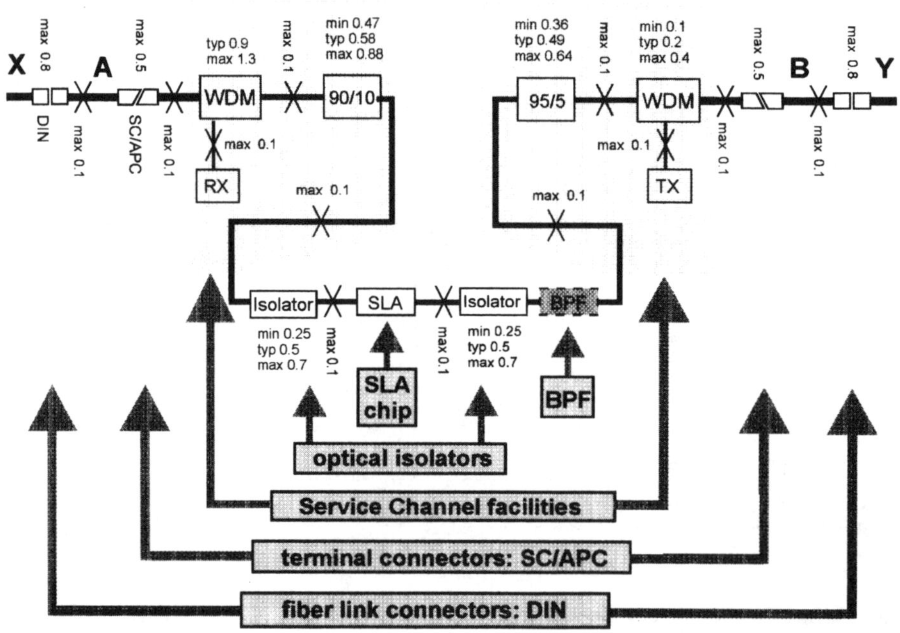

Figure 4. Internal losses of the optical repeater unit

5. Optical System Service. Following ITU-T proposals [3,4] the wavelength for the Service Channel was chosen to be 1510 nm. This Service Channel allows the control and supervision of the remote SOA's located at the in-line ORU's, from the Header PC (see Fig. 1) located at one end of the link (in our case at the CeBIT). In Fig. 5 the link supervision method is indicated. The link is controlled by polling; the Header interrogates each repeater and waits for the answers. The repeaters can also send alarms to the Header even when they are not interrogated.

The SOAC (Fig. 3) monitors the input and output powers of the SOA, drives the Peltier currents, and the Service Channel input and output power. Furthermore the SOAC collects alarms indicating that any of these parameters is out of range. Another task of the SOAC is to change the drive current and output power of the SOA's, when it is required.
The link layer control (LLC, see Fig. 3) organises the data generated at the SOAC, or addressed to it, in a frame. This frame contains information about the origin and destination of the data, its category and allows to detect errors during the transmission. The frame is line coded and transmitted through the Service Channel.
As an example the remote monitoring of the in-line SOA's is illustrated by the repeater status report visible on the screen of the Header PC (Fig. 6). There the optical input and output powers of the SOA's (denoted by SLA's is Fig. 6) and the Service Channel (denoted by CS which stands for control channel) are shown as well as laser and Peltier current.

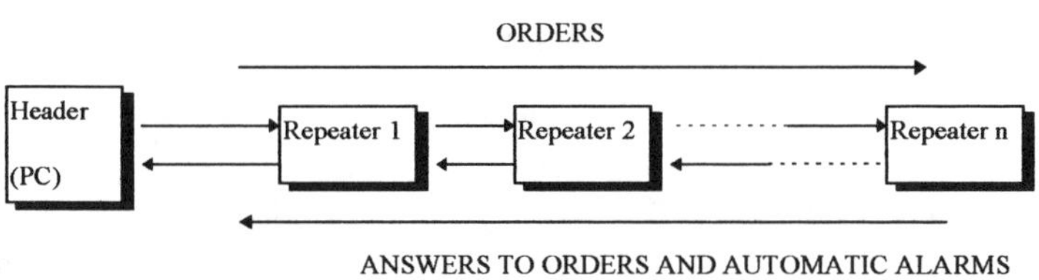

Figure 5. Link supervision method.

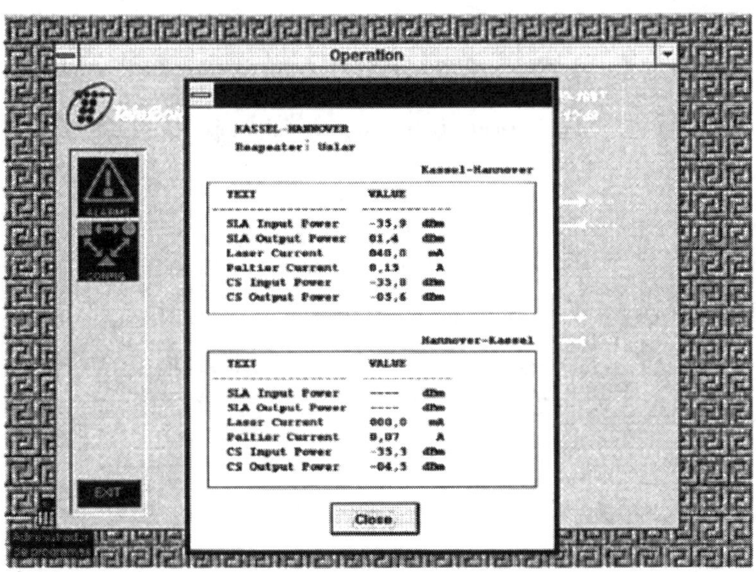

Figure 6. Repeater status report

6. 10 Gbit/s Data Transmission Results. Prior to the CeBIT trial 10 Gbit/s laboratory transmission experiments were performed on a system imitating as realistically as possible the real Kassel-Hannover link. No BER was observed there. On the real link a relatively large PMD of 13 ps appeared at the Bodenburg-Hildesheim upstream fiber. Furthermore it turned out that the attenuation of additional five in-line DIN fiber connectors between the Hannover switching station and the CeBIT stand increased the link loss by more than 3 dB. The system design taking into account some margin in the power budget turned out to be very efficient. Even in the presence of several cascaded fiber connectors with plane facets (which also contribute to markable return losses) the BER was kept in the order of 10^{-11}. Furthermore it was also proved that the 1.3 µm signal channel and the Service Channel do not disturb one another under field conditions.

7. Conclusion. In this paper we reported on the first field trial of 10 Gbit/s data transmission in RZ format at 1.3 µm wavelength over 210 km of installed standard monomode fiber. The link included a total of 9 switching stations at irregular intervals ranging from 10 to 37 km. Optical repeater units were installed in 6 of these stations. These units contained polarisation insensitive semiconductor optical amplifiers. Remote service of the amplifiers was accomplished via a Service Channel supported by an optical signal outside of the working band of the data signal, namely at 1510 nm consistent with ITU-T recommendations. Both, the supervision and traffic channels shared the same optical fiber. Furthermore newly developed flexible TM-64 terminals suitable for RZ and NRZ transmission were employed in this experiment.

This work was carried out in the framework of the European ACTS project UPGRADE. One of the fundamental goals of this project is the introduction of RZ optical soliton transmission at 1310 nm within the European standard single mode fibre optic network. This field test is one of the first major stepping stones to the deployment of an 800 km long system between the capital cities of Madrid and Lisbon.

Achknowledgement. This work is conducted within the ACTS (Advanced Communications Technologies and Services) program sponsored by the European Union. The authors are indepted to the assistance of many colleagues helping to make this field test a success.

References.

[1] L.F. Tiemeijer, P.J.A. Thijs, T. van Dongen, J.J.M. Binsma, E.J. Jansen., J. Lightwave Technol., vol .14, pp. 1524-1533, 1996.

[2] L.F. Tiemeijer, P.J.A. Thijs, T. van Dongen, J.J.M. Binsma, E.J. Jansen, P.I. Kuinersma, G.P.J.M. Cuijpers and S. Walcyzk, IEEE Photon. Technol. Lett., vol .7, pp. 1519-1521, 1995.

[3] ITU-T SG15-D561. Proposal for Out-of Erbium band optical supervisory channel

[4] ITU-T Com 15 Geneva 13-24 November 1995, temporary document 47 (4/15), draft recommendation G.mcs.

Core and ATM Networks
D.W. Faulkner and A.L. Harmer (Eds.)
IOS Press 1997 © AKM

Polarisation Effects in Transmission Fibres

A.S. SIDDIQUI
Department of Electronic Systems Engineering
University of Essex, Colchester, Essex, CO4 3SQ, UK

Abstract. This paper summarises recent work on polarisation effects in optical fibres leading to the development of the first fully polarimetric optical time domain reflectometer (POTDR).

1. Introduction

A major development during the last two years has been the deployment of the first *optically amplified* high bit rate (≥ 5 Gb/s) and long haul (TAT 12/13, TPC 5) optical transmission links. A further parallel development over the last few years has been the design, and implementation upto the demonstrator stage, of large scale optically amplified high bit rate fibre communication *networks* [1,2]. Like the long haul point to point links these networks are also to be optically transparent and their end to end size can be several thousand kilometres. In both cases optical transparency (that is, the absence of 3R regenerators, usually less than 100km apart) results in dispersion and non-linear effects accumulating over the entire length of a link. Dispersion shifted (DS) fibres, together with dispersion management schemes, have now essentially overcome the chromatic dispersion problem. These schemes can also keep the penalties due to non-linear effects to within acceptable limits [3]. However, polarisation mode dispersion (PMD) still remains as a significant and now the main dispersion mechanism which limits the maximum attainable bit rate times distance product for these links. This limitation due to PMD applies not just to the new generation of links and networks using DS fibre. They also apply, again as a highly significant limitation now, to upgrading the transmission bit rate over the vast amount of already installed standard step index fibre where the chromatic dispersion at 1.5μm can now be overcome by one or other of a range of dispersion compensation schemes demonstrated over the last few years [3].

Polarisation effects in transmission fibres are therefore an important current topic for experimental and theoretical investigation and in this talk I will attempt to summarise the recent work in this area at Essex.

The initial stimulus for this work was the development of spun fibre to reduce PMD values in fibre cables for use in optically amplified transoceanic links. The main goal of our work was to understand the interaction between the linear and circular birefringences within the fibre and how this leads to the observed PMD behaviour of the fibre, with and without external twist. It will be seen that the role played by fibre twist is a crucial one in determining the PMD behaviour.

The context of this talk will be confined to fibre *transmission* systems. We should note, however, that polarisation effects also lie at the heart of a, presently steadily growing, range of fibre *sensor* devices.

2. Differential Group Delay (DGD), Polarisation Mode Dispersion (PMD) and the Maximum Bit Rate

Using the principal states model [4] for polarisation propagation in fibres, we define the differential group delay (DGD) in a *short* fibre link as the difference between the transit times, across the link, of the two orthogonal principle states of polarisation for the link. A convenient unit for DGD in real fibres is ps/km. In this context a *short* fibre link is defined as one where there is no mode coupling within the link so that the DGD is a direct measure of the (uniform or average) fibre birefringence. For a given fibre the DGD of a short link therefore increases linearly with length as its unit (ps/km) indicates.

If we now progressively increase the length of the link then it is found that, for link lengths greater than a few kilometres, the measured dependence of DGD on link length reduces to being proportional to the *square root* of the link length. This is due to mode coupling coming into play as the link length increases. In this regime of *long* fibre links the differential group delay over the link is referred to as polarisation mode dispersion (PMD), whose value is usually quoted in units of ($ps/km^{\frac{1}{2}}$).

Now because mode coupling is a random process it is not possible to write down a general deterministic relationship between the DGD and PMD values of a fibre. However, for real fibres a rough rule of thumb appears to be that a long link, formed from a fibre with a DGD of say 1ps/km, is likely to have a PMD lying in the range (0.2 → 0.5) $ps/km^{\frac{1}{2}}$.

Turning now to the limit due to PMD on the maximum practicable bit rate, we note that the design figure taken for the TAT12 link was that the *total* PMD time over the whole link had to be not greater than 7% of the bit period. TAT12 was designed to allow a maximum PMD of 0.15$ps/km^{\frac{1}{2}}$, which would result in a total PMD of 12ps over its 6400 km length. This in turn leads to a maximum practicable bit rate of 5.8 Gb/s which is comfortably above the target figure of 5 Gb/s.

3. The Effect of Fibre Twist on DGD

In [5] the first analytic solution was presented for the DGD of uniform linearly birefringent fibres subject to an external elastic twist. Figure 1 shows the birefringence and dispersion vectors. β_L is the magnitude of the linear birefringence of the untwisted fibre which is here assumed to have an external twist of γ rad/m imposed on it. $\beta_c = g\gamma$, is the twist-induced circular birefringene, where g is the rotation coefficient. Figure 2 which shows the measured and calculated values of DGD against fibre twist also shows that, for a given fibre, there exists a particular twist rate which nulls the DGD due to the linear birefringence in the fibre and that at higher twist rates the DGD is dominated by the twist-induced circular birefringence. These results also elucidate how the DGD, and therefore the PMD, value is lowered in spun fibres.

4. Polarisation OTDR Measurements on Fibres with Twist

In [6,7] an analytical solution is presented for the back-scattered state of polarisation (SOP) from a uniformly linear birefringent fibre subject to elastic twist which shows that the twist significantly influences the periodicity of the observed signal from the POTDR. Therefore values of PMD inferred from the POTDR traces can be significantly in error if even small amounts of twist exists in the fibre. Measurements confirmed these theoretical results (Figure 3) and another, different, set of measurements were made [8] which could be used, by interpreting them using the theory of [6] and [7], to identify fibre sections with and without twist and sections with relatively high and low PMD.

5. Fully Polarimetric OTDR

The polarisation OTDR used in the above measurements, and indicated in Figure 3, consisted of a conventional commercial OTDR instrument with the addition of a polariser (plus a waveplate in some measurements) in front of the detector. It was therefore not a real time, fully polarimetric OTDR arrangement. However, the results obtained with this set up, together with the detailed theoretical analysis carried out in parallel, strongly pointed to a need for a fully polarimetric OTDR set up capable of making real time measurements in the sense of sampling all the Stokes parameters of the back scattered SOP at the same time. Such a

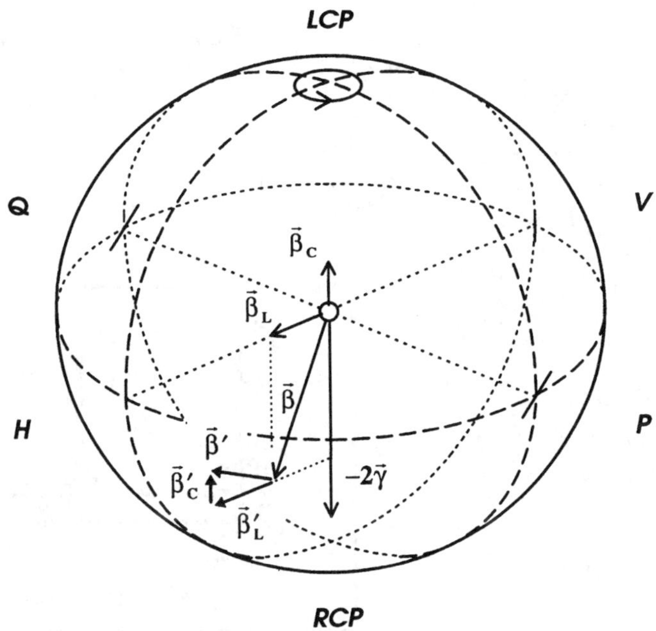

Fig. 1 Birefringence and dispersion vectors in rotating reference frame

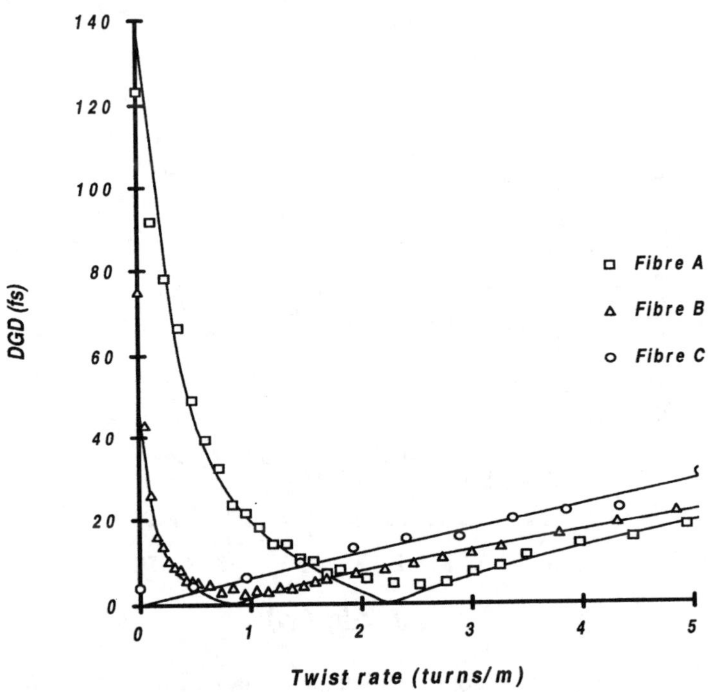

Fig. 2 Measured and calculated values of DGD against twist

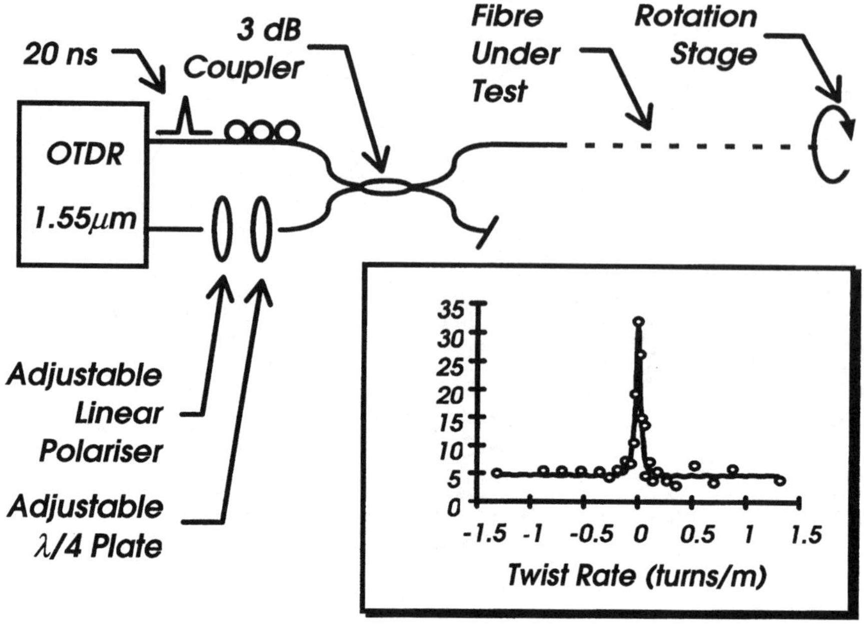

Fig. 3a Experimental POTDR setup. Inset shows experimental and simulated DGD versus twist results for light reflected from far end of the fibre

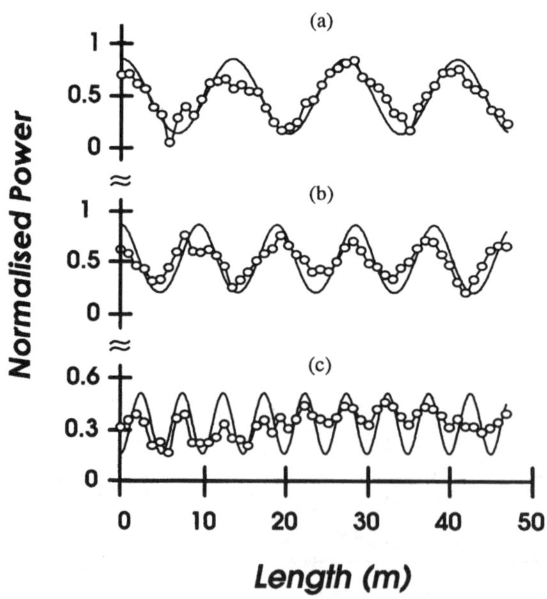

Fig. 3b Normalised output power of POTDR trace versus length, fibre twisted at far end. (a) $\gamma = 0$ turns/m, (b) $\gamma = 0.057$ turns/m, (c) $\gamma = 0.104$ turns/m. The curves are best fit simulations with β_L=0.23 rad/m and input SOP (cos(π/4), 0, sin((π/4)).

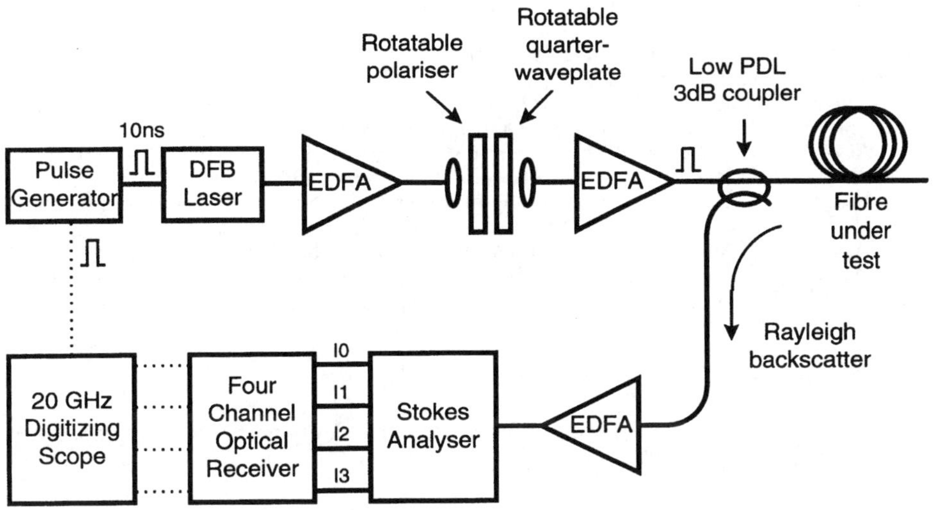

Fig. 4a Schematic diagram of Polarimetric OTDR

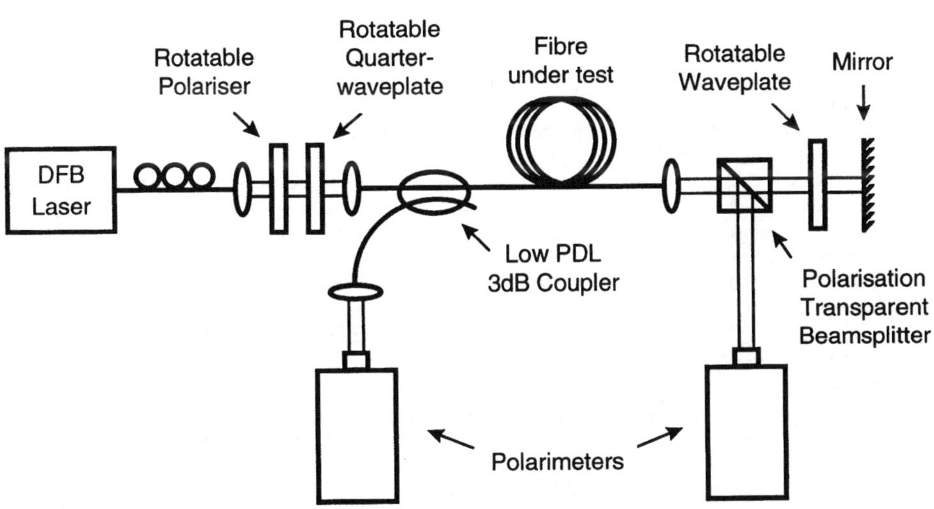

Fig. 4b Forward and return matrix measurement set-up

fully polarimetric optical time domain reflector (POTDR) has been developed recently [9]. Figure 4 shows schematic diagrams of the POTDR set up as well as that of a forward and backward SOP propagation measurement set up.

The POTDR has a spatial beat length resolution of 1 metre with a dynamic range of 30dB. It can therefore measure linear birefringences in fibres too low to be measured, in equivalent terms, by currently available PMD measurement instruments.

At present we foresee two significant applications for the POTDR. The first is in the area of fibre manufacture where the effect of fine tuning the fibre fabrication process in different ways, on the DGD of the resulting fibres, could be determined by examining sample lengths of fibre using the POTDR. The second application could be to examine the birefringence of different sections of an installed standard fibre link to identify sections with abnormally high PMD values. By selectivity replacing these sections only with say DS fibre the overall PMD performance of the whole link could be significantly improved.

6. Conclusion

A summary has been presented of work leading upto the development of a fully polarimetric OTDR and significant potential applications for the POTDR have been indicated.

Acknowledgements

This work was carried out under an EPSRC research grant, Reference GR/K49454. The industrial partners are BT Laboratories, Martlesham Heath and Alcatel Submarine Networks, Greenwich.

At Essex the other workers involved are D.H.O. Bebbington, J.G. Ellison, R.E. Schuh, X. Shan and S.D. Walker. At BT Labs the collaborating workers are L.M. Gleeson, E.S.R. Sikora and N.G. Walker.

References

[1] M.J. O'Mahony et al., The design of a European optical network, *IEEE Journal of Lightwave Technology*, **13** (1995) 817-829.

[2] G. Hill, Optical networks : From myth to reality. In : G. Prati (ed.)., Photonic Networks. ISBN 3 540-76143-8. Springer-Verlag, Berlin, 1997.

[3] R. Tkach, Dispersion compensation in nonlinear systems, *Conference on Optical Fibre Communications*, OFC '96, San Jose, CA, (1996), Tutorial WO, 131-161.

[4] C.D. Poole et al., Phenomenological approach to polarisation dispersion in long single mode fibres, *Electronics Letters*, **22**, (1986) 1029-1030.

[5] R.E. Schuh et al., Theoretical analysis and measurement of the effects of fibre twist on the DGD of optical fibres, *Electronic Letters*, **31** (1995) 1772-73.

[6] R.E. Schuh et al., Theoretical analysis and measurement of the effect of fibre twist on the polarisation OTDR of optical fibres, *Conference on Optical Fibre Communications*, OFC '96, San Jose, California, (1996) paper FA5.

[7] R.E. Schuh et al., Polarisation OTDR measurements and theoretical analysis on fibres with twist and their implications for estimation of PMD, *Electronic Letters*, **32** (1996) 387-388.

[8] R.E. Schuh and A.S. Siddiqui, Measurement of SOP evolution along a linear birefringence fibre with twist using polarisation OTDR, NIST/IEEE/OSA *Symposium on Optical Fibre Measurements*, Boulder, CO, (1996) 159-162.

[9] D.H.O. Bebbington et al., Fully polarimetric optical time domain reflectometer with one metre spatial resolution, *Conference on Optical Fibre Communications*, OFC '97, Dallas, Texas, (1997) paper WL24.

Core and ATM Networks
D.W. Faulkner and A.L. Harmer (Eds.)
IOS Press 1997 © AKM

Characterization of 11000 km of installed standard monomode fibre and statistical analysis in view of network design

H. Gruhl, G. Herchenröder, A. Mattheus, J.Vobian,

Deutsche Telekom AG, TZ, PO Box 64276, D-64295 Darmstadt, Germany

Abstract. The attenuation, chromatic despersion and polarisation mode dispersion of 350 installed standard monomde fibers - with more 11000 km length in total - between the two cities Darmstadt and Hannover have been measured and analyzed in the second and third optical window (around 1.3 μm and 1.55 μm), respectively. At 1310 nm the attenuation and chromatic dispersion had a mean value and variance of 0.37 ± 0.02 dB/km and -0.14 ± 0.26 nm, while we had 0.23 ± 0.02 dB/km and 16.66 ± 0.30 nm for 1550 nm wavelength. The PMD is less than $0{,}5\ \mathrm{ps}/\sqrt{\mathrm{km}}$ in average and up to several $\mathrm{ps}/\sqrt{\mathrm{km}}$ for fibers in old cables while the PMD of new fibers is usually smaller than $0{,}3\ \mathrm{ps}/\sqrt{\mathrm{km}}$.

This set of measured fiber parameters serves as a basis for general all-optical network design.

1. Introduction. In the last few years all-optical network concepts have been investigated in detail world-wide. Laboratory record transmission experiments on standard monomode fibers (SMF) have been performed, such as 10 Gbit/s over 2245 km SMF [1], 40 Gbit/s over 150 km SMF with no in-line amplifier [2], 10 x 20 Gbit/s WDM over 1000 km SMF with 100 km amplifier spacing [3], and others. However, the deployment of operational high capacity systems in real networks has - by far - not yet reached that high technological level. A few 10 Gbit/s SONET terminals and some more N x 2.5 Gbit/s-WDM equipment (with the number of WDM channels starting from N=4 and aiming at N=32 very soon) have been implemented in Japan and in the USA, respectively. In general, most network providers still rely on backbones up to 2.5 Gbit/s data rate. One one hand this is due to commercial reasons regarding data traffic, fiber network infrastructure and prices of very advanced systems. One the other hand there are several technical issues related to ≥10 Gbit/s systems, such as OAM and the physical system behaviour. While laboratory experiments can be optimized with respect to specific conditions operational systems have to cope with the statistical distribution of fiber parameters in the network under consideration.

In this article we report on the measured attenuation, chromatic dispersion (CD) polarisation mode dispersion (PMD) of 350 installed fibers with 11000 km of total length between the two German cities Darmstadt and Hannover. A statistical analysis of the data is presented as well as considerations on advanced optical network design.

2. Fiber Base under Investigation. The parameters of installed fibers were measured along two links, the first one from Darmstadt via Frankfurt and Kassel up to Hannover, and the second one from Frankfurt via Würzburg up to Seesen, both links comprising 33 switching stations in total. The total length of all sections, the mean value and the variance were 922.68 km, 28.83 km and 8.73 km, respectively. In average 12 fibers per station were investigated thus getting a data set for more than 11'000 km of installed SMF.

3. Fiber Attenuation. Fiber attenuation was measured using an OTDR. The main result is shown in Fig. 1. There the number of fibers is plottet versus the attenuation in intervals of 0.01 dB/km for two wavelengths, namely λ=1310 nm and λ=1550 nm. The peaks of the corresponding distributions are located at 0.37 dB/km and 0.23 dB/km. At λ=1310 nm the attenuation of *all* fibers can be found within an interval of 0.1 dB/km in contrast to λ=1550 nm where *most* fibers are within the interval 0.2 - 0.3 dB/km. Along a few links, however, some older fibers exhibit an attenuation between 0.3 dB/km and 0.5 dB/km at this wavelength. The mean value and variance was 0.37 ± 0.02 dB/km for λ=1310 nm while we found 0.24 ± 0.04 dB/km for the other wavelength including all fibers, and 0.23 ± 0.02 dB/km excluding those fibers with more than 0.3 dB/km at 1550 nm.

4. Chromatic Dispersion. Next the chromatic dispersion (CD) of the fiber base was determined by the differential group delay technique, i.e. pulses at different wavelengths were launched into each fiber to measure the group delay time τ versus wavelength. The derivative $d\tau/d\lambda$ is proportional to the CD. In Fig. 2 the distribution of the zero dispersion wavelength λ_0 is plottet versus wavelength for all fibers (see bars in Fig. 2) and for fibers in

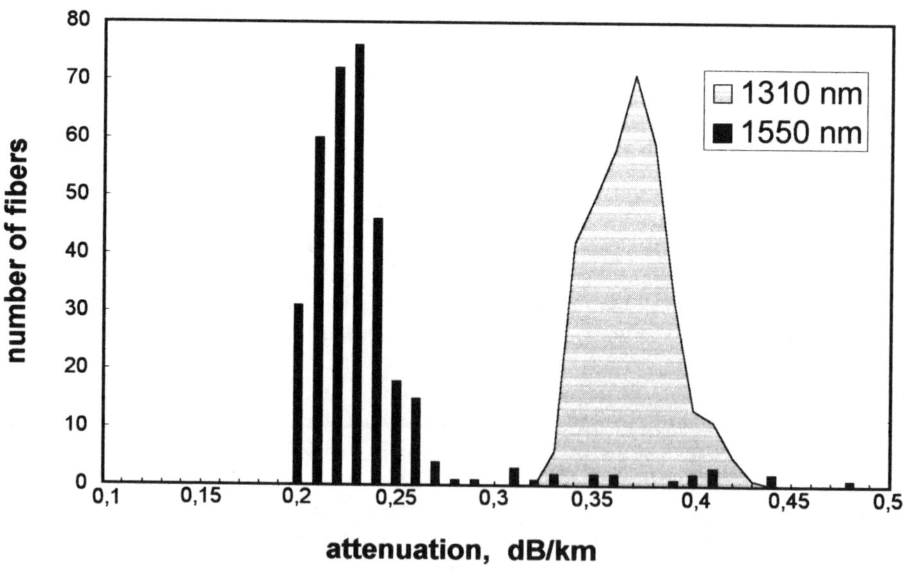

Figure 1. Distribution of fiber attenuation of all measured fibers at 1310 nm and 1550 nm, respectively.

some particular cables (such as cable1-4 in Fig. 2). Fibers within one particular cable (say for example cable 4) exhibit a rather small variance of about $\Delta\lambda_0$=1 nm. However, the zero dispersion wavelength of *all* fibers between Darmstadt and Hannover is scattered over the wavelength range from 1304 nm up to 1318 nm with a mean value of $\overline{\lambda_0} = 1311{,}6$ nm and width of $2 \cdot \overline{\Delta\lambda_0} = 5{,}9$ nm. This histogram can be understood considering that different cables from different manufacturers have been deployed for more than 15 years, and the mean zero dispersion wavelength of the fibers is different in each cable.

Besides the zero dispersion wavelength the distribution of the chromatic dispersion D at characteristic wavelengths λ, in each optical window is of interest. In Fig. 3 the histogram of D(λ) is presented for λ=1310 nm and λ=1550 nm, respectively. The mean values $\overline{D}$, variances $\overline{\Delta D}$, minimum and maximum chromatic dispersion D_{min} and D_{max} are summarized in Table 1 for both wavelengths. As can be seen each of both distributions is limited to an interval of $D_{max} - D_{min} \leq 1.4\ \mathrm{ps}/(\mathrm{nm}\cdot\mathrm{km})$, and the corresponding variances are as small as $\overline{\Delta D} \leq 0.3\ \mathrm{ps}/(\mathrm{nm}\cdot\mathrm{km})$.

wavelength	average disp. $\overline{D}$ $\frac{\mathrm{ps}}{(\mathrm{nm}\cdot\mathrm{km})}$	variance $\overline{\Delta D}$ $\frac{\mathrm{ps}}{(\mathrm{nm}\cdot\mathrm{km})}$	min. disp. D_{min} $\frac{\mathrm{ps}}{(\mathrm{nm}\cdot\mathrm{km})}$	max. disp. D_{max} $\frac{\mathrm{ps}}{(\mathrm{nm}\cdot\mathrm{km})}$
1310 nm	-0.14	0.26	-0.68	0.51
1550 nm	16.66	0.30	15.96	17.36

Table 1. Average chromatic dispersion, variance and minimum and maximum chromatic dispersion of all measured fibers at 1310 nm and 1510 nm wavelength, respectively.

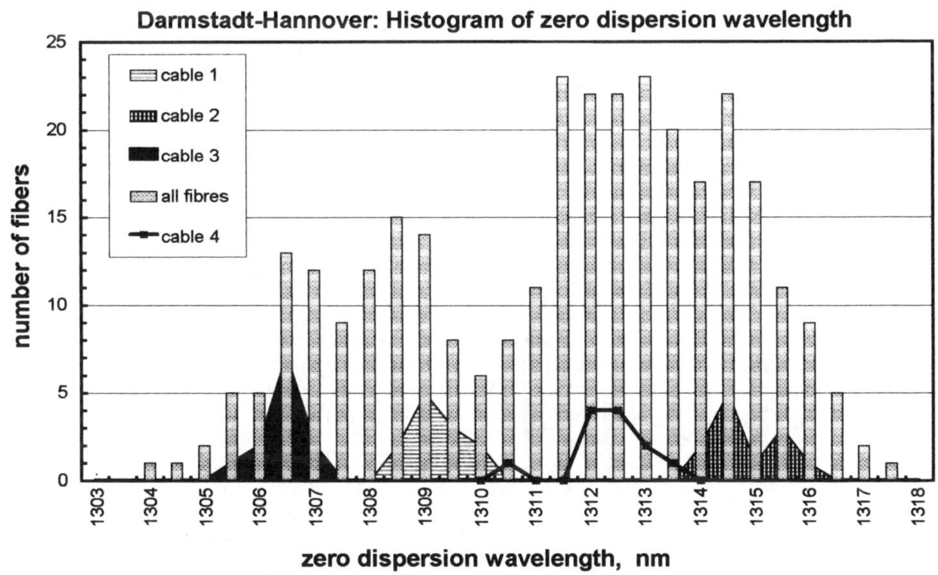

Figure 2. Histogram of zero dispersion wavelength λ_0 for both, all fibers measured between Darmstadt and Hannover, and fibers in selected cables between switching switching stations.

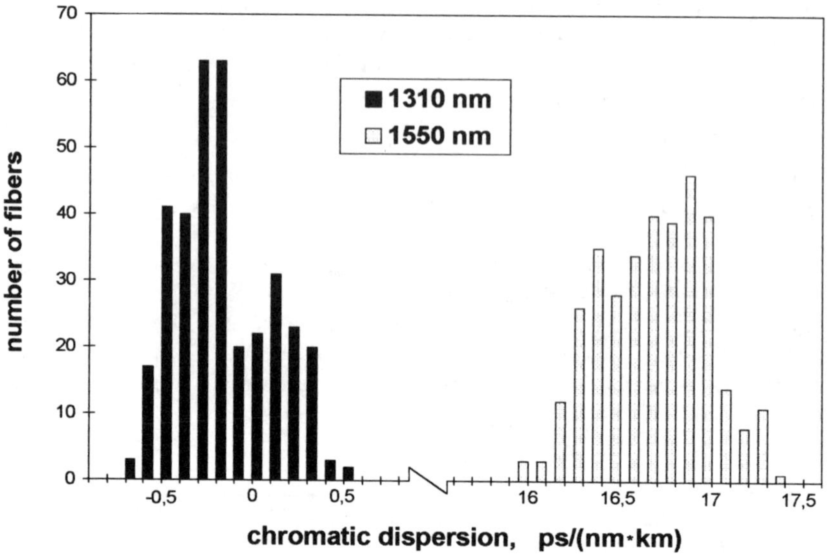

Figure 3. Histogram of chromatic dispersion of all fibers measured at the two wavelengths 1310 nm and 1550 nm, respectively.

The chromatic dispersion at a particular wavelength λ in the vicinity of the zero dispersion wavelength λ_0 is approximately given by $D(\lambda) = (\lambda - \lambda_0) \cdot S$ with the slope S. It turned out that the slope is almost constant in the order of $0.85\ ps/(km \cdot nm^2)$. Thus the large variation of λ_0 (see Fig. 2)is linearly transformed to the variation of the chromatic dispersion in the zero dispersion regime. This phenomenon is illustrated in Fig. 4. An optical signal at a particular wavelength in the 1.3 µm-window (say for example at 1310 nm) experiences a fairly different chromatic dispersion at different fibers.

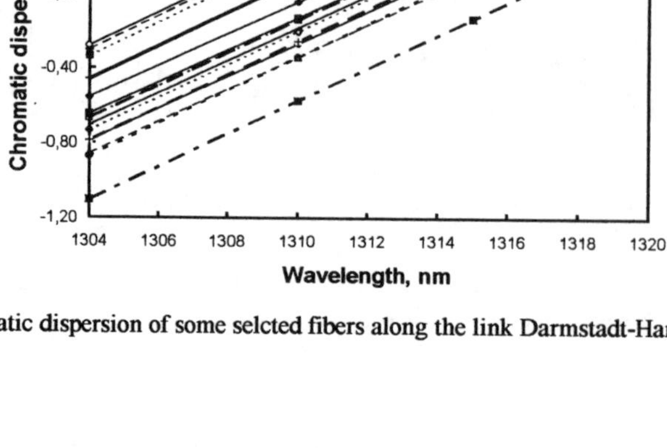

Figure 4. Chromatic dispersion of some selcted fibers along the link Darmstadt-Hannover

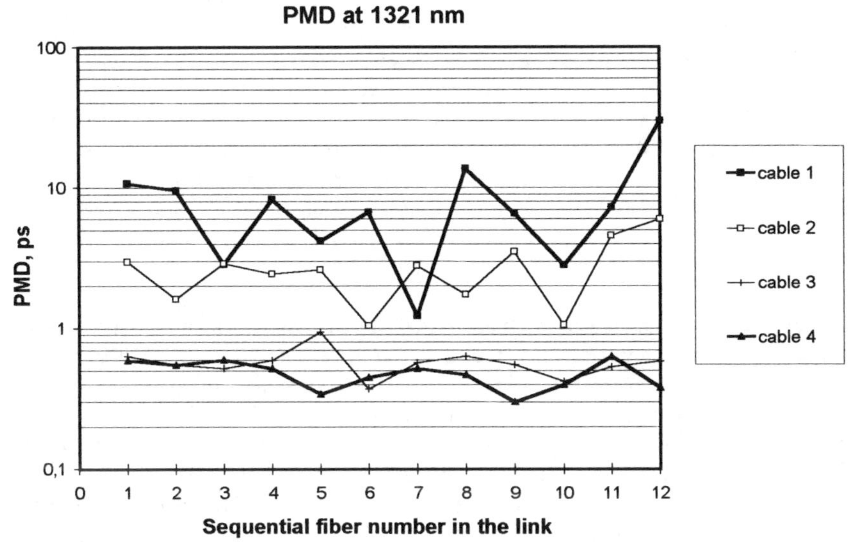

Figure 5. Polarisation mode dispersion (at 1321 nm) of 48 fibers (12 fibers in 4 different cables).

5. Polarisation Mode Dispersion. Polarisation mode dispersion (PMD) measurements were also performed in time domain by an interferometric method [4]. Some characteristic data, measured at λ=1321 nm, are presented in Fig. 5. The averaged PMD (at 1321 nm) of 173 fibers was found to be $0{,}47 \pm 0{,}45 \text{ ps}/\sqrt{\text{km}}$. However, a systematic distribution of the PMD can be observed. This phenomenon is illustrated in Fig. 5. On one hand there are cables where the PMD of all fibers is large, in particular in cable 1, where one fiber exhibits a PMD of 30 ps over 22,4 km SMF (which corresponds to $6{,}3 \text{ ps}/\sqrt{\text{km}}$). On the other hand the PMD is very small for all fibers of other cables (for example in cable 4). The reason for this observation is still under current investigation. However, a correlation can be found between the average PMD of the fibers in a cable, the year of its production and (in some cases) the manufacturer of the fibers, respectively. New SMFs usually exhibit a fairly small PMD (below $0{,}3 \text{ ps}/\sqrt{\text{km}}$) due to improved manufacturing processes avoiding any asymmetry (refractive index, stress, etc.) of the fiber core. It should be noted that the origin and distribution of the PMD in both, fibers on bobbins and fibers in installed cables, is still subject of current research.

6. Aspects of Network Design. The set of measured fiber parameters is a sufficient basis for the design of advanced optical transmission systems. The operation of advanced systems, such as 10 Gbit/s terminals or WDM systems, has to be specified regarding additional fiber parameters besides attenuation, such as fiber chromatic dispersion and PMD.

There is a significant difference whether a transmission system shall be optimized for one particular link or if it shall work at any arbitrary link in the network. The latter approach is of vital interest for a network operator. One has to take the existing installed terrestrial fiber base as it is. System specification must cope with the statistical distribution of all fiber parameters - attenuation, λ_0, CD, PMD - and considerations on the impact of fiber nonlinearities. This task appears to be complicated at first sight. However, the moderate scattering of the attenuation figures (with the exception of very few old fibers) and the CD will most likely not cause fundamental problems for the design of advanced optical networks. It should be noted that the fiber attenuation may gradually increase during the lifetime of a fiber partially due to additional splices - including repair of fiber cuts -, connectors, etc.. PMD might be a more serious problem and is still subject of current research.

7. Conclusion. In this article more than 11000 km of installed standard monomode fibers between the two German cities Darmstadt and Hannover have been characterized in the two optical windows of interest for telecommunication, i.e. around 1.3 µm and 1.55 µm, respectively. The attenuation and chromatic dispersion was found to be in the range of $\alpha = 0.37 \pm 0.02\ \mathrm{dB/km}$ and $D = -0.14 \pm 0.26\ \mathrm{ps/(nm \cdot km)}$ for λ=1310 nm. At 1550 nm we had $\alpha = 0.24 \pm 0.04\ \mathrm{dB/km}$ for all fibers and $\alpha = 0.23 \pm 0.02\ \mathrm{dB/km}$ if a few very old fibers above 0.3 dB/km attenuation were excluded. The chromatic dispersion is as large as $D = 16.66 \pm 0.30\ \mathrm{ps/(nm \cdot km)}$ at 1550 nm.

The zero dispersion wavelength λ_0 may range from 1304 nm up to 1318 nm. However, the variance of λ_0 of the fibers within one cable is much smaller, namely in the order of 1 nm. Thus the zero dispersion wavelength is one characteristic feature of the fibers in a particular cable. Another one is PMD. Fibers in old cables show a rather large PMD of several $\mathrm{ps}/\sqrt{\mathrm{km}}$ in contrast to less than $0.3\ \mathrm{ps}/\sqrt{\mathrm{km}}$ PMD of fibers in all new cables.

Finally the fiber data were briefly evaluated in view of the design of advanced optical communication systems on the installed fiber base. It is evident that existence of PMD and its impact on network design is one of the most important issues.

8. Acknowledgement. This work was carried out in the framework of the European ACTS project UPGRADE. The authors gratefully acknowledge the assistance of the colleagues at the various switching stations.

References.

[1] N. Kikuchi, S. Sasaki, and K. Sekin, Electron. Lett., **31**, pp. 375-377, 1995.
[2] R. Ludwig, W. Pieper, H.G. Weber, D. Breuer, K. Petermann, F. Küppers, A. Mattheus, Electron. Lett., **33**, no. 1, pp. 76-77, 1997.
[3] D. Le Guen, F. Favre, M.L. Moulinard, M. Henry, G. Michaud, L. Macé, F. Devaux, B. Charbonnier and T. Georges, Proc. OFC'97, PD17, Feb. 16-21, Dallas, USA.
[4] J. Vobian, K. Mörl, W. Keilig, Proc. EFOC&N'94, pp. 174-179

Core and ATM Networks
D.W. Faulkner and A.L. Harmer (Eds.)
IOS Press 1997 © *AKM*

Equalization of Bit Distortion Induced by Polarization Mode Dispersion

Henning BÜLOW
Alcatel Telecom Research Division
Dept. ZFZ/NO, Alcatel SEL AG, D-70430 Stuttgart, Germany

Abstract. Electronic polarization mode dispersion (PMD) equalizers as well as a new opto-electonic equalizer are investigated theoretically. All concepts allow to track fluctuating signal distortion due to first order PMD of the fiber link by pure electronic adaptation. However, electronic equalizers not always exhibit low penalties for all possible PMD distortions. The operation of the new opto-elctronic equalizer which is proposed here, is demonstrated with experimentally generated 10 Gbit/s data patterns distorted by PMD.

1. Introduction

The evaluation of ultra-high speed optical transmission systems covering bitrates of 10 Gbit/s and beyond has to face limitations posed by physical properties of the transmission fiber. The limitation given by fiber attenuation and chromatic dispersion are overcome by erbium doped fiber amplifier (EDFA) and dispersion shifted fiber (DSF) or dispersion compensation techniques (DCF). However, signal distortion due to polarization mode dispersion (PMD) of the fiber remain as a major limiting effect for fiber length and bitrate. PMD has the effect of a birefringence which leads, in first order, to dual path propagation and thus to signal distortion. In this paper, electronic PMD equalizers are analyzed with respect to power penalty. Then a new concept of an opto-electronic equalizer is presented which performs the adaptation of the filter parameters also electronically, a theoretical model to describe the operation and the adaptation is derived, and the operation demonstrated using an experimentally generated and distorted 10 Gbit/s data pattern.

2. Signal distortion due to PMD

As it is depicted by Fig. 1 PMD can be regarded as a fiber birefringence with a slow and a fast fiber axis with the principal states of polarisation (PSP) denoted by the corresponding Stokes vectors on the Poincaré sphere $\vec{e}_+$ and $\vec{e}_-$, respectively, and a differential group delay (DGD) $\Delta\tau$ [1]. The signal power splitting between both axis is denoted by the relative powers of γ and $1-\gamma$ transported on the fast and slow PSP, respectively. Fig. 1 illustrates that the DGD between the two optical paths is leading to inter symbol interference (ISI) and thus to signal distortion. The first order PMD distortion is thus quantified by a set of the two parameters $\Delta\tau$ and γ [2]. Due to the statistic nature of fiber PMD this property is commonly

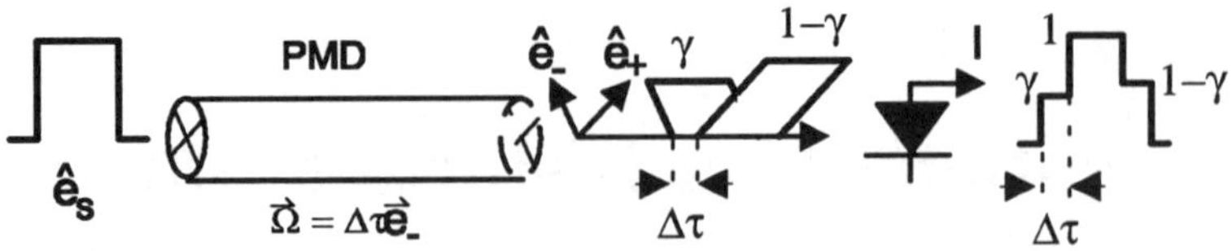

Fig. 1 PMD induced signal distortion and receiver diode signal.

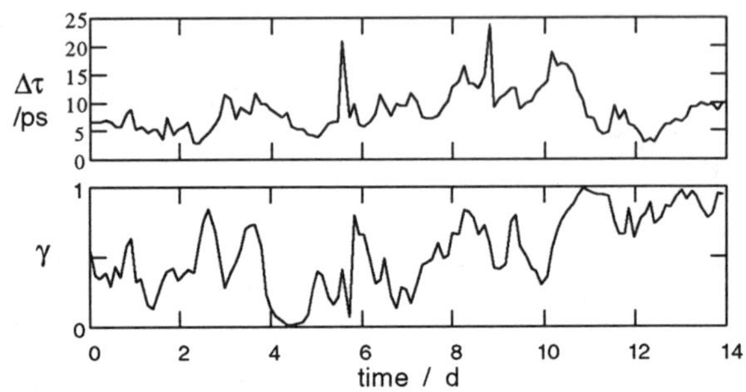

Fig. 2 Temporal variation of DGD $\Delta\tau$ (ps) and fast axis power γ of an installed 246 km field fiber link [6]

quantified by the mean DGD $<\Delta\tau>$ or the length related PMD coefficient $\tau' = <\Delta\tau>/\sqrt{L}$ (L denotes the fiber length). Negligible degradation probability is assumed for a PMD (mean differential group delay DGD) of less than about 10 per cent of the bit period T [3]. Assuming typical PMD coefficients of installed fiber links of $\tau' = 0.1...0.5$ ps/$\sqrt{\text{km}}$ [4], the maximum tolerable link length may exceed 400 km for 10 Gbit/s transmission whereas it can shrink to less than 25 km for 40 Gbit/s.

However, the PMD which is seen by a signal transported on an installed fiber is not constant but it fluctuates due to environmental temperature changes [5]. A typical temporal variation of the two PMD distortion parameters $\Delta\tau$ and γ, which have been measured for a 246 km standard field fiber over a time span of two weeks, is plotted in Fig. 2 [6]. It is clearly visible that significant changes of the PMD take place in the time scale of a few hours. Thus the speed requirements of the distortion tracking electronics of an equalizer seems to be not very high.

3. PMD compensation and equalization

To overcome the link length and bitrate limit posed by fiber PMD, several PMD compensation schemes have been proposed and discussed in the literature. Some of them are listed in Tab. 1. The most straight forward approaches are the optical PMD compensators, which either avoid the dual path propagation by adjusting the transmitter signal state of polarization (SOP) to one PSP by an adaptively tuned polarization controller (scheme 1) [7], or insert an adaptively tuned birefringence opposite to the fiber PMD into the optical path (scheme 2) [8,9]. The adjustable birefringence is built by concatenated polarization controllers (PC) and polarization maintaining fibers (PMF). An opto-electronic scheme (3) overcomes the bottleneck of a variable optical delay line by splitting the slow and fast axis signal with help of a polarization beam splitter (PBS) followed by a conversion of the two undistorted signals into the electrical domain to phase-adjust them with an electrical delay line [10].

<table>
<tr><td>Optical compen-sation</td><td>1.
Signal adaptively launched into one PSP</td><td>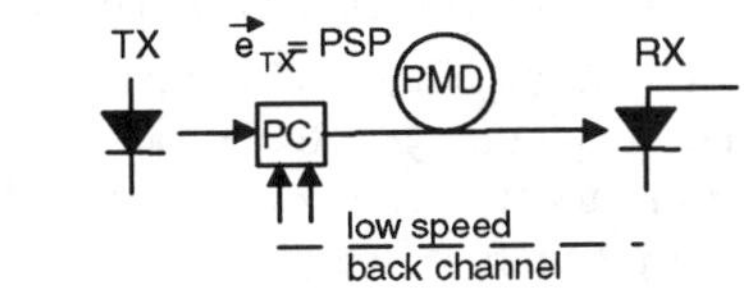
</td><td>[7]</td></tr>
<tr><td></td><td>2.
Adaptive generation of opposite birefringence</td><td>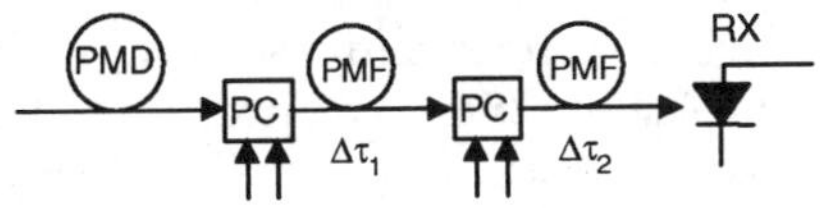
</td><td>[8,9]</td></tr>
<tr><td>Opto-electronic equalization</td><td>3.
Polarization diversity receiver with adaptive optics and electronics</td><td>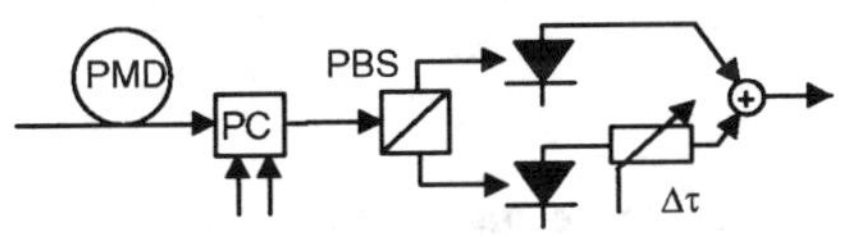
</td><td>[10],
[11]</td></tr>
<tr><td></td><td>4.
Polarization analyzing equalizer (PAE),
fixed optics and adaptive electronics</td><td>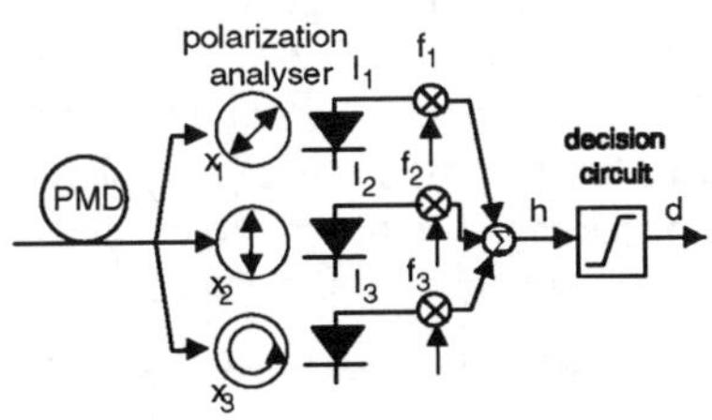
</td><td></td></tr>
<tr><td>Electronic equalization</td><td>5.
Tapped delay line equalizer (linear equalizer, LE)</td><td>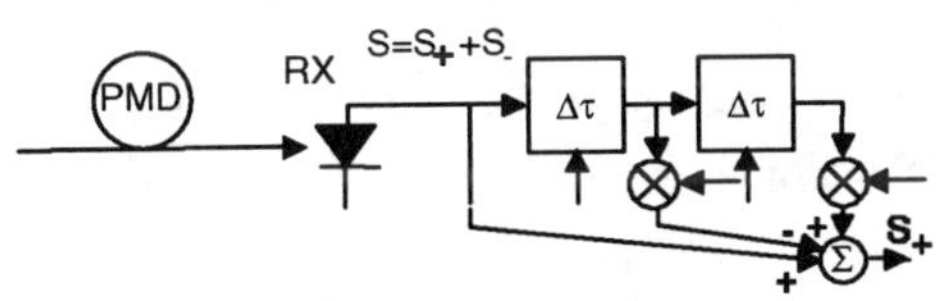
</td><td>[12],
[13]</td></tr>
<tr><td></td><td>6.
Decision feedback equalizer (DFE)</td><td>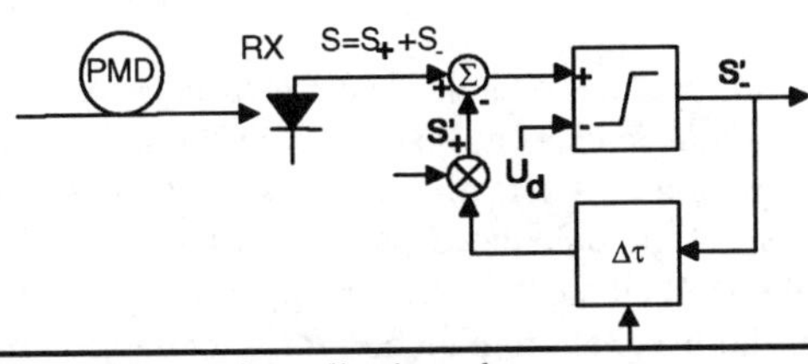
</td><td>[12],
[14],
[15]</td></tr>
</table>

Tab. 1 PMD compensation and equalization schemes

So far, all of the equalizer concepts make use of adaptively tuned polarization controllers which track the fluctuating PSP of the transmission fiber. Here, we propose an opto-electronic PMD equalizer, shown in Tab. 1 as scheme 4, which replaces the "active" optics by branching the distorted signal at the receiver into three parts and directing them through three polarizers onto three photo diodes. Each of the polarizers extract a different signal polarization. The receiver diode signals I_1, I_2, and I_3 are adaptively weighted by weighting factors f_1, f_2, and f_3, then superimposed and finally decided by a decision gate to recover the original transmitted bit stream. The advantage of this Polarization Analyzing Equalizer (PAE) is that no adaptation has to be performed in the optical path but only variable processing of the electrical signals is required, which can be integrated into the receiver and distortion tracking electronics. This PAE concept will be discussed in chapter 5 and 6 in detail.

A third family of PMD equalizers proposed in the literature are the electronic equalizers. They recover the data from the distorted signal detected by the receiver photo diode [12]. Beside tapped delay line filters, referred to as linear equalizer LE (scheme 5) [12,13], also nonlinear decision feedback equalizers (DFE, scheme 6) have been discussed in the literature [12,14,15]. These two concepts also process the signal in the electrical domain and, in contrast to the PAE, they require only one receiver diode. They will be discussed in the next chapter.

4. Penalty of electronic equalizers

The most common linear equalizer (LE) is a FIR filter in the form of a tapped delay line filter as shown in scheme 5 of Tab. 1. PMD distortion equalization has already been demonstrated experimentally [13]. In order to obtain a rough estimate of the power penalty on a LE matched to a PMD distortions ($\Delta\tau,\gamma$), we derive the transfer function of a realization based on variable tap spacing. The photo diode signal s(t) generated by the distorted optical signal can be written as a superposition of two signals s_- and s_+ which are undistorted data signals d(t) with a mutual group delay of $\Delta\tau$ and modified amplitudes γ and $1-\gamma$, respectively:

$$s(t) = s_-(t) + s_+(t) \quad \text{with} \quad s_-(t) = \gamma\, d(t+\Delta\tau/2) \quad \text{and} \quad s_+(t) = (1-\gamma)\, d(t-\Delta\tau/2) \tag{1}$$

In eq. 1 s_- (or s_+) can consecutively be replace by a delayed s_+ (or s_-) with modified amplitude, yielding the weighting factors $a_\pm^i$ and delays $i\Delta\tau$ shown in Tab. 1:

$$s_\pm(t) = \sum a_\pm^i\, s(t \mp i\Delta\tau) \quad \text{with} \quad a_\pm = \begin{cases} -\dfrac{\gamma}{1-\gamma} \\ -\dfrac{1-\gamma}{\gamma} \end{cases} \quad \text{and} \quad i = 0....N\text{-}1 \tag{2}$$

In this equation N denotes the number of taps. Assuming a distorted "1010..." input pattern and a signal independent noise density at the LE input, it is straight forward to calculate eye diagram opening and effective noise amplitude at the LE output and thus signal-to-noise ratio. In Fig. 3 the calculated penalties of a LE with N = 3 taps are drawn over the plane span by the PMD distortion $\Delta\tau$ and γ. $\Delta\tau$ is related to the bit period T. In the upper half of the figure the weights are matched for s_- extraction whereas the lower half is governed by a filter extracting s_+. It is obvious that the LE exhibits "blind spots" for distortions with $\gamma \approx 0.5$.

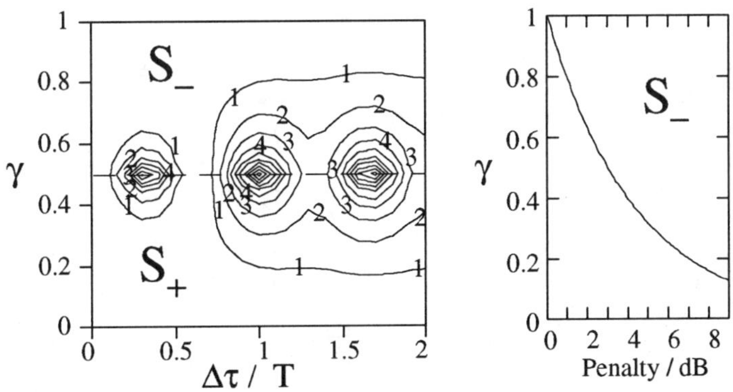

a) Penalty of LE (3 taps) over plane of PMD distortion (Δτ,γ)   b) Penalty of DFE (on x-axis) vs PMD distortion (γ on y-axis)

Fig. 3 Power penalty (in dB) of LE and DFE

The second electronic equalizer shown in Tab. 1 is the decision feedback equalizer (DFE). Experimental verification of PMD compensation as well as concepts for high-speed realization have been discussed in the literature [14,15]. Again, feedback loop weighting factor and feedback delay can be derived from eq. 1. Under the assumption that the fast axis signal is correctly decided at the decision circuit output yielding s_-', it is possible to generate the slow axis signal s_+' with help of s_-':

$$s_+'(t) = \frac{1-\gamma}{\gamma} s_-'(t - \Delta\tau) \quad . \tag{3}$$

This operation is performed by the loop elements (delay plus weight). Regarding s_+ as an echo within the distorted signal s, we cancel the echo by subtracting s_+' from s at the decision circuit input. Since the fast axis signal selected by the DFE has the amplitude γ, we can roughly estimate the power penalty by the ratio 1/γ. In Fig. 3 b the penalty (in dB, x-axis) is drawn versus the relative power γ (y-axis). In comparison to the penalties of the LE shown on the left hand side of the figure, the DFE exhibits a moderate penalty of approximately 3 dB for distortions at which the LE exhibits blind spots. However, with decreasing power in the fast axis the penalty of the DFE strongly increases.

5. Polarization analyzing equalizer (PAE)

A possible way to keep the advantage of equalizer adaptation in the electronic domain and to avoid penalty poles at specific PMD distortions, is the PAE. In order to analyze the operation principle of the PAE and to elaborate a strategy to optimize the weighting factors f_i we recall the basic setup shown as scheme 4 in Tab. 1. The three pass states-of-polarization (SOP) are denoted by the three unity Stokes vectors $\vec{x}_1$, $\vec{x}_2$, and $\vec{x}_3$. They should form a triple of orthogonal vectors, e.g. linear (0°) SOP, linear (45°) SOP, and left circular SOP. For the analysis we will write the three diode currents and the weighting factors as vectors: $\vec{I} = (I_1, I_2, I_3)$ and $\vec{f} = (f_1, f_2, f_3)$.

Fig. 4 shows the polarization states of the distorted signal (unchirped). Three SOPs appear at the input port of the PAE: SOP_+ and SOP_-, the principal states of polarization (PSP) of

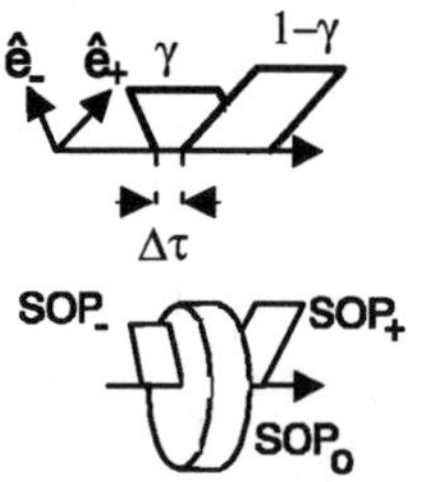

Fig. 4 States of polarization (SOP) at PAE input

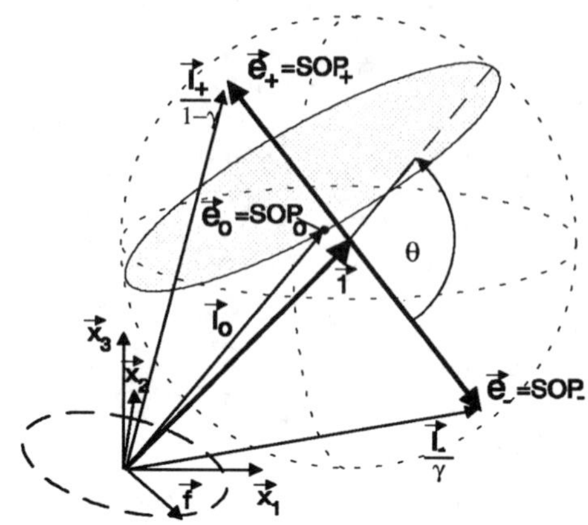

Fig. 5 Diode currents I and weighting factors f for the different input SOPs (+,-,0)

slow and fast fiber axis, and SOP_O, denoted by $\vec{e}_0$, the state of polarization appearing when both axes simultaneously transport a signal. Since the transmission of the polarizer can be calculated by $T = 0.5(1+\vec{x}\,\vec{s})$ where $\vec{s}$ denotes the Stokes vector of the signal SOP [16], we can write the diode signals $\vec{I}_i$ for all possible input SOPs (+,-,0) as follows

$$\vec{I}_i = P_i \begin{pmatrix} 1+\vec{x}_1\vec{e}_i \\ 1+\vec{x}_2\vec{e}_i \\ 1+\vec{x}_2\vec{e}_i \end{pmatrix} \quad , \text{ with } P_i = 1-\gamma,\ 1,\ \gamma \quad \text{for } i = +,-,0 \tag{4}$$

The vector representation of eq. (1) is shown in Fig. 5 where the Poincaré sphere is shifted from the origin to the point $\vec{1} = (1,1,1)$. SOP_+ and SOP_-, together with the corresponding diode currents $\vec{I}_+$ and $\vec{I}_-$ are also visible at opposite points of the sphere. The circle with a shaded filling represents all possible states of polarization SOP_O. The circle is perpendicular to $\vec{e}_+$ and $\vec{e}_-$ and its position on this axis is a function of the power splitting determined by θ with $\gamma = 0.5(1+\cos\theta)$. The actual SOP_O and the corresponding current $\vec{I}_0$ are sensitive to the optical phase difference of the signals transported in SOP_+ and SOP_-.

The weighting vector $\vec{f}$ is also shown in Fig. 5. In general, it has to be oriented perpendicular to a current vector $\vec{I}_+$ (or $\vec{I}_-$) which yields a blanking of the superposition signal h when the actual signal polarization is only SOP_+ (or SOP_-). E.g. in Fig. 5 the plane perpendicular to $\vec{I}_+$ is marked by a dashed circle around the origin. The threshold of the decision circuit has to be chosen to recover the undistorted signal d from the superposition h. In the case of Fig. 5 it is the fast axis signal s_- transported in SOP_-.

Thus we can conclude that the operation principle of the PAE is not to extract an undistorted signal transported in a PSP but to blank or, at least, to minimize the signal of a PSP and to restore the data using the remaining signal. In general, the weighting factors have to be adapted to remain orthogonal to one of the two diode signals $\vec{I}_+$ or $\vec{I}_-$ of a PSP and to maximize the eye opening.

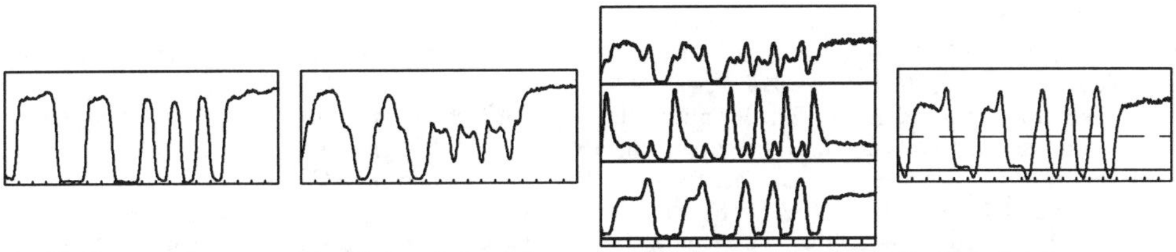

a) Undistorted data pattern.

b) Distorted signal S leaving high PMD lab fiber.

c) Photo diode signals I_1, I_2, and I_3.

d) Calculated superposition signal h for weights of f = (0.6, -0.3, 0.75).

Fig. 6 Measured 10 Gbit/s pattern

6. Experimental demonstration of the PAE operation

In order to demonstrate the operation principle of the PAE with realistic 10 Gbit/s patterns distorted by PMD, a laser diode was externally modulated with a lithiumniobate Mach-Zehnder modulator (MZI) driven by a pattern generator (Fig. 6 a). After transmission through a high PMD lab fiber ($\langle\Delta\tau\rangle = 60$ ps) [3] the distorted signal ($\Delta\tau \approx 100$ ps, $\gamma \approx 0.5$, see Fig. 6 b) was then passed through a polarizer with help of a fiber beam expander and, finally, detected by a photo diode (26 GHz bandwidth) and monitored with a sampling oscilloscope. Three different polarizers (linear 45°, linear 90°, circular) were inserted into the optical path and the oscilloscope traces stored. The corresponding three diode signals are drawn in Fig. 6 c. Finally, the weighted superposition was simulated in a computer by adding the three diode signal with weighting factors $\vec{f} = (0.6,-0.3,0.75)$ yielding the signal h (see Fig. 6 d). It is obvious that with a threshold adjusted to about half amplitude (dashed line) a decision gate would recover the slow axis signal of the fiber. The fast axis signal is nearly blanked.

7. Conclusion

Electronic filters as well as a new concept for an opto-electronic equalizer have been analyzed theoretically with respect to the equalization of first order PMD distortion. All equalizers, which are supposed for implementation in the receiver, are based on adaptive electronic signal processing and they thus have no need for polarization controllers in the optical path, as it is common for optical equalizers. Electronic equalizers, like tapped delay line filter or decision feedback equalizer, exhibit a high penalty (>5 dB) for specific PMD distortions. A new polarization analyzing receiver (PAE) has been proposed which branches the optical input signal to three receiver diodes and hence avoids penalty poles. A theoretical analysis of this concept and a demonstration of the operation with experimentally generated and distorted 10 Gbit/s patterns has been presented.

8. Acknowledgement

This work has been partly supported by the German Government (BMBF) under the national "Photonik 2" program (contract 01BP435).

References

[1] N.Gisin, J.P.Pellaux, „Polarization mode dispersion: time versus frequency domains", Optics Communic., vol.89, no.2,3,4, 1992, pp. 316-323

[2] H.Bülow, "Operation of digital optical transmission system with minimal degradation due to polarisation mode dispersion", Electron. Lett., 1995, vol.31, no.3, pp. 214-215

[3] H.Bülow, " Polarisation Mode Dispersion (PMD) Sensitivity of a 10 Gbit/s Transmission System ", proc. ECOC 96, Oslo, TuD.3.6, 1996, pp. 2.211-2.214

[4] J.Vobian, K.Mörl, W.Keilig, "Different aspects of polarization mode dispersion measuring technique", proc. EFOC&N'94, 1994, Geneva, pp. 174-179

[5] T.Takahashi, T.Imai, M.Aiki, "Time evolution of polarisation mode dispersion in 120 km installed optical submarine cable", Electron. Lett., vol.29, no.18, 1993, pp. 1605-1606

[6] H.Bülow, G.Veith, "Temporal Dynamics of Error-Rate Degradation Induced by Polarisation Mode Dispersion Fluctuation of a Field Fiber Link", submitted for publication

[7] J.H.Winters, Z.Haas, M.A.Santoro, A.H.Gnauck, "Optical equalization of polarization dispersion", SPIE Vol. 1787 Multigigabit Fiber Communications, 1992, pp. 346-357

[8] T.Takahashi, T.Imai, M.Aiki, "Automatic compensation technique for timewise fluctuating polarisation mode dispersion in in-line amplifier systems", Electron. Lett., vol.30, no.4, 1994, pp. 348-349

[9] T.Ozeki, M.Yoshimura, T.Kudo, H.Ibe, "Polarization-mode dispersion equalization experiment using a variable equalizing optical circuit controlled by a pulse-waveform-comparison algorithm", proc. OFC'94, TuN4, 1994, pp. 62-64

[10] T.Ono, S.Yamazaki, H.Shimizu, D.Emura, "Polarization Control Method for Suppressing Polarization Mode Dispersion Influence in Optical Transmission Systems", J. of Lightwave Technol., vol.12, no.5, 1994, pp. 891-898

[11] B.W.Hakki, "Polarization Mode Dispersion Compensation by Phase Diversity Detection", Photon. Technol. Lett., vol.9, no.1, 1997, pp. 121-123

[12] J.H.Winters, R.D.Gitlin, "Electrical Signal Procressing Techniques in Long-Haul Fiber-Optic Systems", Trans. on Commun., vol.38, no.9, 1990, pp. 1439-1453

[13] J.H.Winters, M.A.Santoro, "Experimental Equalization of Polarization Dispersion", Photon. Technol. Lett., vol.2, no.8, 1990, pp. 591-593

[14] S.Kasturia, J.H.Winters, "Techniques for High-Speed Implementation of Nonlinear Cancellation", J. on Select. Areas Commun. vol.9, no.5, 1991, pp. 711-717

[15] J.H.Winters, S.Kasturia, "Adaptive Nonlinear Cancellation for High-Speed Fiber-Optic Systems", J. of Lightwave Technol., vol.10, no.7, 1992, pp. 971-977

[16] C.D.Poole, D.L.Favin, „Polarization-Mode Dispersion Measurements Based on Transmission Spectra Through a Polarizer", J. of Lightwave Technol., vol.12, no.6, 1994, pp. 917-929

Core and ATM Networks
D.W. Faulkner and A.L. Harmer (Eds.)
IOS Press 1997 © AKM

Crosstalk evaluations of optical add/drop multiplexer in WDM-ring networks

C. Caspar, A.Gladisch*, H.-M.Foisel, B.Strebel

Heinrich-Hertz-Institut für Nachrichtentechnik Berlin GmbH
Einsteinufer 37, D-10587 Berlin, Germany
Tel. + 4930 31002 586/587, Fax. + 4930 31002 213
email: caspar@hhi.de
**Deutsche Telekom AG Technologiezentrum*
Ringbahnstraße 130, D- 12103 Berlin, Germany
Tel. + 4930 75744881, Fax. + 4930 7529957
email: gladisch@fz.telekom.de

Abstract. The influences of power and wavelength (frequency) uncertainties as well as crosstalk were investigated by means of a recirculating loop experiment with 10 Gbit/s NRZ formatted signals. Two general kinds of optical add/drop multiplexer (OADM) were evaluated experimentally by introducing into the loop arrayed-waveguide grating (de)multiplexer (AWGM) or fibre Bragg grating OADM. The measurement of cascading effects due to OADM based on arrayed-waveguide grating have shown an acceptable frequency uncertainty of ±3.5 GHz for both transmitter and AWGM, respectively. The measurement of the relation between OADM crosstalk and gain variations have shown a tolerable gain detuning of approximately ±3 dB and at least the measurement of relation between crosstalk and detuning of an OADM based on fibre Bragg grating have given a range of stable operation within ±20 GHz or 17°C to 37°C, respectively.

1. Introduction

The use of optical frequency and wavelength multiplexing is a method that allows, besides utilising the technically feasible transmission bandwidth of an optical fibre, also the construction of flexible configurable, scalable and modular networks [1][2]. For the different network levels solutions were proposed which make use of optical wavelength multiplexing, where an application in the sector of transport networks appears to be highly advisable under present-day aspects. The typical network elements in such an optical transport layer are the optical cross connector (OXC), based on a combination of space and wavelength multiplex stages, and the optical add/drop multiplexer (OADM). The network structure that can be designed with OADM is the optical ring network or an optical bus. This network type is primarily employed in the local transport area. Compared with SDH ring networks being equipped with electric add/drop multiplexers (ADM), they have, in addition to an improved capacity utilisation, the benefit that the introduction of an optical network level allows new network elements to be inserted without the necessity of reconfiguring the network. Furthermore, the OADM allows non-relevant traffic to bypass an electrical ADM and independent logical networks to be built up on a physical network. Moreover the introduction of optical network elements allows also an enhanced optical protection switching and a transport of signals with no SDH- format.

The optical add/drop multiplexers are the key-elements for multi-wavelength ring networks and many different technologies to realise add/drop function have been reported. Some examples are fibre Bragg gratings, accousto-optics, Fabry-Perot-filters, arrayed-waveguide gratings, dielectric filters and Mach-Zehnder- interferometers [3].

Independently of the technology the general physical requirements on OADM are

- low insertion loss
- flat pass- and stop- band
- high crosstalk suppression
- high frequency stability

In the following sections we investigate the basic requirements to be met by an OADM in ring networks with respect to optical power and frequency variations, while neglecting the effects of noise accumulation due to Er-doped fibre amplifiers and optical nonlinearities of the transmission. The behaviour of different OADM in ring networks was studied by means of a recirculating loop experiment [4].

2. Classification of OADMs

The general behaviour of an OADM is to drop one ore more channels of an incoming data stream and to insert also some channels carrying a new information (Fig. 1). OADMs applied in optical ring networks can be classified into two categories. At first one can consider OADMs based on a multiplexer/demultiplexer structure, that demultiplexes and multiplexes completely the incoming WDM-signal. Such an OADM has a periodical transfer function between the network ports input 1 and output 1. For such structure it can be expected that cascading effects and wavelength uncertainties are the most critical limiting factors because the overall transfer function is narrowed and distorted by that effect. Therefore the number of OADMs that can be passed by a channel is limited [5].

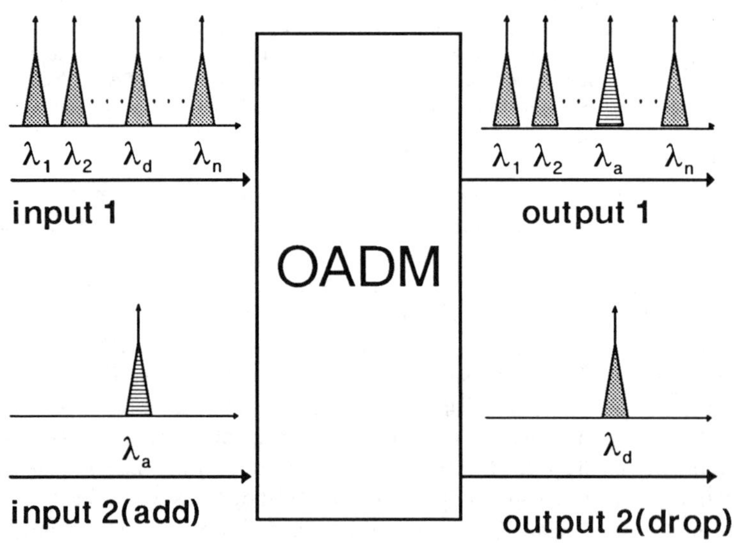

Fig. 1: Functionality of an OADM.

The second type of OADM we distinguish is characterised by a notch of the transfer function between the network ports (input 1; output 1). OADMs based on fibre Bragg grating or Fabry- Perot- interferometer are examples for that type. For the second type we expect no critical cascading effects as for type one but due to the narrow stopband the requirements on the frequency stability are more critical.

3. Description of the loop experiment

3.1. Experimental investigation on cascading effects

The set-up of the loop testbed is shown in Fig. 2. An externally modulated 10 Gbit/s ASK NRZ signal was fed into a loop which consisted of 75 km dispersion shifted fibre (D = 0.2 ps/[nm km] at λ = 1555 nm), two EDFAs and as device under test a commercially available 8x8 arrayed wavegude grating multipexer (AWGM). After each round trip a part of the data signal was coupled out to a receiver for bit error ratio (BER) measurements.

The 8x8 AWGM had a channel spacing of 100 GHz, a bandwidth B_{3dB} = 38 GHz and B_{1dB} = 22 GHz and an insertion loss of 4.5 dB. The 3-dB-bandwidth of a cascade of (de)multiplexers amounts to B_{10} = 13 GHz and B_{15} = 10.6 GHz. For the theoretical evaluations the AWGM passband characteristic was fitted with a Bessel filter.

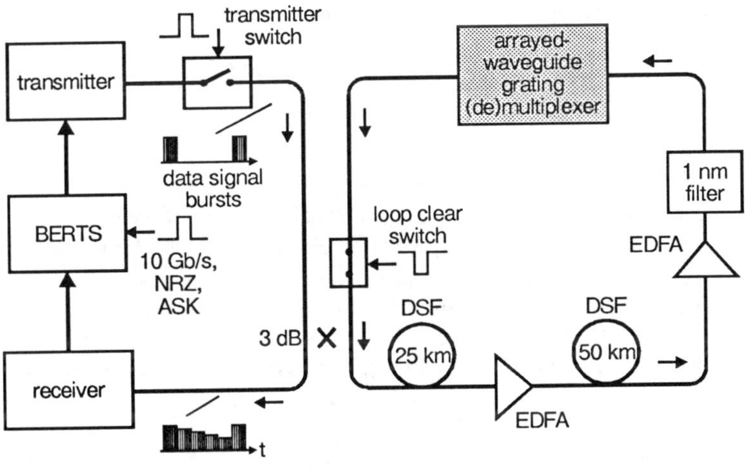

Fig. 2: Loop testbed for testing the cascadability of optical arrayed-waveguide grating (de)multiplexers.

The penalty for a constant BER (for measurement simplification BER = 10^{-7}) depends on the number of round trips and on the frequency detuning Δf of the AWGM centre frequency from the optical carrier frequency (Fig. 3). As shown in Fig. 3 a rapid decrease of the bandwidth have been observed and therefore the requirements on stability of carrier frequency as well as of centre frequency of AWGM increases. In the example for 10 round trips the stability must be better than ± 7 GHz and if one assume statistically independent processes for the transmitter and for the centre frequency of the AWGM the resulting stability requirement is ± 3.5 GHz for the AWGM and for the transmitters as well. Therefore it seams to be realistic to apply a frequency monitoring or control for the transmitter because long term measurements have shown that 5 to 10 percent of the lasers exhibit an ageing or drift of more than 10 GHz [6].

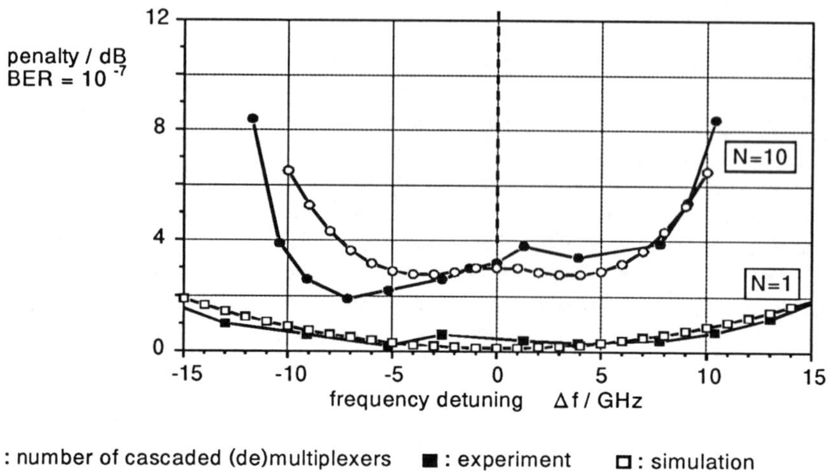

Fig. 3: System penalty as a function of carrier frequency detuning for different numbers of round trips.

3.2. Experimental investigation on crosstalk and power variations

An OADM has the task of guiding a given channel of the wavelength λ_d from a group of wavelengths with the channel spacing $\Delta\lambda$ to an output 2 and to offer the network at output 1 the remaining channels and a new channel on λ_a (normally $\lambda_a = \lambda_d$) carrying new information (Fig.1). A complete ring network consists of several OADMs and Er-doped fibre amplifiers as well as of a central hub where all channels are separated and, via a cross connector (either optical, electrical or combined), routed to exchanges or other network levels.

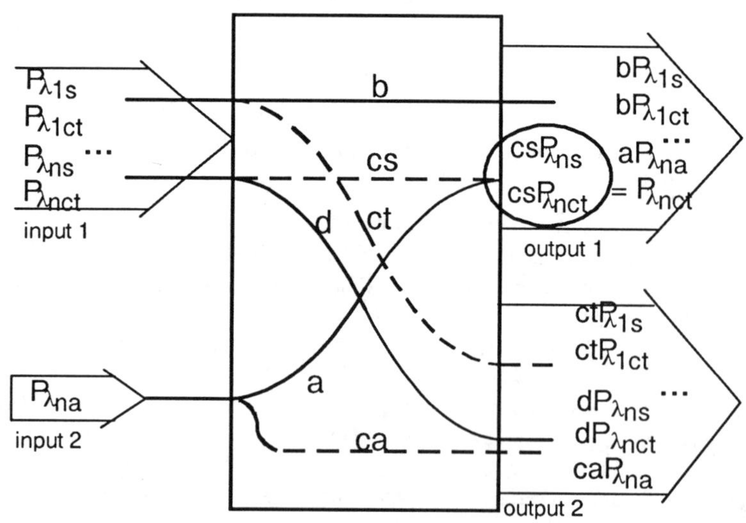

Fig. 4: Different types of crosstalk in an OADM:
cs - crosstalk of the droped channels to the ring ports;
ct - crosstalk of all channels which are not dropped to the drop port;
ca - crosstalk of the added channel to the drop port.

Besides the wanted drop and add function of the multiplexer, technical and physical limits entail unwanted crosstalk effects, i.e. a small portion of the power (ca*$P_{\lambda a}$) of the add-channel can also be detected at output 2. Moreover, there are crosstalk effects (ct) from all channels except λ_d on output 2 and λ_d of the network-input (input 1) on output 1 (cs) (Fig.4). For simplification let us assume that the factor for all wavelengths be c (cs = ca = ct = c). In the following, all investigations will be limited to the deterioration of a system due to crosstalk from unwanted channels, i.e. nonlinear effects and disturbance caused by the noise of optical amplifiers are neglected. In principle, crosstalk can occur in two different ways:

1. In the case that wavelength of all channels involved in the crosstalk are widly spaced to the wanted signal channel, i.e. that all conceivable optical signal differences are much larger than the electrical receiver bandwidth, the disturbance is proportional to the sum of crosstalk powers. The theoretical limit for a single disturbing channel is given by [7],[8]

$$p = -10 \cdot \log(1-c).$$

2. In the case that mutual interference exists between the individual channels or also with the wanted channel in the range of the receiver bandwidth, a carrier beat noise is caused, which may considerably reduce the system sensitivity [7]. For instance, the beat noise of the wanted channel and the crosstalk of the inserted channel causes, in the most unfavourable case, a noise current, which results for a Gaussian approximation in a system penalty [7] of

$$p = -5 \cdot \log(1 - 4 \cdot q^2 \cdot c).$$

Here c is the factor of the power crosstalk between input 2 (add) and output 2 (drop) (Fig. 4) and q is 5.9 for a bit to error ratio (BER) of 10^{-9}.

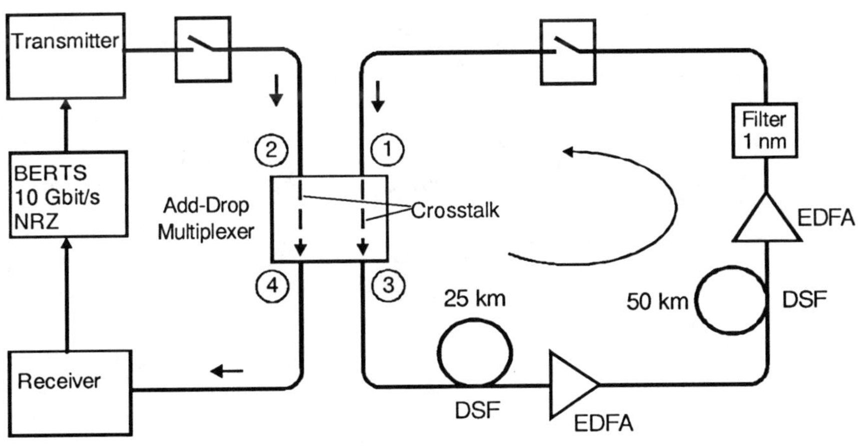

Fig. 5: Application of the loop experiment to measure the crosstalk behaviour of OADM.

In an optical ring network with wavelength multiplexing it is quite possible that a great number of such interferences between the crosstalk components occur on one wavelength, for instance the last OADM in a ring network "sees" on its drop wavelength λ_d also signals resulting from the crosstalk of previous OADMs on λ_d. In order to measure the influence of the different kinds of crosstalk an OADM and a model of an OADM were been inserted in the recirculating loop (Fig. 5). The BER and the penalty for a BER of 10^{-7} were measured, while the round trip gain of the ring and also the "in-band" crosstalk value (cs; ca) were modified. In order to measure the influence of both kinds of crosstalk, from add to drop (ca) as well as in the ring (cs), the time slot of measurement was set after the second round trip.

The experimental results for two different crosstalk values and two theoretical estimations are shown in Fig. 6. The remarkable "tub"-form of the curves is due to the fact that an increased ring gain decreases the relative power ratio between the crosstalk power in the ring and the wanted signal power and moreover that a decreased ring gain decreases the relative power ratio between the crosstalk form add to drop and the wanted signal power. In our experiment the tolerable gain variation was ±3 dB for an extra penalty of 1 dB and an network with OADMs that have a crosstalk of -25 dB. A tolerable gain variation of ±3 dB is a practicable value but on the other hand 3 dB variation for a channel could be induced by switching of some channels in a multi wavelength network [9] .

Fig. 6 shows also two simulated curves representing two different receiver types and a Gaussian noise model. The measured curve is within the boundaries given by the two theoretical curves.

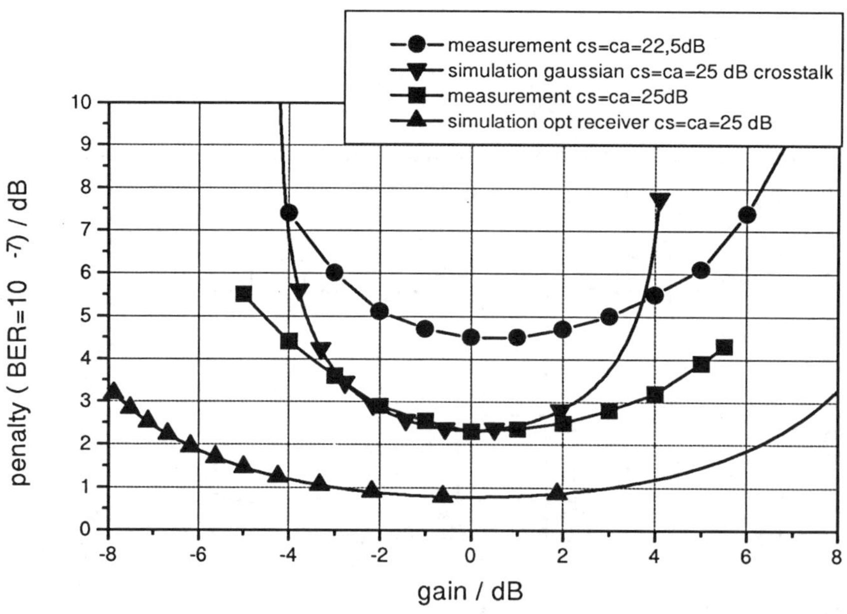

Fig. 6: System penalty due to detuning ring round trip gain.

3.3. Experimental investigations of notch type OADMs

With the experimental set-up shown in Fig. 5 the influence of transmitter frequency instabilities was measured for OADM with one notch ("type 2") as well. The OADM was build up by a fibre Bragg grating and 2 optical circulators, while the 3 dB width of the fibre Bragg grating was 80 GHz. The extinction ratio of the OADM was better than 30 dB. The BER and the system penalty was measured while the centre frequency (wavelength) of the

OADM was tuned by varying the temperature of the fibre Bragg grating. Again bursts of a 10 Gbit/s datastream were fed into the loop by the OADM and the BER after the second round trip was measured. Measuring after the second round trip gave a realistic experimental model for a ring with wavelength reuse, because crosstalk of the round trip remained in the loop and disturbed the second round trip (crosstalk cs) and the third round trip models the crosstalk from add to drop. The results which are shown in Fig. 7 indicates a broad region of stable operation among ±20 GHz and an increase of the system penalty du to the increase of crosstalk for higher detuning of the fibre Bragg grating. On the other hand the temperature region of the Bragg grating for stable operation is in between 19°C and 37°C which is not enough for an uncontrolled operation.

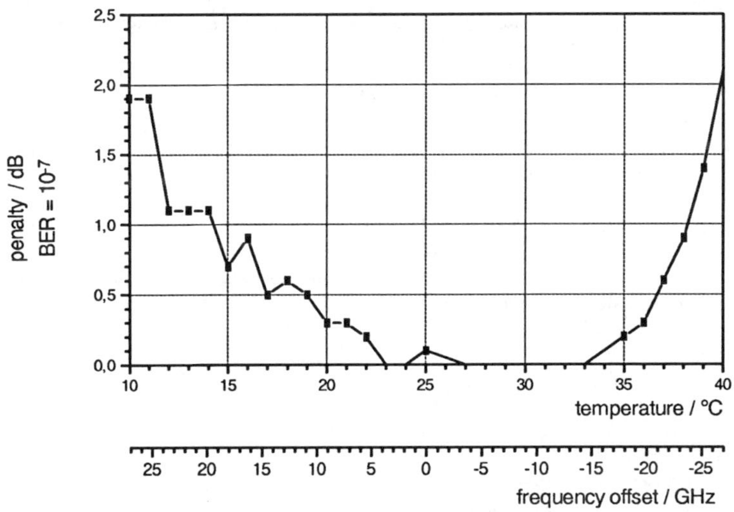

Fig. 7: System penalty versus the frequency detuning of a fibre Bragg grating OADM

4. Conclusions

Different kinds of OADM were tested in a recirculating loop experiment while three different measurement series were carried out. The measurement of cascading effects due to OADM based on arrayed-waveguide grating have shown an acceptable frequency uncertainty of ±3.5 GHz for both transmitter and AWGM, respectively. The measurement of the relation between OADM crosstalk and gain variations have shown a tolerable gain detuning of approximately ±3 dB and at least the measurement of relation between crosstalk and detuning of an OADM based on fibre Bragg grating have given a range of stable operation within ±20 GHz or 17°C to 37°C, respectively.

All the practical results indicates the need to design simple supervision or control strategies for optical ring networks with OADMs. On the other hand a more favourable way in order to reduce network costs and complexity, especially in WDM based passive optical networks, will be the development of highly stable OADMs with flat pass- and stop- bands.

References

[1] C.Brackett, *et al*: A scalable multiwavelength multihop optical network...;IEEE Journal of Lightwave Technology 5(1993)5,pp 736-753

[2] S. Johansson.: Transparent optical multicarrier networks; ECOC 92;We A9.2

[3] H.Venghaus, *et al*: Optical add/drop multiplexers for WDM communication systems; OFC 1997 Dallas; Technical Digest THJ1; pp 280-281

[4] C. Caspar, H.-M. Foisel, R. Freund, U. Krüger, B. Strebel: Cascadability of arrayed-waveguide grating (de)multiplexers in transparent optical networks; OFC Dalles 1997

[5] R. S. Vodhanel, et al.: Performance of an 8-wavelength 8-node WDM ring network experiment with 80 Gbit/s capacity OFC`96, San Jose, USA, Technical digest, postdeadline paper PD28

[6] Y. C. Chung, J. Jeong, L. S. Cheng: Aging-induced wavelength shifts in 1.5-μ DFB lasers; IEEE Photonics Technology Letters, vol. 6, No. 7, pp. 792-795, 1994

[7] E.L.Goldstein; L.Eskildsen; A.F.Elrefaie: Performance implications of component crosstalk in transparent lightwave networks; IEEE Journal of Lightwave Technology 6(1994)5 , pp 657-660

[8] R.J.Hess.: Fibre optic handbook, Prentice Hall 1990

[9] J.L.Zyskind, *et al*: Fast link control protection for surviving channels in multiwavelength optical networks; ECOC 96 Oslo; Technical Digest Volume 5 ThC. 3.6; pp. 49-52

Abbreviations

AWGM	Arrayed Waveguide Grating Multiplexer
BER	Bit Error Ratio
EDFA	Erbium Doped Fibre Amplifier
OADM	Optical Add Drop Multiplexer
OXC	Optical Cross Connector
SDH	Synchronous Digital Hierarchy
WDM	Wavelength Division Multiplexing

Core and ATM Networks
D.W. Faulkner and A.L. Harmer (Eds.)
IOS Press 1997 © AKM

Design of wavelength branching units in long haul WDM networks

Andrew Lord
BT Laboratories,Martlesham Heath, Ipswich, UK IP5 7RE

Abstract. This paper discusses some of the key physical design issues relating to long haul WDM networks. These networks require a degree of optical pre-emphasis to equalise the performance of each WDM channel. This is required at each point in the system where a new channel is added. The design of the branching unit should allow for a pre-emphasis control facility, and measures to counteract ageing and failure mechanisms. Consideration is also be given to ensuring that channels don't unduly affect the performance of other channels.

1. Introduction

WDM technology may be applied to submarine systems to both multiply the capacity and allow remote branching by wavelength add drop multiplexing (WADM). There are many advantages in adopting architectures which involve the use of wavelength add drop multiplexers (WADMs) to add and drop traffic at intermediate / remote locations. Unfortunately these WADMs lead to overall system design problems arising from the differing power levels on the branches. The aim of this paper is therefore to address the problems of wavelength independence by considering wavelength channel pre-emphasis and use of optical amplifiers near the WADM.

2. Pre-emphasis

2.1 Background

Point to point long haul WDM systems usually require pre-emphasis of the launch powers, since optical amplifiers do not provide perfectly flat optical gain. Therefore, channels receiving less gain need to be boosted, relative to the others at the transmitter.

Dropping a wavelength channel before the other channels in the system means that with the same launch power, it will perform better, since it has less far to travel. Also, the replacement channel will perform better, if it is multiplexed at the same power level as the others. An optimised (and hence cheaper) system will introduce pre-emphasis at the branching unit (BU) to equalise the signal to noise ratios of all the channels. Since each BU is uniquely positioned, the required pre-emphasis will be different for each one. Unfortunately, changes to the system configuration, such as repairs, loss of spurs etc., will change the pre-emphasis required at the remote BUs. Consequently, the variation in performance of the various channels can be large. The BU must therefore be designed to take account of these effects, as discussed in this section.

2.2 Modelling a simple four channel WADM system

Fig.1 shows a schematic of a typical four channel system:

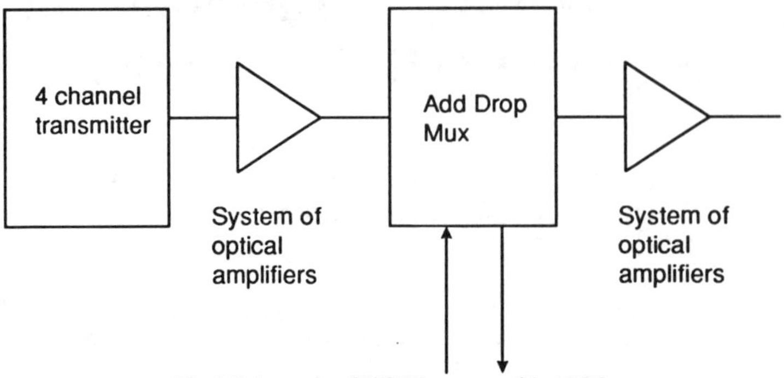

Fig.1 Schematic of WDM system with ADM

In this system, three channels are passing straight through, with the fourth being dropped and replaced. The system has amplifier outputs of 8mW, total length of 6300km with ADM located after 3500km. The repeater spacing is 70km. A simple linear model of noise propagation and amplifier saturation is used to analyse this system. We assume that an initial pre-emphasis is used to counter any amplifier gain variations, and this study focuses on the effects of power variations at the WADM.

Fig.2 shows the signal to noise ratios of the various channels in the system, calculated at the WADM.

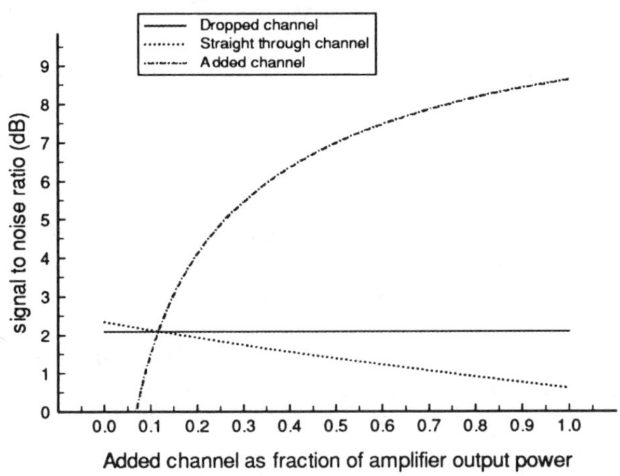

Fig.2 Signal to noise ratio of various channels

The figure shows, not surprisingly that the dropped channel is unaffected by the power of the added channel. It also shows the location of the optimum operating point, where all SNRs are the same. As the added channel power increases, the performance of this channel improves at the expense of the straight through channels. If the added channel is removed, then the other channels increase in power and hence

performance. Similar performance variations are obtained if we vary the trunk channel input powers and keep the added channel power fixed.
It is therefore essential to have some form of pre-emphasis (relative channel power) control at the WADM as well as the launch. Indeed, variations in power on either trunk or spur, caused by repairs, changes in amplifier output powers etc., could impact quite heavily on performance. The above results have neglected effects such as nonlinearity.

2.3 Amplifiers inside the BU

By incorporating optical amplifiers in the BU, their automatic output power control can be used to regulate the powers entering and leaving the BU, and thus achieve pre-emphasis control. Fig.3 shows a possible scheme in which amplifiers are present at all entry points to the BU.

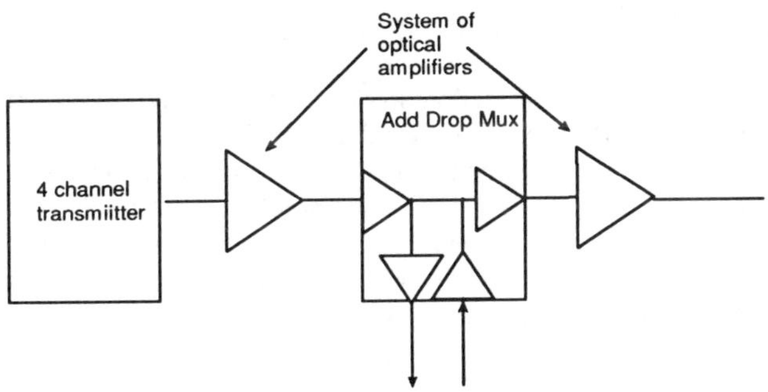

Fig.3 Possible BU internal configuration

This ADM model automatically controls the power of all aggregate signals entering the BU. This configuration provides a buffer against incoming power variations. The amplifier output power of one of the two input amplifiers will need to be adjustable to optimise the pre-emphasis.

Inclusion of optical amplifiers in the BU results in a robust design, although it makes the BU a more complex component. An alternative approach is to use main line amplifiers which have output power control (rather than relying on automatic control from saturation).

3. WDM systems with Add Drop Muxes

One interesting point about the use of WADMs in long haul WDM systems is that the ADM function can actually help system performance. There are several related issues here:

i. Wavelengths do not necessarily travel the full length of the system, because they can be re-launched on the way (i.e. a noisy channel can be dropped and a clean channel added). Thus the total system length is not necessarily the key factor determining performance. What is important is the longest distance travelled by a channel before it is dropped. Of course the noise sitting between the channels may never be dropped, and will build up. This noise will not directly affect the signal to noise ratio as it is out of the channel bands, but it could affect amplifier saturation.
ii. Dropping a wavelength removes both the signal and a portion of noise (the amount contained within the multiplexer bandwidth). The signal that is re-injected is

noise-free. This removes any four wave mixing products that may have been building up in this bandwidth.

iii. A key parameter indicating the robustness of the system design is the required pre-emphasis at various points on the system. Adding a channel at some distance along the system permits it to have a lower launch power, since it has no noise associated with it. Use of saturating optical amplifiers results in an increase in the power levels of the other channels, hence improving their performance.

iv. In a system containing 4 channels on each of two fibres, it would make sense to drop channels from both fibres, rather than having one fibre completely committed to 'express' traffic. Dropping from both fibres alleviates the system penalties from both fibres allowing a bigger repeater spacing. In the system, the repeaters for each fibre will be located in the same housing, and so the repeater spacing will be determined by the performance of the worse of the two fibres. An optimum configuration, under these circumstances, would involve equalising the performance of the channels from both fibres.

It is therefore concluded that ADM functionality and long haul WDM are mutually beneficial. The task of network design (i.e. which channel goes where) is not just driven by logical connectivity requirements, but also by transmission issues. It is of course possible to produce a network design, based purely on topological considerations, and then to assign wavelengths to give the best performance for this particular network. However, a network optimised for performance, connectivity and cost, requires simultaneous logical and physical modelling.

4. Functionality of WADMs

At first sight it is a fairly innocuous fact that a system has two trunk fibre pairs taking channels in opposite directions. However, these fibres will drop channels at the same WADM. In terms of maintaining a simple wavelength management scheme, it might make sense to drop the same wavelength from both directions - thus allocating each spur with a specific wavelength (or more if needed). However, this restriction already imposes a design limitation as shown in the following example. Fig.4 shows a simple schematic of a system with two fibre pairs, and ADMs operating on all trunk fibres.

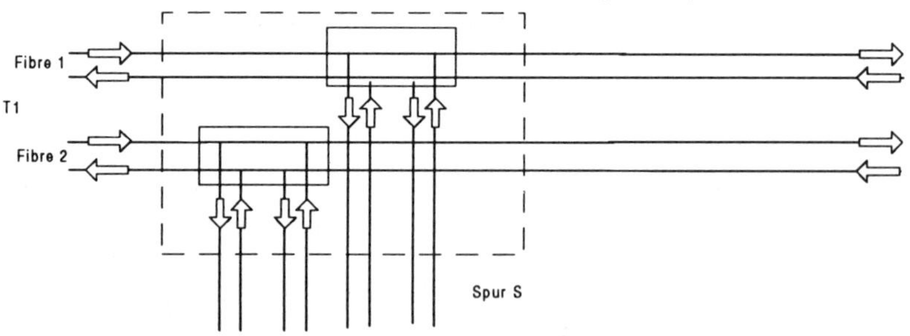

Fig.4 General BU dropping channels from two fibre pairs

The figure shows that there could be as many as eight spur fibres connecting a remote spur to the main trunk system. If on a specific fibre pair, the wavelength dropped from both directions is the same (this may be desirable from a management point of view), then clearly two fibres will be required, as this is the only means of separating the channels. Two separate fibres are also needed to add channels at the same wavelength.

Clearly it would be beneficial to reduce the spur fibre count. Various possibilities arise. The add and drop from each channel could be combined with a coupler at the ADM, cutting the fibres to 4 (or only 2 if just one of the fibre pairs has an ADM). Problems with bi-directional transmission include cross-talk problems - particularly near-end cross-talk if the two wavelengths are the same (as they are in this case). Operation of amplifiers bi-directionally also causes a problem - not least because they would have to be different from trunk amplifiers which would automatically have isolators in them. Not all spur links would need amplifiers necessarily. The following figure shows a novel implementation (although simpler schemes not using four port circulators are possible).

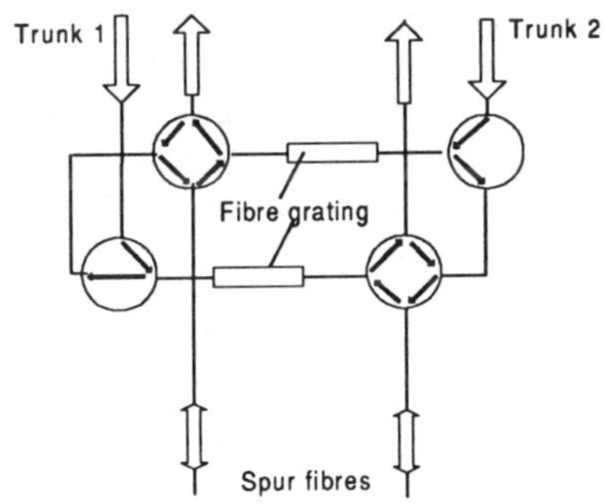

Fig.5 ADM design achieving spur multiplexing

The above figure provides a neat way of multiplexing add and drop channels from a trunk fibre pair. The two internal 4 port circulator routes, whilst transmitting any reflected light coming their way, will only direct the light to a dead end at the subsequent 3 port circulator. As usual in this type of circuit, if the spur attempts to add the wrong wavelength, it simply reappears at the spur's drop.

Another way of reducing spur fibre count would be to combine fibres carrying channels in the same direction, but this requires different wavelengths. This is feasible when the spur needs more than one wavelength drop. It is simple to combine two dropped channels onto a single spur fibre - the two channels could either both originate from one fibre pair, or alternatively, there could be one channel from each fibre pair. Dropping several channels from one fibre simply requires the addition of gratings in series, thus automatically dropping channels on the same fibre. If the channels were dropped from different fibre pairs, then a coupler would be needed to multiplex them together. It may be more common, however, to drop just one channel from one fibre pair at the majority of the spurs. The resulting four fibres can be combined into two bi-directional fibres as already discussed.

The use of four port circulators and fibre gratings provides a wide range of wavelength drop functionality.

5. Effect of optical transients on DLS independence

As reported by Zyskind et al at OFC 1996 [1], sudden changes to optical power give rise to transient ripples on the output power of erbium amplifiers. For example, in a four wavelength system, if two of the wavelengths are instantly switched off, then the other two will increase in power until they have increased by 3dB. The power increase however will tend to overshoot the new quiescent level, before settling back. The overshoot can be several dB, and the time scale decreases with increasing numbers of amplifiers. Many amplifiers will give a rise time of the order of 100 ns. Conversely, if the two wavelengths are switched back on, the power in the other two channels, before settling back at the original level, will undershoot by up to several dB. Transient effects of this nature can give rise to error bursts.

5.1 Decision threshold penalty

Not only do the signals over / under shoot, but any noise in the system will do the same. This means that the signal to noise ratio remains constant during the transience. However, the receiver decision threshold sits at a fixed voltage and this level will remain unchanged. The receiver will be ac coupled, with a specific low pass frequency cut off. Up until a time delay of order the reciprocal of this cut off frequency, the data will pass through the receiver with the zeroes at a fixed voltage, and with the ones changing due to the transience. So, for example, if a signal overshoots, the decision threshold will be closer to the zero level than the one level. Because the SNR is constant, the optimum decision threshold will rise as the level rises. This means that a fixed decision threshold will no longer be optimal. Fig.6 shows a plot of Q penalty vs. under / over shoot in dB. (Q is related to bit error rate via BER = Erfc Q).

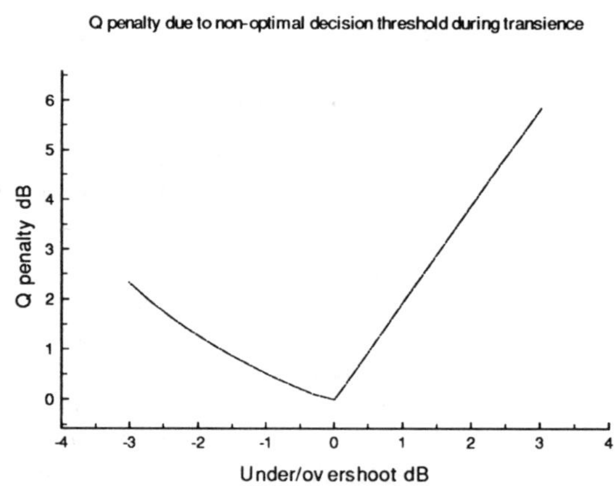

Fig.6 Q penalty as a function of transience

As is seen in the figure, large penalties can arise from under/overshoots of only a few dB (2-3dB is quite possible for a transience effect). There is an asymmetry to the figure, which arises from an interplay between spontaneous noise on 'ones' and 'zeros' and the decision threshold level.

5.2 Receiver ac coupling

After a time equivalent to the receiver time constant, the electrical data will adjust its mean level back to zero. The decision threshold will not have regained the optimum position, but the error performance will be different from the time before the receiver ac coupling came into play. If the new configuration (i.e. more or less channels than before) is to remain for any length of time, this decision threshold will need to be re-optimised to take account of the new signal levels entering the receiver. Fig.7 shows the Q penalty variation with under/overshoot, before the decision threshold has been re-optimised. The asymmetry in fig.7 is in the opposite sense to fig.6.

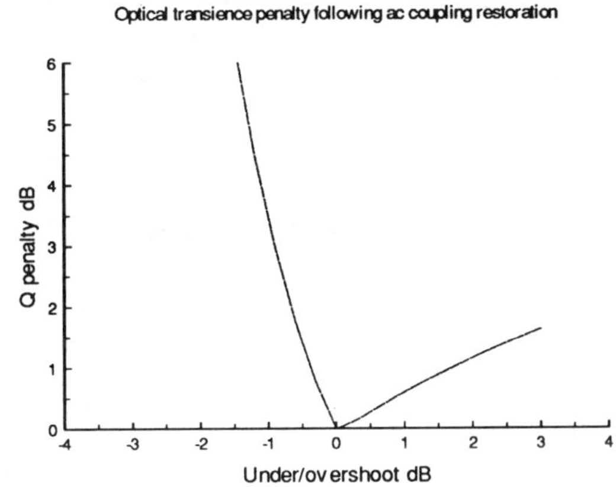

Fig.7 Q penalty with transience after receiver coupling

5.3 FEC recovery

FEC is able to correct not only random errors but also bursts of errors (it may, depending on the exact implementation be able to cope with several hundred ns - i.e. 1000 bits - of burst errors before losing alignment). However, system Qs (pre FEC) are below the 10^{-9} BER level, and FEC is used to recover the error rate to be better than 10^{-9}. In other words, FEC is being used, largely to permit extended repeater spacing and hence lower line Q. The actual Q produced during a transience could easily be lower than that which the FEC is able to deal with. As the system ages and the Q degrades, the erosion into the system margin will require more of the FEC gain, hence leaving less available for optical transient effects.

6. Conclusions

This paper has discussed key issues relating to the introduction of wavelength branching units to long haul systems. In particular, it has shown that care has to be taken to design BUs that are robust enough to cope with power fluctuations. The BU should also provide a facility for optimising the channel pre-emphasis of the system. The most comprehensive way of achieving this is by using optical amplifiers as an integral part of the BU design. It is also important that a BU is universal, so that multiple spares are avoided. Use of amplifiers in BUs aids this requirement.

The paper also considered the design of branching units and showed that use of fibre gratings and optical circulators (particularly four port circulators) can give rise to a wide variety of functionality.

Finally, the paper warned of the effects of optical transients, which could occur of individual channels were switched on or off. These transients can in principle affect the performance of other channels.

References

[1] Zyskind JL, Sun Y, Srivastava AK, Sulhoff JW, Lucero AJ, Wolf C and Tkach RW, 'Fast Power Transients in Optically Amplified Multiwavelength Optical Networks', OFC '96, PD31

Core and ATM Networks
D.W. Faulkner and A.L. Harmer (Eds.)
IOS Press 1997 © AKM

Numerical analysis of transparency limitations in optical core networks

R.Freund, B.Strebel

Heinrich-Hertz-Institut für Nachrichtentechnik Berlin GmbH
Einsteinufer 37, D-10587 Berlin, Germany
Tel.: +49 30 31002 418, Fax: +49 30 31002 213, E-mail: freund@hhi.de

Abstract. Transparent optical core networks offer the possibility to use the optical fibre amplifier bandwidth by optical frequency division multiplex (OFDM) techniques, whereas routing within transparent optical crossconnects in conjunction with linear and non-linear fibre effects leads to significant decrease of transmission length. These effects can be illustrated by the transparency length diagram. To overcome the problems with the combined analytical investigation of transparency limits a software package, *BroadNeD*, which allows the numerical analysis of transparency limitations in optical networks has been developed. Results of typical simulation scenarios are presented.

1. Transparent optical networks

Future core networks of a global telecommunication system shall transmit high information capacity using multichannel optical frequency multiplex (OFDM) techniques with bitrates of up to 10 Gb/s per channel. The OFDM channels are switched in the network nodes with respect to space and frequency domain in a circuit switching manner. So only a low channel switching speed is needed compared to the data bitrate. In the OFDM trunks connecting the nodes, all channels are commonly amplified by fibre amplifiers (EDFA). Therefore wide areas of a core network could be designed optically transparent [1, 2]. In this case the signals have to be switched optically transparent in the crossconnects [3]. Fig.1 shows the scheme of an transparent optical core network. In this network the signals pass no opto-electronic interfaces between transmitters and receivers. The network is said to be optical transparent, if the bandwidth of the channel is not affected by the optical switching and routing procedure. The networks are invariant against codes, transmission modes and protocols.

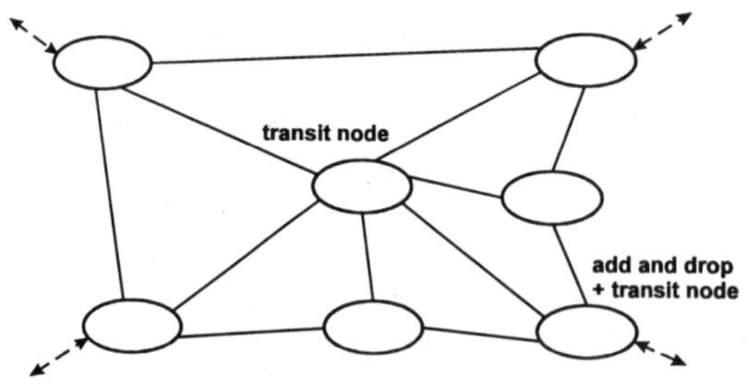

Fig.1: Transparent optical network

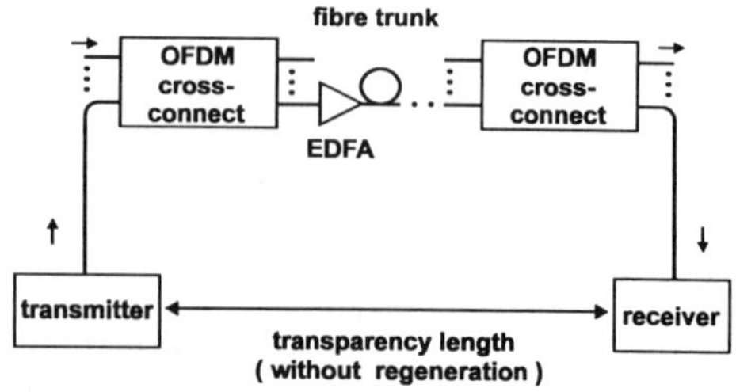

Fig.2: Transparent optical signal path

In Fig.2 a transmission path in a transparent optical core network is depicted. The signal passes a series of OFDM crossconnects and fibre trunks. The routing of the information in the nodes of a transparent core network is performed in optical crossconnects [3]. The signals are switched optically in the frequency and space domain without crossing opto-electronic interfaces. The network is characterized by the optical transparency length. It is defined as the transmission length without time regeneration of the signal, over which a predetermined signal quality can be guaranteed.

2. Transparency limitations

A complete transmission path in a transparent optical network (Fig.2) consists of OFDM crossconnects and fibre trunks including fibre amplifiers. The transparency length in the fibre trunks is limited by:

- chromatic and polarization dispersion,
- selfphase and crossphase modulation,
- four wave mixing,
- amplifier noise,
- polarization noise, and
- reflections.

In the crossconnects transparency limitations are given by:

- amplitude and phase noise in amplifiers and frequency converters,
- crosstalk in space and frequency stages,
- filter bandwidth narrowing, and time delay distortions.

The transparency range of a transmission path can be illustrated by a transparency range diagram (Fig.3), in which the limiting parameters of a photonic network are depicted [2]. This diagram shows the maximum transmission length as a function of the number of crossconnects in a transmission path. Dashed lines describe the case of multichannel systems. The marked area represents the transparency range of the network. Within this area a signal can be transmitted with a given penalty, e.g. 2 dB.

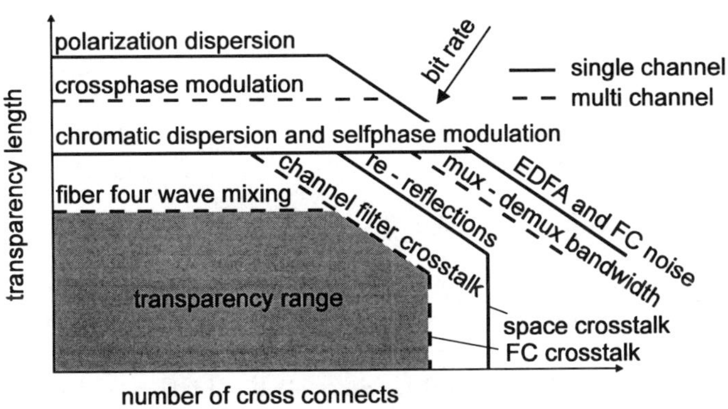

Fig.3: Transparency range diagram

3. Numerical analysis of optical transparency limits

The influence of effects limiting optical transparency can be determined analytically [2]. However, in most cases it is difficult to combine the several effects since a closed theory for a description of different effects like fibre nonlinearity, amplifier noise, and crosstalk does not exist. In contrast, numerical analysis provides an efficient and sometimes the only applicable method for combined consideration of more than one effect [3,4].

The numerical simulation results presented here were calculated using *BroadNeD*, a self designed simulation package based on *Ptolemy* [5]. Each optical system-component is modeled mathematically by a specific subroutine. Some examples of these modules are shown in Fig.4. The simulation software includes a large variety of modules like: transmitters, receivers, fibres, EDFAs, filters, frequency converters, and others. These modules can simply be combined for the analysis of very complex signal path configurations.

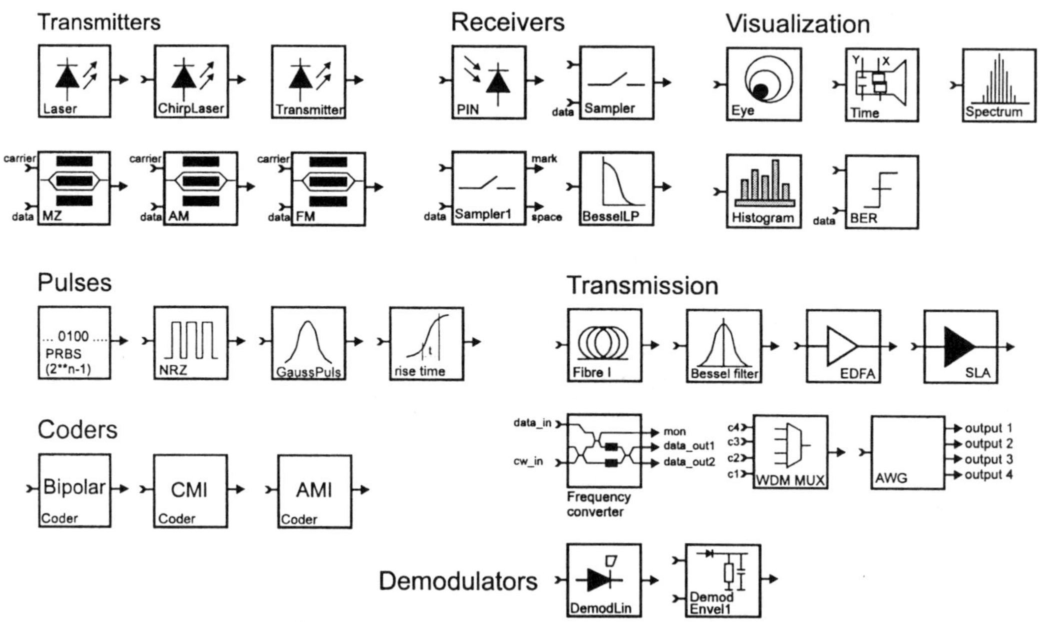

Fig.4: Examples of BroadNeD modules for simulation of transparent optical signal paths

4. Simulation of transparent optical signal paths

The (de)multiplexing function of OFDM data streams may be performed by arrayed-waveguide gratings (AWG) within optical crossconnects. Fig.5 shows the bandwidth narrowing effect of the non-flat, parabolic filter shape of a cascade of AWGs with 100 GHz channel spacing. The resulting bandwidth of a cascade of N = 10/15 round trips amounts to 13 GHz and 10.6 GHz, respectively. For the numerical analysis the AWG transfer function is fitted with a Bessel-filter.

The penalty for a 10 Gb/s NRZ-signal at a constant bit error rate (BER = 10^{-7}) depends on the number of round trips N and the frequency detuning Δf of the AWG center frequency from the optical carrier frequency. Fig.6 shows the good correspondence between experimental and simulation results [6], even in the case of quasi single sideband transmission at $\Delta f = \pm 5.5$ GHz and

N = 15 cascaded AWGs. For the simulation a pseudo random bit sequence (PRBS = 2^{10}-1) was used. The BER was estimated by a Gaussian approximation.

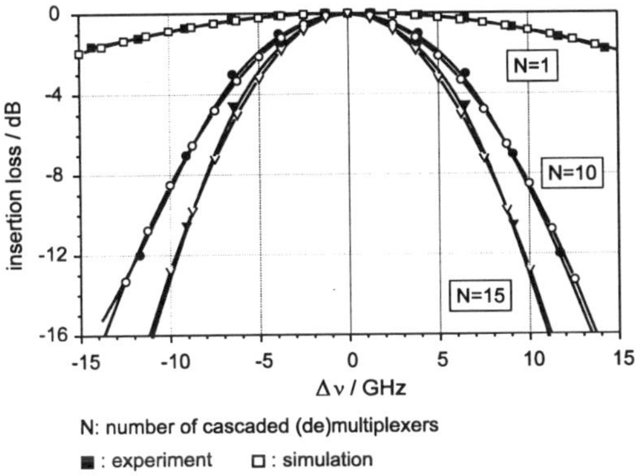

Fig.5: Passband characteristic of N cascaded AWGs

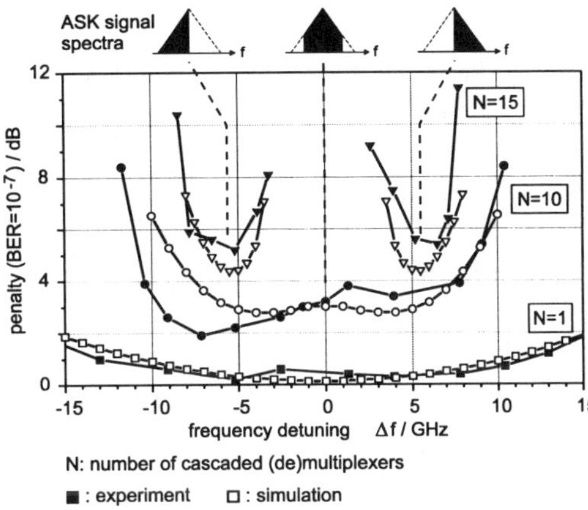

Fig.6: Sensitivity penalty due to frequency detuning of N cascaded AWGs

Another application of *BroadNeD* is the calculation of transparency range diagrams for a given transmission path (Fig.7), taking into account a number of effects, e.g.: chromatic dispersion, fiber nonlinearity, ASE of the EDFAs, and crosstalk occurring in the crossconnects. Tab.1 gives an overview of the relevant system parameters used for this simulation example.

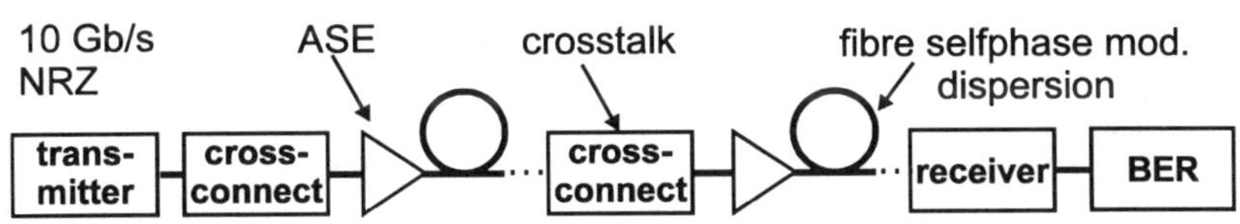

Fig.7: Transparent optical transmission path configuration

Fig.8 shows, how the interaction of effects acts on the transparency range for a sensitivity penalty less than 2 dB. Horizontal lines indicate the transparency limits due to fibre trunks, whereas vertical lines the limits due to optical crossconnects. The transparency range itself is limited by all these effects as shown in the marked area of the diagram.

For investigation of linear effects (dispersion, ASE, crosstalk) a data stream of 4096 bits was used to determine the transparency limits. The transparency limit due to fibre dispersion occurs near 10.900 km. The EDFA noise limits the transmission length to about 12.100 km. For 10 crosstalk sources a tolerable crosstalk per source of about -33.5 dB was determined. The interaction of fibre dispersion, EDFA noise and crosstalk (-33.5 dB) in the crossconnects leads to a transparency limit of about 5.600 km for 4 crossconnects. A data stream of 1024 bits was used to determine the transparency ranges due to fibre nonlinearity (signal power 0 dBm). The interaction off ASE, dispersion and nonlinearity leads to a transparency limit of near 4.400 km. In all cases the BER was estimated by a Gaussian approximation. The precision of this kind of system analysis depends upon the time and frequency resolution of the sampled signals, the accuracy of the used mathematical model as well as the number of calculated bits for the BER estimation.

module	parameter	value
transmitter	bitrate wavelength data stream	10 Gb/s 1550 nm PRBS
Mach-Zender modulator	extinction ratio 3 dB bandwidth	> 50 dB (ideal) 10 GHz
fibre	attenuation dispersion dispersion slope nonlinearity coefficient length	0.2 dB/km -0.1 ps/nm/km 0.08 ps/nm/nm/km 2.6 10^{-16} cm^2/W 75 km
EDFA	gain ASE bandwidth noise figure	15 dB 120 GHz 4 dB
PIN receiver	conversion factor noise 3 dB bandwidth	1 A/W thermal (20 pA/√Hz), shot noise 10 GHz

Tab.1: Relevant system parameters for the simulation

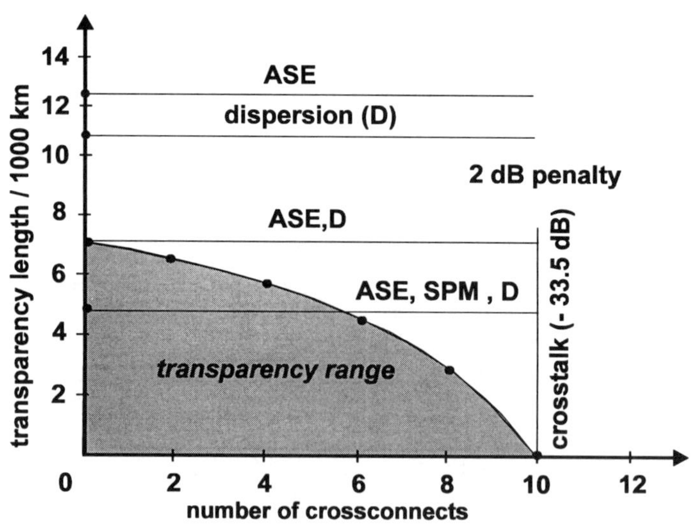

Fig.8: Transparency range diagram for the investigated signal path (Fig.7, Tab.1)

5. Conclusion

The transparent optical core network contains frequency tubes with a defined bandwidth, which are routed by OFDM crossconnects in the frequency and space domain. The network can be characterized by the transparency range diagram. The limits in this diagram depend on the crossconnect components and the fibre trunk characteristics. The presented simulation software *BroadNeD* is an efficient tool for the investigation of transparency limiting effects in optical networks.

Acknowledgement

This work was carried out under a research contract with the Deutsche Telekom AG.

References

[1] P. Cochrane, R. Heckingbottom, D. Heatley: Optical transparency, Optics & Photonics News 1994, Vol.5, No.8, pp.15-54

[2] B. Strebel, C. Caspar, H.-M. Foisel: Transparency limitations in optical core networks. Proc. of the NOC ´96, Broadband Superhighway, pp.193 - 199

[3] B. Strebel, C. Caspar, H.-M. Foisel, R. Freund: Optical transparency in future core networks. Photonics in Switching 1997, paper PThA1

[4] F. Matera, M. Settembre: Performance evaluation of optical systems operating on long fibre links. Optical and Quantum Electronics, 27 (1995), pp. 831-846

[5] J. Buck, S. Ha, E. A. Lee and D. G. Messerschmidt: Ptolemy: A Framework for Simulating and Prototyping Heterogeneous Systems. Int. J. Computer Simulation, special issue on Simulation Software Development", January, 1994

[6] C. Caspar, H.-M. Foisel, R. Freund, U. Krüger, B. Strebel: Cascadability of arrayed-waveguide grating (de)multiplexers in transparent optical networks. OFC ´97, Technical Digest, Vol.6, pp. 19- 20

Core and ATM Networks
D.W. Faulkner and A.L. Harmer (Eds.)
IOS Press 1997 © AKM

Design Techniques for SDH Mesh-Restorable Networks

F. Poppe*, M. Pickavet*, P. Arijs and P. Demeester,
University of Gent - IMEC, Department of Information Technology,
Sint-Pietersnieuwstraat 41, 9000 Gent, Belgium.
Tel. no. +32 9 264 34 47
Fax. no. +32 9 264 35 93
e-mail {poppe,mpick,arijs}@intec.rug.ac.be

Abstract. We studied the design of SDH mesh-restorable networks. We propose three different algorithms for solving the design problem: an integer linear programming (branch-and-cut-and-price) algorithm, a zoom-in approach which combines a genetic algorithm with deterministic optimisation routines and a GRASP based heuristic algorithm. We compare the numerical results we obtained by using these algorithms. For each of the algorithms, we study the cost of the designs which it produces and the time and memory requirement.

1. Introduction

The design of low cost reliable telecommunication networks is a notoriously difficult problem. Choosing the implementation technology and the network architecture are typically tasks for human experts. Computer-aided design procedures can be used to determine low cost reliable designs once the implementation technology and the network architecture have been chosen and reliable node-to-node demand data have been gathered. We develop and compare several design procedures for telecommunication networks based on the Synchronous Digital Hierarchy (SDH) with a mesh-restorable architecture (see [1]). An SDH mesh-restorable network is a cost-efficient solution in many cases. When a failure occurs in such a network, the connections which are affected are recovered by a restoration mechanism, which relies on the presence of spare capacity on the links of the network, and uses this spare capacity to reroute the affected connections. Spare capacity is not dedicated to one particular failure scenario or to one particular connection but is shared among all failure scenarios and all connections.

We studied the problem of designing an SDH mesh-restorable network. Given is a set of node locations and a VC-4 node-to-node demand. Asked is to determine between which nodes links should be installed and which capacity should be assigned to each link. The capacity which is installed has to be sufficient to carry the demand, not only when all network components are functioning properly, but also in case a single link fails. We assume that the restoration mechanism is a link restoration mechanism (see [1], [2]). This

* Research Assistant of the FWO-V (Fund for Scientific Research (Belgium)).

means that in case of a link failure, the affected connections are rerouted between the endpoints of the failed link.

The installation of a link with a certain VC-4 capacity between two nodes involves the installation of a cable between those nodes, and a sufficient number of line systems, each line system using up one fibre of the cable. Furthermore, since the line system signal level degrades with transmission distance, regenerators have to be placed at regular intervals along a cable. Accordingly, we considered three different costs: the cable installation cost, which is roughly proportional to the distance between nodes, the line system installation cost, and the regenerator cost.

We considered STM-1 (155 Mbit/s), STM-4 (622 Mbit/s) and STM-16 (2.5 Gbit/s) line systems. Economies of scale play a role in the selection of line systems. For instance, an STM-4 offers a capacity which is four times as high as that of an STM-1 at less than double the cost (a similar relationship holds between STM-16 and STM-4). As a result, when the VC-4 demand offered to the network is relatively high, one may expect a predominant use of STM-16 line systems, STM-4's and STM-1's only being installed where STM-16's would be severely underused.

It is appealing to handle this problem in a sequential fashion because algorithms for solving parts of the problem already exist. We could for instance start by designing a low cost VC-4 link topology on the nodes which satisfies certain connectivity constraints in order to ensure reliability, using the algorithm described in [3]. Routing the VC-4 demand and assigning initial capacities to the VC-4 links in this VC-4 link topology can then be done using the algorithm described in [4]. Once the working capacity of each of the VC-4 links is known, the algorithm described in [5] can be used to determine the amount of spare capacity which has to be added on the VC-4 links in order to enable link restoration in case of VC-4 link failures.

A simple sequential approach generally does not result in an acceptable design because at each stage, too many problem aspects are discarded. Therefore, we developed several new algorithms for solving the SDH mesh-restorable network design problem. In this paper, we compare three different algorithms for solving the design problem: a branch-and-cut-and-price algorithm, a zoom-in algorithm and a GRASP based heuristic algorithm. For each of the algorithms, we study the running time and the cost of the designs which it produces.

2. An integer linear programming algorithm

Formulating the SDH mesh-restorable network design problem as an integer linear programming (ILP) problem has the advantage of eliminating all possible misconceptions about the design problem. It encompasses identifying the decision variables (columns of the ILP problem) and parameters of the problem, the relationships (rows of the ILP problem) between these variables and parameters, and the objective function which has to be minimised. This formulation has been explained in much detail in [6].

In order to lower the number of decision variables, the ILP formulation we developed only considers STM-16 line systems. This is justified - as a first approximation - when the demand estimations are high because the network cost reduction which can be realised by replacing underused STM-16 with STM-4's and STM-1's will then be relatively small.

Once the design problem has been formulated as an ILP problem, it can in principle be solved to optimality using a general ILP algorithm (such as a general branch-and-bound algorithm or a general cutting-plane algorithm, see for instance [7]). In practice however, this is only possible for very small problem instances (for instance when the network only contains 8 nodes and only 13 candidate link locations are considered, see [6]). For larger problem instances, which have more practical importance for designers of telecommunication networks, the use of such a general ILP algorithm simply would take too much time, and would require too much memory. Therefore, in order to solve larger problem instances, we developed a specialised branch-and-cut-and-price algorithm, which is explained in much detail in [6]. This specialised algorithm is especially effective in reducing the amount of memory needed to solve the ILP problem (reducing the number of columns and rows in the ILP formulation of the problem). However, even with the specialised algorithm, it is still not possible to solve the ILP problem to optimality in a reasonable amount of time for larger networks. We chose to limit the number of nodes of the branch-and-bound tree which the algorithm explores. Therefore, the solution produced by this algorithm is not necessarily the optimal solution to the ILP problem. However, in general the algorithm produces solutions which are satisfactory and can not be improved in a trivial way.

The ILP algorithm has its limitations in that it can not be used to consider a very large number of candidate link locations. The network designer has to choose carefully which links he wants to consider as possible locations for installing a cable. On the other hand, the ILP algorithm performs very well when an existing cable infrastructure can be used, and the only thing left to do for the designer is to assign a capacity to each link. In that case, the ILP algorithm produces near-optimal results.

3. A zoom-in approach

A second method to design SDH mesh-restorable networks uses a zoom-in technique, which means that the topology and capacity assignment are constructed rather roughly in the beginning and the cost model and search focus are refined as the algorithm proceeds. This strategy is induced by the difficulty of the overall design problem, which implies that tackling the problem as a whole (i.e. considering the entire cost model from the beginning) would result in enormous running times for practical problems of bigger dimensions.

In a first phase, a genetic algorithm (see for instance [8],[9],[10]) produces a topology and an estimate of the needed link capacities (we refer to [11] for a detailed description). The cost model in this phase considers the cable installation cost (proportional to the length of the link), a linearisation of the line system cost (proportional to the capacity of the link) and a linearisation of the regenerator cost (proportional to both length and capacity). The link capacities (working and spare capacities) are estimated by using a bi-routing strategy which routes an adapted demand matrix.

In a second (optional) phase, we try to improve the result of the genetic algorithm by a local optimisation technique, still using the same cost model.

The third phase considers the real cost model (as described in Section 1) and clearly distinguishes between working and spare capacity. In order to reduce the calculation time needed to evaluate a topology, the working traffic is routed along the shortest paths. The spare capacity is assigned to satisfy full link restorability in case of link failures. Using this extended cost model, we search for a better topology using a local optimisation technique. The solutions in this phase are all feasible solutions of the problem described in Section 1, since the capacity assignment allows full routing of the demand and ensures complete link restorability in case of single link failures.

Finally, the fourth phase concentrates on a further improvement of the capacity assignment without changing the network topology. Here bifurcation of the working traffic is allowed, as long as the demand is divided in integer VC-4 parts.

The algorithm returns the final topology, the working and spare capacity assignment on each link, the number of STM-1, STM-4 and STM-16 line systems on each link, the number and type of regenerators on each link and the paths of the working traffic.

The impact of each of the four phases depends on the relative importance of cable installation cost versus line system costs. If the former is big enough in comparison with the latter (which is a realistic assumption in most practical cases), the topology found in the first phase (genetic algorithm) usually cannot be improved by the second phase (local optimisation using simplified cost model) and the third phase (local optimisation using correct cost model). This shows that generally the choice of the topology does not depend very much on the exact capacity allocation procedure (routing strategy, restoration or protection strategy) as long as the overall capacity cost remains roughly the same. In extreme situations with high demands in a small area however, the local optimisation techniques might improve the topology of the first phase.

4. A GRASP based heuristic algorithm

The previous two methods focused on the topological aspect of network design: which links to use in order to meet the goal of the planning problem. In this section we describe a traffic based heuristic approach. This approach is aimed towards finding the appropriate path for each demand pair. The algorithm is based on greedy randomised adaptive search procedures (GRASP) as described in [12].

A GRASP is an iterative process, with each GRASP iteration consisting of two phases: a construction phase and local search phase. During the construction phase a feasible solution is iteratively constructed, by adding one element at a time. The choice of the next element to be added at each construction step, is determined by a selection criterion (based on a 'fitness', which measures the benefit of selecting each element) which is greedy but also includes an element of randomness by not always selecting the best next element but sometimes also the second or third best one. Furthermore the fitness of each element is adapted after selecting an element, to reflect the changes brought on by this selection. During the local search phase the solution of the construction phase is optimised by looking for better solutions in the neighbourhood of this solution. Once a local optimum is reached a new GRASP iteration is started. The best solution found during all GRASP iterations is kept as the final solution.

The philosophy behind this approach is to rapidly generate good initial solutions, which serve as a starting point for the local search phase. Indeed, the key towards success for a local search technique consist not only of the suitable choice of a neighbourhood structure, but also of a good initial solution. By using the initial solutions obtained in the first phase of the GRASP iteration we hope this local search will converge rapidly to a good local optimum.

We apply the GRASP approach to our problem by setting up one path at a time during the construction phase. For each possible path we calculate the additional cost to set it up. Then one path is selected among the paths with lowest additional cost and it is set up. We use the same approach for setting up paths between the source and target nodes of the links that have to be protected against fibre breaks. In this way a good initial network is constructed. During the local search phase we iteratively try to reroute some paths (both working and spare capacity paths are considered) to further decrease the overall network cost.

The implementation of this algorithm was still in progress at the time of writing, however a simplified version of the algorithm was already implemented in order to present some preliminary results. One of the main simplifications is that only one path was used for each demand pair as well as for the path which protects a link. Besides, the local search phase was not fully developed yet, but it did allow to produce some first results.

5. Numerical results

The different algorithms described above were tested extensively on several quite diverse problem situations: we varied the scale of the network (international network - regional network - metropolitan network), the number of nodes and the density of the demand matrix. The results mentioned here apply to the following problem instances. We consider only STM-16 line systems, the cable installation cost is 2 per km, the STM-16 line system cost is 3.5, and the regenerator cost is 0 (all costs are expressed in million Belgian francs). We concentrate on two 20-node problem instances with exactly the same demand matrix but implemented on a different scale (a regional and a metropolitan network, the network diameter is 300 km and 30 km, respectively). All calculation times are measured on a Pentium Pro 200 Mhz.

The solutions obtained by the integer linear programming algorithm are shown in Figure 1. The number of candidate links was restricted, in order to reduce the memory requirement of the algorithm. The selection procedure for the candidate links was based on the solution of a weak linear relaxation of the travelling salesman problem on the 20 nodes. We will show that this selection procedure is usually adequate by showing that the cost savings which can be obtained by considering a much larger number of links are relatively small. The solutions shown in Figure 1 were obtained within 28 seconds for the regional network (with a cost of 3038, considering 26 candidate links) and within 327 seconds for the metropolitan network (cost 649, 37 candidate links).

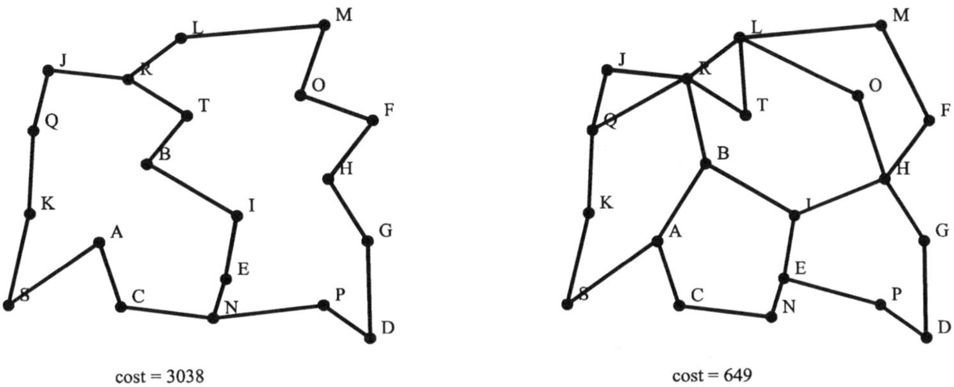

Figure 1 : regional and metropolitan network (ILP algorithm)

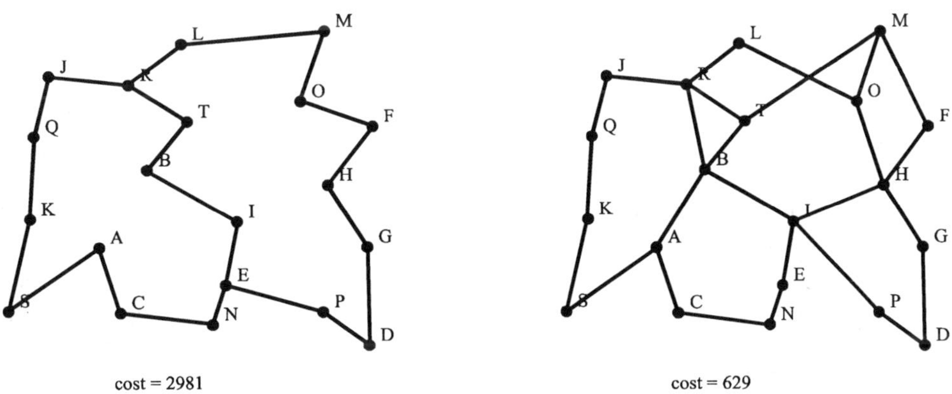

Figure 2 : regional and metropolitan network (zoom-in algorithm)

Figure 2 shows the results of the zoom-in approach. The solutions were obtained within 58 seconds for the regional network (with a cost of 2981, considering 71 candidate links) and within 912 seconds for the metropolitan network (cost 629, 71 candidate links). At first sight, the crossing links in the metropolitan network are unexpected. However, when examining the demand matrix, high traffic between node M and node S on one hand and between node L and node D on the other hand is noticed. As a consequence, introducing crossing links leads to a considerable reduction of the line system costs and a (small) reduction of the overall cost.

The results of the GRASP approach were obtained using the same candidate link positions as in the ILP algorithm. For the regional network the cost of the solution which was found is 2-3% above the results of the first two algorithms, for the metropolitan problem, the gap is bigger: about 15-20%. The calculation times needed to produce these results are in the same order of magnitude as the zoom-in algorithm. The significant gaps with respect to the cost of the found solutions are mainly introduced by the restriction on the spare capacity assignment (only one path is used to protect a link failure) and the preliminary stage of the local search phase. The relatively good solution found for the regional network indicates that the algorithm has the potential to produce good results when it is further refined.

The results of the first two algorithms stress the importance of the network scale on the topology: the regional network is very sparse while the metropolitan network is clearly meshed. Indeed, the cable installation cost severely depends on the scale of the network, the line system cost on the other hand is not influenced by the network scale. The cable installation cost, which represents the driving force towards a sparse network, has an enormous impact in the regional case, whereas the influence of the line system cost, driving force towards dense networks, is very clear in the metropolitan case.

When comparing the results for the regional network, we notice a remarkable resemblance: the difference in cost (1.2 %) is due to one link connecting node P with node N or E respectively. The solution found with the zoom-in algorithm, could not be found by the ILP algorithm, because of the absence of link P-E in the set of candidate links. On the other hand however, the severe restriction of the number of candidate links (only 26 links for a 20-node problem) does not severely decrease the quality of the solution, which indicates the intrinsic power of the candidate link choosing heuristic based on the Travelling Salesman solution.

More differences, both with respect to the cost (3.2 %) and the chosen links, are noticed when comparing the results for the metropolitan network. Now the overall network cost is divided almost fifty-fifty between cable installation and line system cost. This leads to a much tougher problem than the regional problem - which can also be noticed from the calculation times.

A comparison of the three algorithms with respect to the time and space requirements indicates that the ILP approach uses less CPU time, whereas the zoom-in approach and the GRASP algorithm consume less memory (e.g. even for a 50 node problem the zoom-in approach did not exhibit any memory problems).

6. Conclusion

We studied the design of SDH mesh-restorable networks. We formulated a design problem involving the design of the physical layer (cable infrastructure) and the design of the SDH VC-4 layer (both working and spare capacity). The use of heuristics to solve this design problem seems justified, because attempts to solve it to optimality using an ILP approach failed. Three heuristics were developed to tackle the design problem: an integer linear programming (branch-and-cut-and-price) algorithm, a zoom-in approach and a GRASP approach. The first two methods yield comparable results, whereas the implementation of the third method is still in a preliminary phase.

Acknowledgement

This work has been supported by the Flemish Government through the IWT project ITA/950214/INTEC

References

[1] T.H. Wu, *Fiber Network Service Survivability* (Artech House, London, 1992).

[2] K.-I. Sato, *Advances in Transport Network Technologies, Photonic Networks, ATM and SDH* (Artech House, Inc., Norwood, 1996).

[3] C.L. Monma and D.F. Shallcross, Methods for Designing Communications Networks with Certain Two-Connected Survivability Constraints, Operations Research 37 (4) (1989) 531-541.

[4] Gavish B. and Neuman I., A System for Routing and Capacity Assignment in Computer Communication Networks, IEEE Transactions on Communications 37 (4) (1989) 360-366.

[5] Lee K., Park K. and Park S., Spare Channel Assignment for DCS Mesh-Restorable Networks, *Proc. of the Third International Conference on Telecommunication Systems, Nashville* (1995) 296-307.

[6] F. Poppe and P. Demeester, The Design of SDH Mesh-Restorable Networks: Problem Formulation and Algorithm, *Proc. of the Fifth International Conference on Telecommunication Systems* (1997) to be published.

[7] G.L. Nemhauser and L.A. Wolsey, Integer Programming, in: *Optimization*, G.L. Nemhauser, A.H.G. Kan, M.J. Todd, eds., Handbooks in Operations Research and Management Science 1 (North-Holland, Amsterdam, 1989).

[8] J.H.Holland, Adaptation in Natural and Artificial Systems (University of Michigan Press, 1975).

[9] L.Boullart, A.Krijgsman and R.A. Vingerhoeds, Application of Artificial Intelligence in Process Control (Pergamon Press, 1992).

[10] E. Falkenauer, Tapping the Full Power of Genetic Algorithms through Suitable Representation and Local Optimization: Application to Bin Packing, Evolutionary Computation 2 (1995).

[11] M. Pickavet, F. Poppe, J. Luystermans and P. Demeester, A Genetic Algorithm for Solving the Capacitated Survivable Network Design Problem, *Proc. Of the Fifth International Conference on Telecommunication Systems* (1997) to be published.

[12] T.A. Feo and M.G.C. Resende, Greedy Randomised Adaptive Search Procedures, *Journal of Global Optimization* 6 (2) (1995) 109-133.

Core and ATM Networks
D.W. Faulkner and A.L. Harmer (Eds.)
IOS Press 1997 © AKM

A Comparison of 2 and 4 Fibre MS SPRings

M. Robinson and I. Wright
Fujitsu Europe Telecom R&D Centre Ltd. Uxbridge UK

Abstract: This paper presents the results of a reliability and cost comparison between two transmission networks employing different ring protection schemes. The two schemes under study are the 2 fibre and 4 fibre Multiplexer Section Shared Protection Ring (MS SPRing). The impact of network size, in terms of total number of nodes and span between nodes, as well as the time to repair for any equipment or fibre failure within the network are considered in the reliability analysis. The cost analysis compares the hardware and installation costs of the two networks in a greenfield site where two types of fibre routing approaches are considered.

1. Introduction

SDH networks employing ring architectures are being extensively deployed throughout the world. There are a number of ring protection schemes available, one of which is a shared protection scheme. The main advantage of a shared protection scheme is its ability to re-use transmission capacity around the ring. Capacity re-use can lead to bandwidth efficient network solutions specifically for networks where the traffic distributions are predominantly between adjacent nodes on that ring, Tier 1 networks for example. Currently two shared protection schemes are being specified by the standards bodies, the 2 fibre and 4 fibre Multiplexer Section Shared Protection Ring (MS SPRing).

This paper compares the two mechanisms in terms of network reliability and relative network cost per STM-1 tributary.

Detailed explanations of how these two mechanisms operate are given in [1] and the following sub-sections provide a brief summary of their operation and the key differences between the two.

1.1. The 2 fibre MS SPRing

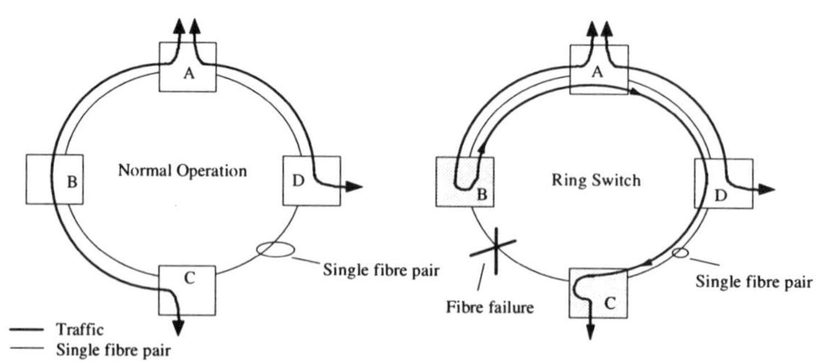

Figure 1 The 2 fibre MS SPRing operation

Figure 1 shows the 2 fibre MS SPRing in normal operation and under fibre failure condition. Traffic is transported bidirectionally between terminating nodes, in this example between nodes A and C and nodes A and D, normally over the shortest path. Half of the transmission capacity between any two nodes is dedicated for working traffic whilst the other half is dedicated to protection traffic under fault conditions. An STM-16 MS SPRing will therefore have a maximum of 8 x STM-1 protected working traffic capacity between any two nodes.

Under fault condition, for instance an optical card failure at node B or as in this example a fibre failure between nodes B and C, the nodes adjacent to the failure, nodes B and C, will detect a loss of traffic. Node B will loopback the traffic that was intended for the failed path onto the protection path in the opposite direction around the ring utilising the dedicated spare capacity of the fibre pair towards the destination node, node C. Node C will bridge the traffic from the protection channel of the other fibre pair to the dropped tributary port so as to maintain traffic connectivity.

1.2. The 4 fibre MS SPRing

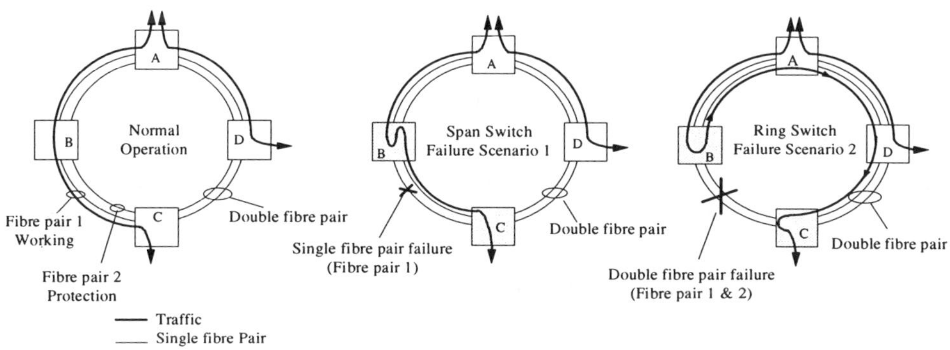

Figure 2 The 4 fibre MS SPRing operation

Figure 2 shows the 4 fibre MS SPRing in normal operation and under a single and double fibre failure condition. Traffic is transported between terminating nodes in the same manner as the 2 fibre MS SPRing however the working traffic is transported between nodes on a dedicated fibre pair and a second dedicated fibre pair is used for protection traffic under fault conditions. The maximum protected working traffic capacity between any two nodes on a STM-16 4 fibre MS SPRing will therefore be 16 x STM-1, this is twice the capacity of the 2 fibre MS SPRing.

The 4 fibre MS SPRing has two forms of protection switching, a span switch and a ring switch. Referring to Figure 2, in failure scenario 1 a single fibre pair carrying traffic, fibre pair 1, between nodes B and C is broken. The Span switch mechanism switches the traffic onto the protection fibre pair, fibre pair 2, between these two nodes. In failure scenario 2 a double fibre pair, working and protection fibres, between nodes B and C are broken. Under these circumstances a ring switch occurs as with the 2 fibre MS SPRing the main difference being that the working traffic is rerouted onto a dedicated protection fibre pair. Although only fibre failure scenarios are discussed in these examples the same holds for optical interface failure where the term fibre failure can be interchanged with the term interface failure for example.

2. Network Reliability

The two networks were modelled and assessed in terms of their reliability. This was achieved by calculating the total downtime in minutes/year of a single STM-1 tributary being transported over the longest working path under normal operation of a ring of N nodes. The basic model used for our network reliability analysis is given in Figure 3.

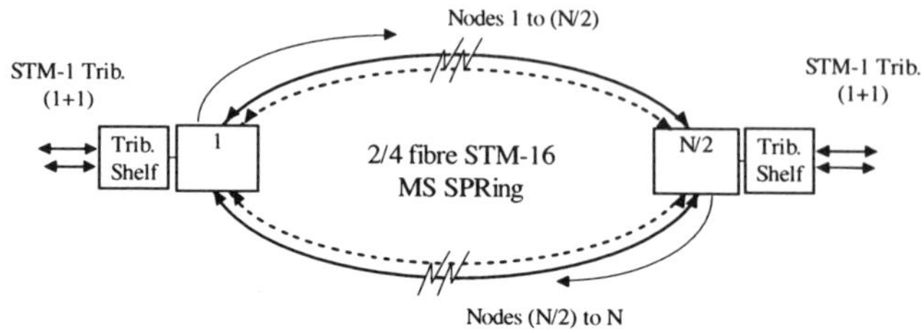

Figure 3 Network Reliability Architecture

2.1. *Network Assumptions*

With reference to Figure 3 the following assumptions were made:

⇒ The STM-1 tributaries will employ Multiplexer Section Protection (MSP)

⇒ The tributary shelves for the 2 fibre and 4 fibre networks are common

⇒ 2 and 4 fibre MS SPRing Network Elements (NEs) are common with the exception of additional aggregate optics and span switching cards for the 4 fibre MS SPRing nodes.

⇒ The maximum number 'N' of NEs will be 16 as defined in [1].

In the analysis the following three network types were modelled:

⇒ A 2 fibre MS SPRing

⇒ A 4 fibre MS SPRing without diverse fibre routing

⇒ A 4 fibre MS SPRing with diverse fibre routing

The 4 fibre MS SPRing was modelled for a Diversely Routed (DR) and Non Diversely Routed fibre network (NDR). A DR fibre network will have the fibre pairs between NEs laid in different trenches and therefore different ducts. For a NDR fibre network the fibre pairs will lie in the same ducting.

Reference [2] considered the reliability of 2 fibre and 4 fibre MS SPRing networks for NDR fibre paths only and fixed Mean Time To Repair (MTTR) figures for equipment and fibre plant. This paper considers the impact of varying MTTR figures for equipment and fibre as well as employing fibre route diversity for the 4 fibre MS SPRing network.

2.2. *Network Variables*

The parameters that were varied for the three networks were :

⇒ **Number of Nodes** - Networks comprising 4, 8 and 16 nodes were modelled.

⇒ **Node Spacing** - The internodal spans, and therefore fibre lengths, between NEs were varied between 20, 40, 60 and 80km. This directly impacts on the overall network reliability as network downtime is predominantly attributable to fibre failure and increases as a function of the fibre length. The maximum distance between nodes will be 80km which corresponds to a L16.2/3 interface as defined by [3]. As the fibre failure rate is measured in failures in time per km then 80km will give rise to the maximum fibre failure rate, for example a fibre span of 80km will have 4 times more failures in a given period than a 20km fibre span.

⇒ **Mean Time To Repair** - The Mean Time To Repair (MTTR) figures were varied for each model, the figures used were :

⇒ MTTR Cable Failure :	12 & 24 hours
⇒ MTTR Card Failure :	5 & 24 hours

2.3. Reliability

Network reliability was assessed in terms of the number of minutes per year that a network would not meet its required quality of service. Network models were generated and the down time in minutes per year for a single STM-1 tributary were calculated for the three networks discussed over a system lifetime of 15 years. A fibre failure rate value of 250 failures per km in 10^9 hours as presented in [4] was used in the models.

2.3.1. Fibre Failure

The predominant cause of a fibre failure between two nodes is due to an underground duct being accidentally dug up, therefore only failure figures for these dig up incidents are considered in the reliability model as a fibre failure mechanism. This will impact on the NDR 4 fibre MS SPRing as it is fair to assume that all the fibres in a given duct will fail when the duct has been dug up. Therefore in terms of reliability it is equally likely that downtime will arise due to fibre failure for a single or double fibre pair in the same duct. Down time due to fibre failure will occur for the following two cases :

⇒ **Case 1. A 2 fibre MS SPRing or Non Diversely Routed 4 fibre MS SPRing** - In the event of a fibre failure in path A ***and*** path B, see Figure 4.

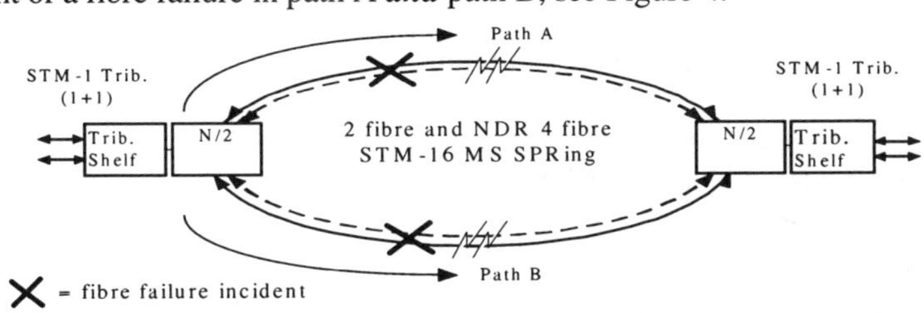

Figure 4 2 Fibre and Non Diversely Routed 4 fibre MS SPRing Fibre Failure

⇒ **Case 2. A Diversely Routed 4 fibre MS SPRing -** Referring to Figure 5, a double fibre failure on any single span, where a span is a point to point fibre pair link between two NEs, over path A (node 1 to node 2 in this example) ***and*** any fibre failure over path B. The same would be true if the fibre failure incidents occurred in the other paths, two fibre failures in path B and a single fibre failure in path A.

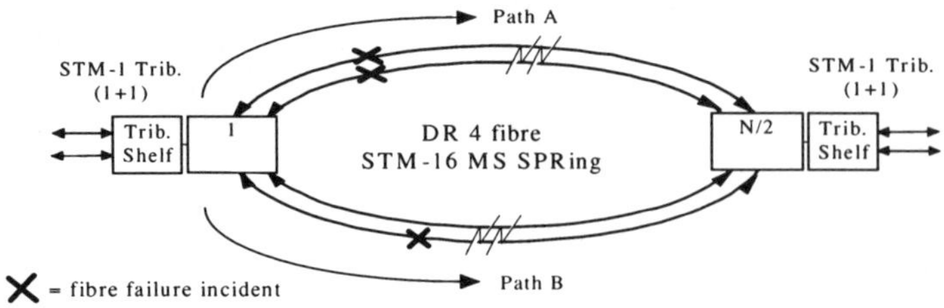

Figure 5 Diversely Routed 4 Fibre MS SPRing Fibre Failure

2.3.2. Equipment Failure

The models were also developed to include the effects of equipment failure on network downtime for the different network architecture.

2.4. Simulation Results

The following graphs show the downtime in minutes per year of an STM-1 tributary being transported over three networks of 4, 8 and 16 nodes with NEs separated by 20, 40, 60 and 80km of fibre for different figures of MTTR for card and cable failures.

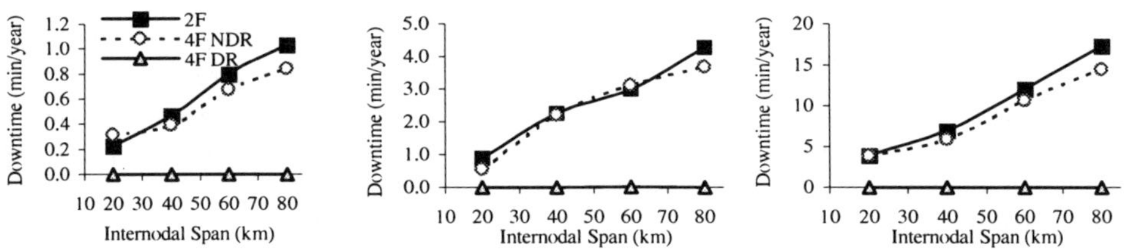

4 Node Rings 8 Node Rings 16 Node Rings

Figure 6 Network Downtime, MTTR Card - 5h, Fibre - 24h

Figure 6 shows the network downtime for the three networks understudy with MTTR figures of 5 hours for a card replacement and 24 hours for a cable fault repair. The optimum choice between the 2 fibre and NDR 4 fibre network in terms of network reliability is not clear as the downtime figures for both are very similar. For the DR 4 fibre network no downtime was measured over the simulation period used for all scenarios.

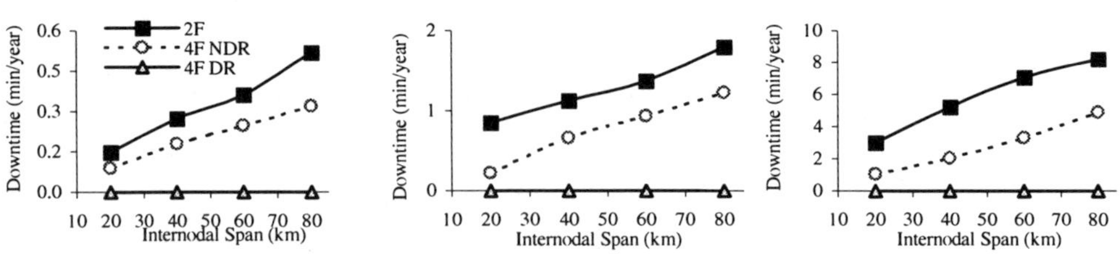

4 Node Rings 8 Node Rings 16 Node Rings

Figure 7 Network Downtime, MTTR Card - 5h, Fibre - 12h

The graphs of Figure 7 show network downtime for the three networks understudy for MTTR figures of 5 hours for a card replacement and 12 hours for a cable fault repair. In comparison to Figure 6 it can be seen that improving the MTTR for a cable failure approximately halves the network downtime. In this scenario the difference in terms of downtime between the 2 fibre and NDR 4 fibre network is more apparent. For the DR 4 fibre network no downtime was measured over the simulation period used for all scenarios.

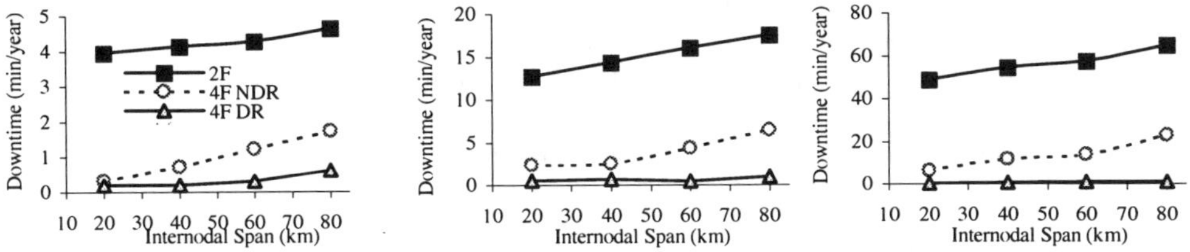

4 Node Rings | 8 Node Rings | 16 Node Rings

Figure 8 Network Downtime, MTTR Card - 24h, Fibre - 12h

The graphs of Figure 8 show network downtime for the three networks understudy for MTTR figures of 24 hours for a card replacement and 12 hours for a cable fault repair. Again the difference in terms of downtime between the 2 fibre and NDR 4 fibre network is more apparent than the previous results. In this example downtime for the DR 4 fibre network was measured and low in comparison to the other two networks.

2.4.1. Discussion

Figure 6 to Figure 8 show that network downtime will vary significantly depending upon network size, i.e. number of NEs and the span between NEs, as well as the MTTR figures for both equipment and cards. It is clear that the MTTR of both fibre and equipment will directly impact on the network downtime for the different network architectures in different ways.

The 4 fibre DR network clearly has superior reliability characteristics to the other two networks. The downtime measured for the MTTR parameters of Figure 6 and Figure 7 were negligible, in fact only significant downtime was measured for the MTTR parameters presented in Figure 8. The DR 4 fibre network will always provide improved reliability when compared with the other two network scenarios discussed.

In reference to Figure 6 to Figure 8 for the 4 fibre NDR network and the 2 fibre network the following can be concluded :

⇒ As the network downtime associated with card failure increases with respect to fibre failure then the 4 fibre NDR network will provide increased reliability in comparison to the 2 fibre network.

⇒ As network downtime associated with card failure decreases with respect to fibre failure then the downtime associated with the 4 fibre NDR network and the 2 fibre network will tend towards one another ultimately providing equally reliable network solutions.

For any of the previous networks of a fixed size, in terms of number of nodes and span between nodes, it should be noted that downtime due to fibre or card failure will obviously increase as the MTTR value increases.

Although there are more components associated with the 4 fibre NDR NEs than the 2 fibre NEs the vastly improved reliability achieved by the span switching capabilities of the

4 fibre MS SPRing shadow any increased likelihood of card failure once the protection path is used.

It is clear from the above that deployment of either a 2 fibre or 4 fibre MS SPRing will provide an extremely resilient network which in the models simulated provide downtime figures for an STM-1 path which equate to an average availability greater than 99.98%. As might be expected the 4 fibre MS SPRing provides improved protection capabilities when diversely routed paths are used. In the NDR case the optimum choice between the 4 fibre or 2 fibre network is not clear in that the network operators maintenance/repair strategy will directly impact on each networks performance differently.

3. Relative Cost Analysis

A simplified cost analysis has been performed to compare the installation and equipment costs of a 2 fibre and 4 fibre MS SPRing network at a greenfield site. A greenfield site will have no existing fibre infrastructure or transmission equipment. The cost analysis was carried out to determine the complete installation costs of ducting, equipment and fibre as well as the relative cost per STM-1 tributary for the following networks:

⇒ A 2 fibre MS SPRing

⇒ A 4 fibre MS SPRing without diverse fibre routing.

⇒ A 4 fibre MS SPRing with diverse fibre routing.

3.1. Cost Breakdown

Ducting costs - This includes the hardware and civil costs involved in laying new ducting, a fixed cost per km is assumed for all ducting scenarios.

Fibre costs - This includes hardware and civil cost of installing a fibre bundle in existing ducting. For Diversely Routed fibre networks, fibre and ducting costs increase two fold. For 2 fibre MS SPRing and Non Diversely Routed 4 fibre MS SPRing networks only a single fibre bundle in a single duct will be required between two equipments.

Equipment costs - The cost of an STM-16 2 fibre and 4 fibre MS SPRing ADM with 4 protected (1+1) STM-1 tributaries is to be considered. The 4 fibre ADM will cost approximately 1.5 times more than the equivalent 2 fibre ADM due to the additional aggregate interfaces and switching cards for the 4 fibre equipment.

Costs associated with buildings are not considered here and are assumed to be the same for both 2 fibre and 4 fibre architectures. Fibre, ducting and equipment costs are given in the following Table 1:

Table 1 Hardware Costs

Hardware	Unit Cost
Fibre Bundle and Single Duct	0.75/km
2 fibre ADM	1/node
4 fibre ADM	1.5/node

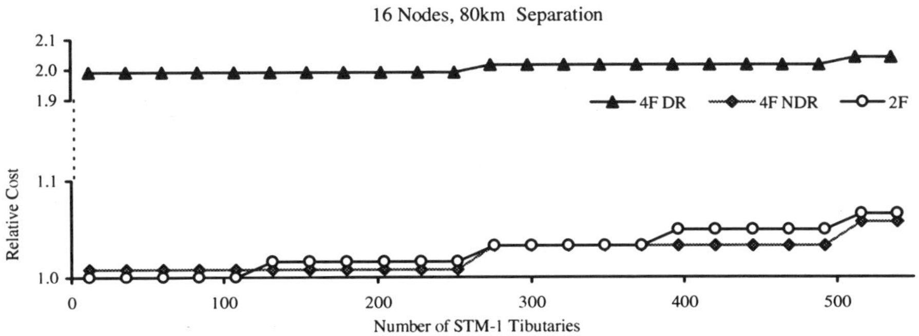

Figure 9 Tributary Access Vs Relative Network Cost - Greenfield Site

Figure 9 shows the relative cost per STM-1 tributary for the various networks, note the Y axis has a discontinuous scale. The network costs are normalised against the cost of installing a 2 fibre MS SPRing. After the networks are in place we assume that there is unlimited access to spare free fibre. When the capacity demand exceeds that of the networks capabilities an additional set of sixteen nodes is added. This assumption implies an evenly distributed traffic demand pattern increase, localised traffic demand pattern increases are not considered here. As there is spare fibre then only the additional cost of the nodes is added to that of the existing network. It is clear to see that although the maximum capacity of a 2 fibre MS SPRing is half that of the 4 fibre MS SPRing it will remain a considerably cheaper solution per STM-1 tributary compared to the Diversely Routed 4 fibre MS SPRing network due to the lower initial installation costs of the 2 fibre MS SPRing network. The 4 fibre NDR MS SPRing network is initially more expensive than the 2 fibre MS SPRing due to the increased equipment cost, however, when a network upgrade is required for the 2 fibre MS SPRing, the 2 fibre MS SPRing becomes more expensive.

4. Conclusion

The above discussions have shown that the optimum network protection architecture will depend on a number of factors: the number of nodes, node separation, number of tributaries, route diversity, MTTR parameters and cost of implementation. Although the Diversely Routed 4 fibre MS SPRing gives improved results, the worst case 2 fibre MS SPRing and Non Diversely Routed 4 fibre MS SPRing network, comprised of 16 nodes and a ring circumference of 1280 km, still has an average availability figure of greater than 99.98 %. The 4 fibre MS SPRing will provide increased transmission capacity and the Diversely Routed network will always provide improved availability at a significantly increased cost.

References

1. ITU-T Recommendation G.841, DIGITAL NETWORKS "TYPES AND CHARACTERISTICS OF SDH NETWORK PROTECTION ARCHITECTURES" 07/95

2. M. To and P. Neusy, "Unavailability Analysis of Long-Haul Networks", IEEE JSAC, vol. 12, No. 1, pp. 100-109, Jan. 1994.

3. ITU-T Recommendation G957, DIGITAL SECTIONS AND LINE SYSTEMS "OPTICAL INTERFACES FOR EQUIPMENTS AND SYSTEMS RELATING TO THE SYNCHRONOUS DIGITAL HIERACHY" 07/95

4. SG15-Delayed Contribution, Reliability of fibre optic links in U.S. Telecoms. networks D.169, May 94

Core and ATM Networks
D.W. Faulkner and A.L. Harmer (Eds.)
IOS Press 1997 © AKM

A set of metrics for the evaluation of SDH transport networks

A. Antonopoulos, J. J. O' Reilly, P. Lane
Department of Electronic & Electrical Engineering
University College London
Torrington Place, London WC1E 7JE, UK

Abstract. Continuing advances in optical communications coupled with the widespread adoption of the SDH standards, is defining the future of telecommunication networks. The advent of Synchronous Digital Hierarchy (SDH) transmission has sparked many debates about network architecture and the implementation of transport networks. The variety and the differences between the candidate network architectures resulted in significant network planning complexity. The difficulty in defining an optimum transport network solution is apparent in nearly all current planning cases. An increasing demand to understand the "goodness" of the various network solutions is evolving. This paper presents a new framework to evaluate the performance of SDH transport network solutions. This framework uses a set of proposed metrics and contributes to the network planning process, providing enhanced insight to guide the network planner when selecting the type and the characteristics of network architectures for practical deployment. Most useful though, is for network operators who need to understand the real potential of the candidate network solutions and how this potential matches their specific requirements.

1. Introduction

As telecommunication networks become more complex, the demand to clearly asses the network performance of different network solutions increases [1,2]. Due to the characteristics of new networks, e.g. variety and complexity of new services, mixing of them and differences in the type and the scope of the different telecommunication operators around the world, the task of meeting this demand is not trivial.

Traditionally, network planners were using tools aimed at designing transport network solutions at minimum cost. Today, with the demands of modern networks, this is no longer the case. Operators require survivability, manageability, simplicity, growth abilities and various other characteristics along with costs issues for their networks. Finding the optimum solution, balancing all these requirements, requires a firm usage of metrics that will produce optimum or near optimum solutions on a per planning case basis. To achieve this, the ability to recognise the "goodness" of each proposed network solution is necessary both for the network planner in order to be able to design the network and for the network operator to evaluate and compare the various proposals.

Network planning has so far focused on a rigorous evaluation of all the factors that impact overall performance. Here, we argue that for the complex processes underlying planning a more pragmatic approach needs to be adopted. For example, the routing of traffic through a network impacts the distribution of spare capacity available to support unexpected growth in traffic demand and the network costs (by best utilising your trunk routes). Therefore the quality of routing could be used as a metric to assess the performance of a candidate solution.

Instead, the real issue is the provision of spare capacity at low-cost; the quality of routing is not really an issue that interests the operator. As long as the equipment vendor can supply the required flexibility to respond to the changes in traffic then the operator is not interested in knowing if this is achieved at low-cost by the adoption of efficient routing that minimises equipment requirements or if it achieved at low-cost due the use of lower-cost equipment. Of course, efficient routing would normally minimise cost, but since the routing is hidden from the operator behind the cost of the network solution the operator does not have to made aware directly of the routing quality as a metric.

Defining the performance of the various SDH network solutions with the complexity and requirements of modern networks is therefore a task that needs to be explored [3,4].

2. Defining the Network Performance of a Transport Network Solution

Since the network costs are not the only requirement for efficient transport network planning any more, and numerous other requirements have been developed, a simple and single evaluation of the proposed network solution is not possible. When different transport network solutions are examined and need to be compared, some commonly accepted metrics must be defined, in order to evaluate the network performance of each solution. For defining network performance we need to go back to the fundamental question why do we built telecommunication networks.

Telecommunication operators want to build a network to offer communication services to their present and future contracted customers. The network must curry the traffic of its customers both at present time and in the future. This traffic must be carried with a desired, according to the operator, continuity and quality. After building this flexible and desired network and offering a service to its customers, the operator wishes to have the ability to maintain, operate and manage its network in an optimum, again according to the operator, way. Providing a network operator with a network solution satisfying its demands, implies an overall network cost for the present time and the future that must fit the business plans of the network operator. Therefore, the performance of the network solution needs to be separately evaluated according to some major areas of interest and importance that contribute to different network requirements [4]. Metrics for assessing the network performance of each area of interest must be defined separately. These metrics can be used to comment on the network performance of each area of interest of a given network solution. Considering these, we could define some areas of interest to examine transport network performance as follows:

- Network performance in terms of network capacity
- Network performance in terms of resilience
- Network performance in terms of maintenance, operation and management
- Network performance in terms of costs

A set of metrics for each of these areas of interest has been developed. Even with a clearly defined set of metrics though, the optimality of each candidate solution will remain case dependent. An overall performance evaluation is not possible as it is the operators' decision to weight, according to its priorities, the different network performance areas of interest and their different metrics.

3. Transport network performance in terms of capacity

Each network solution aims to carry the present and future offered traffic of the present and future customers of the operator. The parameters that can be used to characterise the performance metrics of a transport network solution in terms of capacity are:

- The offered total traffic in the network (T).

- The installed capacity of the network which is not reserved for protection (I).
- The working capacity of the network (W).

The latest is defined as the capacity carrying traffic or being reserved for protection of traffic. The installed capacity in the network is a direct contributor to the network solution costs. A number of metrics could be defined, describing how efficient a network solution can be, in terms of capacity.

3.1. Response abilities against unexpected traffic changes

The ability of the network solution to respond to changes of traffic characteristics in terms of:

i) amounts of offered traffic

ii) traffic pattern trends

Also, an indication of the scale of the changes (if necessary) in architecture of the network, required to respond to these types of traffic changes, could be another metric.

In order to develop some metrics describing this ability two scenarios must be examined. The first scenario assumes that traffic changes occur from unpredicted or predicted growth only. This implies that the traffic characteristics change due to amounts of traffic additional to the ones provisioned for the initial deployment of the network. The second scenario suggests that even originally provisioned final year traffic, on which the design was based, can change radically (complete churn) from fundamental changes in the future role of the network and in the type of services it will finally carry. The latest questions the carrying traffic abilities of a network solution in the absence of any advance information on the characteristics and amounts of traffic to be carried.

3.1.a. Traffic changes occurred from unpredicted or predicted growth only

In this case, the response ability can be demonstrated by the following four metrics :

i) The spare capacity (S). The idle capacity of the network. This can be calculated simply as follows:

$$S = (I - W)$$

It can be defined as the amount of idle capacity that can be used for any extra traffic not provisioned in the planning case, without affecting the traffic already provisioned.

ii) The average percentile spare capacity ($\bar{s}$). If c_i is the installed capacity of the multiplex section i, s_i is the spare capacity of the multiplex section i and n is the total number of multiplex sections in the network solution, then

$$\bar{s} = \frac{\sum_{i=1}^{n} \frac{s_i}{c_i}}{n} \times 100$$

iii) Distribution of spare capacity (D). It indicates how the spare capacity is distributed in the network. If again c_i is the installed capacity of the multiplex section i, s_i is the spare capacity of the multiplex section i and n is the total number of multiplex sections in the network solution, then

$$D = \frac{\sum_{i=1}^{n} \left(\left| \frac{s_i}{c_i} - \bar{s} \right| \right)}{n} \times 100$$

If the distribution of the spare capacity is calculated on a per subnetwork basis it should be noted that it has not the same impact for SNCP subnetworks as for MS-SPRing subnetworks, since in a MS-SPRing subnetwork, routing flexibility can ease the disadvantages of a high distribution of spare capacity [5].

iv) Extra traffic carrying abilities. This indicates how much more traffic could be carried by the network in addition to the provisioned traffic. Since this is very much dependant on the characteristics of this extra traffic a worst and a best case of the amount of extra traffic that could be carried can be defined, ET_w and ET_b respectively. These can be used as crude metrics to quantify the limits in the amounts of extra traffic that could be carried by a proposed network solution.

When producing network solutions with SNCP-rings only, the spare capacity of each ring subnetwork equals the spare capacity of any multiplex section of the ring [6,7]. In this case, ET_w equals the spare capacity of the ring with the smallest amount of spare capacity. This follows the assumption of the worst scenario that all growth traffic will originate/terminate within this subnetwork. ET_b equals the sum of spare capacities of each ring subnetwork by assuming the best case scenario of intra-ring growth traffic only.

When MS-SPRings form a candidate network solution, ET_w and ET_b are more difficult to calculate since the extra traffic that can be carried depends on its traffic pattern trends [7]. ET_w is calculated by assuming a future pure hub traffic pattern trend of the growth traffic in the MS-SPRings of the solution. The hub node in each ring is located between the MS-SPRing's two multiplex sections of which the sum of spare capacities is the lowest. ET_w equals the sum of this sum of spare capacities across the MS-SPRings of the architecture. ET_b is simply calculated by adding the sum of the spare capacities of the multiplex sections of each MS-SPRing. This is after assuming a pure node to adjacent node traffic pattern trend in the growth traffic.

When the network solution is a mesh architecture, which employs 1+1 protection, ET_w and ET_b are very complex to calculate. For finding a theoretical lower limit for ET_w we assume that the working or the protection trail of all growth traffic traverses the multiplex section with the smallest amount of spare capacity. Therefore, the theoretical lower limit for ET_w equals the smallest amount of spare capacity found on a multiplex section of the mesh. A theoretical upper limit for ET_b is calculated by assuming only single hop growth traffic trails which will ideally require only 2-hop protection trails. Thus, this limit equals one third of the available spare capacity in the architecture.

In network architectures formed by combinations of SNCP-ring, MS-SPRing and mesh architectures the appropriate combinations can help the planner calculate ET_w and ET_b.

3.1.b All possible traffic changes can occur

In the case which all possible traffic changes can occur the only two metrics that can be used are the minimum and maximum total traffic that the proposed network solution can carry. These can be calculated using similar assumptions to the ones used in calculating the metrics for the extra traffic carrying abilities. Let SCS_i be the capacity equivalent to the ring system level of each of the t SNCP-ring in the networks solution and SCM_i be the capacity equivalent to the ring system level of each of the m MS-SPRing with n_i nodes in the network solution. Then the minimum total traffic equals the amount of traffic that can be carried by the SNCP-ring with the lowest ring system level and the maximum total traffic is:

$$\text{Maximum total traffic} = \sum_{i=1}^{n} SCS_i + \sum_{i=1}^{m} \frac{n_i SCM_i}{2}$$

Again if the network solution comprises of a mesh architecture, it is very difficult to define the minimum and maximum total traffic that can be carried by the architecture. Using similar assumptions to the ones used in calculating the metrics for the extra traffic carrying abilities we can calculate that the capacity that can be carried by the multiplex section with the least installed capacity produces a theoretical lower limit to the minimum total traffic that can be carried by the architecture. A theoretical upper limit to the maximum total traffic that can be carried by the architecture equals the installed capacity divided by three.

The above metrics will in any case produce really rough estimates of the mesh architecture's ability to respond to future traffic changes since they correspond to really extreme situations. From all the metrics proposed in these section the most important are the spare capacity, it's average value and it's distribution.

3.2. The network utilisation (U)

Utilisation of the links of the network (link load):

$$U = \frac{W}{I}$$

It indicates the efficiency in the use of the installed capacity. It should be balanced with the previous metrics in the response abilities against unexpected traffic changes. It could be an indirect way of indicating investment efficiency although a good utilisation does not necessarily implies lower costs (e.g. if more complex and expensive equipment is used).

3.3 Network architecture and routing performance (N)

The ability to route the traffic optimally:

$$R = \frac{T}{W}$$

This can be used as a metric for the capacity utilisation goodness of the network and indirectly can indicate cost and spare capacity efficiency.

3.4 Importance of the proposed metrics

The metrics proposed in 3.1 are the most important and are clear indications of the networks' performance in terms of capacity. The operator is particularly interested on how flexible and resilient to future changes is the network. The metrics of 3.2 and 3.3 are calculated easier but are less important. They produce theoretical results on the "goodness" of the network in terms of capacity that are not crucial in the real world evaluation of it. They could be indirect indications of the response abilities against unexpected traffic changes or of the network costs though.

4. Transport network performance in terms of resilience

Any network solution can be related to different characteristics defining the quality of the services that the operator wishes to offer to its customer. Metrics can be defined to assess the network performance in terms of quality of offered services:

4.1 Levels of protection

Ability of a network solution to provide/support different levels of protection on the network, depending on the service, the geographical area or even on the end customer. Protection, no protection or mixture of the two define this metric.

4.2 Availability

Network availability is a measure of the degree to which the network is in an operable state called upon to perform [8]. Availability could be defined differently for the network operator and the end customer. For the network operator, availability can be defined as the proportion of time his network is entirely available. E.g. Single points of failure in a network with automated protection, or failures non-affecting traffic, result in unavailability for the network operator. This is because although the operator is able to offer its services with apparent continuity to the end customer, operable state of the entire network is not present.

For the end customer network availability can be defined as the proportion of time he can use the network. It should be noted that in a protected network, single points of failure still add to the total network unavailability with the time it takes for the network to self-heal. The network availability figures used to describe network solutions are usually customer facing availability figures.

The best approach to assess the overall transport network solution availability is to use the following four metrics [8]:

i) An histogram of the availabilities or unavailabilities of all the paths of the network.

ii) The expected availability/unavailability.

iii) The worst case availability/unavailability.

iv) The distribution of the path availabilities/unavailabilities

4.3 Survivability

Network survivability, defined for the services offered to the end customer, is the proportion of time a service is survivable [8]. In other words, the proportion of time a service is available to the end customer. It is a measure of goodness for the network and it is more difficult to capture than availability. Survivability should be examined under various outage scenarios. It can be defined accordingly under each scenario [8,9].

5. Transport network performance in terms of maintenance, operation and management

A network solution proposed must demonstrate its abilities for maintenance, operation and management from the operator, from the moment it is initially deployed onwards [10]. Metrics to quantify this large and complex area are not easily defined. As metrics the following could be used:

5.1 The number of systems and subnetworks to be managed

A direct indication of the management complexity. This is probably the most important metric, indicating to the network manager the number and the relation of the network entities to be managed.

5.2 Routing complexity

The way the routing of the traffic in the network is done and its complexity.

5.3 Service provisioning and flexibility

The ability of the network to introduce new services and manage the existing ones. Also the ability to support a mixture of a large number of them.

5.4 Path set-up

The ability of the network to set-up the paths required.

5.5 Maintenance abilities

The ability of the network operator to maintain any of the components of its network, without disrupting the running services. E.g. 4-fibre MS-SPRings can have maintenance work in the ducts without disruption of services [7,11].

6. Transport network performance in terms of costs

One of the usually most important issues describing a network solution are the costs related to it. One clearly important metric for networks is the following:

6.1 Network costs

The overall network costs of the final year solution.

6.2 Cost distribution

How the costs are distributed during the different phases of the network planning and deployment from the initial to the final year.

6.3 Maintenance and operation costs

What are the expected maintenance and operation costs of the network during its deployment and during its operation later.

7. Overall transport network performance

Quantifying an overall transport network performance for the purpose of commonly comparing network solutions is not feasible. As already mentioned, each operator will weight very differently, and according to its scopes, the different areas of network performance and the different metrics of them. It is possible and vital to commonly quantify most of the metrics proposed, but we cannot define how these are going to contribute to the overall transport network performance of a network solution. This lies entirely on the special characteristics of each case. E.g., The network performance definition of a voice network in China will be very different from the one of a military data network in the US and both will be very different from the one of an ATM network in Europe.

New tools to explore near optimum solutions on a per planning case basis should use metrics to produce performance evaluations for the various performance categories defined above. A degree of "goodness" of the solution for each separate performance type should be able to be produced using a set of metrics.

The solutions produced by tools should achieve good performance in the areas of interest that the operator is judging to be the most important, and should optimally balance the performance with the other areas of interest. This requires feedback information from the operator in order to produce the most applicable solution. As a final draft, a number of candidate solutions should be produced, all fitting closely the requirements of the operator, in order for the operator to do the appropriate weighting and choosing the best solution.

Conclusions and Future Work

A framework has been proposed based on a metric platform that can help evaluate the candidate transport network solutions produced in real network planning cases. The importance and usage of the metrics proposed has been discussed in details. Usage of these metrics, if properly embedded in future planning tools, could significantly help the network planner to achieve custom- made competitive proposals.

Future work will initially focus on examining case studies to demonstrate the applicability of this framework and later on the development of tools that use the proposed framework in order to produce competitive network solutions.

References

[1] I. Hawker and P. Cohraine, "Future Scenarios for Telecommunication Networks", European Transactions on Telecommunications, Vol. 4, No. 6, pp. 71-78, Nov.- Dec. 1993.

[2] S. Ablett, "Identifying the Real Economic Benefits for Network Operators", Proceedings of IIR Conference on SDH, London, 1994.

[3] M. Sexton and A. Reid, "Transmission Networking : SONET and the Synchronous Digital Hierarchy", Artech House, 1992.

[4] A. Antonopoulos, J. J. O' Reilly, P. Lane, "Metric Dependant Tool Solutions for SDH Transport Networks", IEEE Symposium on Planning and Design of Broadband Networks, Montebello, Quebec, Canada, October 1996.

[5] C. C. Shyur, Y. M. Wu and C. H. Chen "The Capacity Comparison and Cost Analysis for SONET Self-Healing Ring Networks", IEICE Transactions on Communications, vol. E77-B, No 2, pp.218-225, Feb.1994.

[6] T. H. Wu and R. C. Lau "A Class of Self-Healing Ring Architectures for SONET Network Applications", IEEE Transactions on Communications vol. 40, pp. 1746-1756, 11 Nov. 1992.R.

[7] A. Antonopoulos, J. J. O' Reilly, P. Lane, "Analysis of Factors Affecting the Capacity Performance of SDH Shared Protection Rings", European Conference on Networks & Optical Communications (NOC '96), Heidelberg, Germany, June 25-28 1996.

[8] A. Antonopoulos, J. J. O' Reilly, P. Lane, "A Framework for the Availability Assessment of SDH Transport Networks", Accepted for publication at the IEEE ISCC '97, Cairo, 1-3 July 1997.

[9] T. H. Wu, "Fiber Network Service Survivability", Artech House, Norwood, 1992.

[10] R. F. Holter, "SDH/SONET - A Network Management Viewpoint", IEEE Network Magazine, Nov. 1990.

[11] R. Lau "An Architect for a SONET Self-Healing Ring", Bellcore Contribution to T1X1.5/89, May 10, 1989.

Core and ATM Networks
D.W. Faulkner and A.L. Harmer (Eds.)
IOS Press 1997 © *AKM*

Distributed-Controlled Multiple-Ring Architecture with Classified Path Restoration for Nation-Wide Trunk-Transmission Networks

Masahito Tomizawa, Shinji Matsuoka, Yoshihiko Uematsu, and Kazuo Hagimoto
NTT Optical Network Systems Laboratories, 1-1 Hikari-no-Oka,Yokosuka-shi, 239 Japan

Abstract. This paper provides an architectural study of optical multiple-ring trunk-transmission networks using high-speed Time Division Multiplexing (TDM), and proposes two algorithms for distributed control environments. We propose a path-setup algorithm that uses a Token protocol over Section Overhead (SOH) bytes by which network-nodes communicate with each other to reserve bandwidth. A classified path restoration algorithm is also proposed that offers 3 path classes in terms of restoration performance. Class A paths, the most reliable, never lose any bit even against unpredictable disaster. They are realized by path-duplication at the source node, route diversity, and hitless switching at the destination node. Class B paths are restored by re-routing, where the original path-setup algorithm is reused. Class C paths are the most economical because a failed path is restored by maintenance action.

1. Introduction

In the near future, many different kinds of "multi-media" service will require quite different levels of Quality of Service (QoS). Existing trunk-networks handle "paths", each of which is constructed by multiplexing a variety of circuits, in a uniform manner even though they have a wide variety of quality requirements i.e., survivability or transmission performance. For example, a path accommodating phone-call circuits from residences (bit-loss for several seconds is accepted) and a path in a leased-line (even a single bit-loss may cause significant damage) are handled in the same way in the trunk-network. The total cost of such a network is determined by the paths requiring the highest quality, even if such paths are rare. In the future, the difference in required qualities will be much larger than it is at present. Therefore, in trunk-transmission networks it is important to operate and administrate the paths in a flexible and cost-effective manner in terms of survivability and/or transmission performance. This paper considers a classified path operation scheme, and proposes algorithms for path-setup and path-restoration in high-speed Time Division Multiplexing (TDM) ring-based networks.

2. Network Architecture

The architecture we consider is as follows.

• multiplexing: high-speed TDM at the physical layer

Recent trunk-transmission networks have link-speeds as high as 10 Gbit/s through the use of Synchronous Digital Hierarchy (SDH): they are feasible as confirmed in the Broadband backbone network testbed in Japan [1]. To form more simple and reliable networks, the path is assumed to be defined as a single and direct connection between a node-pair for simple path-administration [2]: for example, a concatenated Virtual Container (VC) in SDH format, which may include Virtual Paths (VPs) having the same source-destination node-pair in the trunk-network. It should be noted that path capacities should be freely

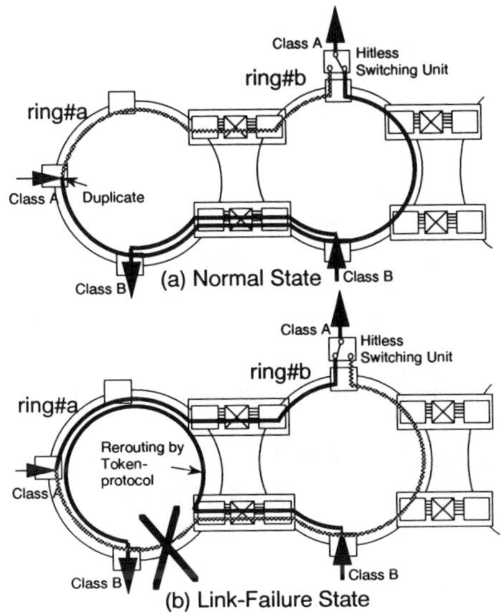

Fig. 1: Multiple-ring architecture. (a) Normal state, (b) Link-failure state.

selectable unlike the conventional path definition. This is a definition issue for network administration, physically there is no major change.

• physical topology: multiple-ring

The ring is a fascinating physical topology because it achieves lower cost due to its reduced number of fibers, its simplified node structures (Add Drop Multiplexers: ADMs), and higher survivability through self-healing [3]. To construct a trunk-transmission network covering wide or global areas, the multiple-ring topology, rings connected in parallel, is more suitable than a single ring because it offers shorter transmission delay. Fig. 1 shows the multiple-ring structure, where pairs of rings are connected via two bridge nodes.

• Network control: distributed control

Distributed control requires smaller Operation systems (OpS) on each node, achieves higher survivability against node-failure than centralized control, suppresses the size of the database needed for network administration, and enables faster response. In a distributed control network, each node has information only about its adjacent links and the paths to added/dropped to/from the network. Note that details of the database structure and up-date mechanism remain to be clarified.

3. Path Set-up Algorithm

In such a network, we prepare 3 classes of paths [2], each of which has different restoration performance. The path restoration classes are summarized in Table I.

Table I. Classified path restoration scheme

Class	Restoration scheme
A	Hitless switching at the destination
B	Re-connection by path-setup algorithm
C	Maintenance action

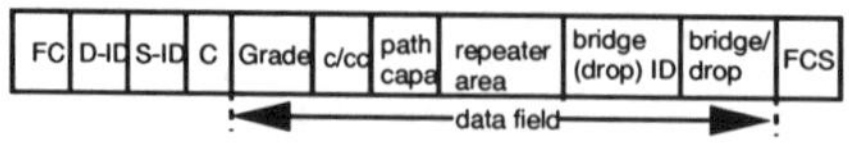

Fig. 2: Token protocol for path-setup

• Class A paths, which never lose one bit even against unpredictable disaster, employ path-duplication at the source node, complete diversity routing, and phase-adjusting (hitless) switching at the destination node. We have already developed hitless switching equipment [4].
• Class B paths are established by single-route path-setup in the normal state, while they are restored by re-connection upon failure using the path-setup algorithm.
• Class C paths are most economical because failure is only eliminated by manual maintenance of equipment or fibers.
In order to realize the described classified scheme in distributed control environments, a new path-setup or routing algorithm that handles all classes is necessary.

For distributed path-setup control, nodes communicate with each other using the Token protocol constructed on Section Overhead (SOH) bytes. Note that main-signal transmission employs high-speed TDM. The reason for using the Token protocol for network control is that in trunk-networks, demands for path-connection change are not as frequent as circuit-connection changes in subscriber networks or LANs. The SOH bytes carrying the Token protocol are likely to be Data Communication Channels (DCC). ITU-T G.784 specifies that the DCC bytes are formed by the OSI 7 layer full-stack protocol for network management inside the SDH-Management Subnetwork (SMS) site [5]. Accordingly, the Token protocol in DCC is inconsistent with the specification in this point. It should be noted, however, that in management packet transfer by the full-stack protocol, the connection set-up to transfer control packets and the error control mechanism employed in the Link Access Protocol on D-channel (LAPD) in the data link layer can cause transfer delay in path setup/restoration. The error control mechanism seems to be dispensable because of the very high quality of the physical layer; SDH optical transmission lines typically achieve Bit Error Rates (BERs) better than 10^{-12}. Hereafter, we consider the Token protocol only for DCC.

Three kinds of packets are used for path-setup: routing packet, response packet, and routing information packet. As can be seen in Fig. 2, the data field in the Token protocol is customized for the proposed path-setup, which includes information of the class of the path, the direction of packet-flow, path capacity, and the flags set by each intermediate node between source and destination stamps. The major characteristics of the proposed algorithm are summarized hereafter: the ring configuration refers to Fig. 1.
• The source node in ring #a, which wants to establish a path to a node in ring #b, gets the Token in ring #a, writes path-class and capacity in the corresponding area in the packet,

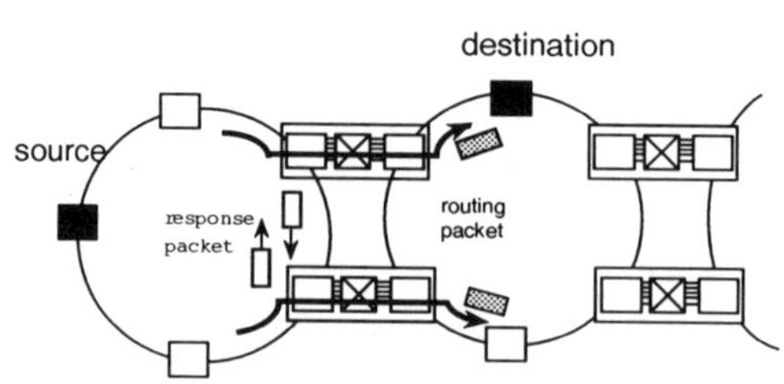

Fig. 3: Packet transfer in the bridge nodes.

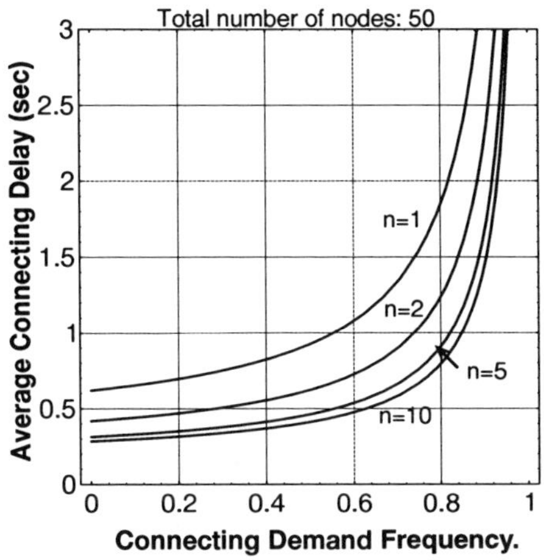

Fig. 4: Average connecting delay versus connecting demand frquency.

and sends the packet in both directions as a "routing packet" for double route searching.
• Each repeater node stamps its own ID in the 'repeater node area' in the received routing packet if the node has enough capacity to pass the indicated path in the direction the received packet is traveling. Otherwise, the node stamps NG.
• The bridge node receives the routing packet, stamps ID or NG in the corresponding area of the routing packet, duplicates the packet and stores it in buffer memory, and sends the original packet as "response packet" to ring #a. The bridge node seizes the Token in ring #b, then sends the stored routing packet in the direction of received packet-flow, clockwise or counterclockwise. This restriction of packet-flow realizes complete route diversity; the two path-routes do not share a single link or repeater node. Fig. 3 illustrates the packet transfer scheme in the bridge nodes.
• When the destination node confirms path-setup by reading the information stamped in the routing packets, it sends a "routing information packet" towards the source node in ring #a. If the path of interest requires class A, two routing information packets are transmitted via the two routes. If the path requires class B or C, a single routing information packet is sent via the shorter route.
• For a class B path, each repeater node in a route other than the established one reserves protection path-capacity, or confirms that its reserved protection path-capacity can restore the path indicated in case of failure. This is done by watching the response packet which is transferred from the bridge (or source) node upon receiving the routing information packet. When the routing information packet arrives at the source node, path-routing is completed.

Another important issue is the timeslot mapping of each routed path. Under distributed control, each node is responsible for determining the bandwidth utilization of its two adjacent links, and has no path-timeslot mapping table of the other links. For efficient use of transmission resources in this distributed control situation, each node needs to reorder its handling slots by itself. Against this, we have already proposed a path-accommodation algorithm for high-speed TDM [3]. The scheme reuses bandwidths by sorting timeslots at each node based on the node-ID of each paths. The details are described in Ref. [3]. Timeslot mapping is determined after receiving and transferring the routing information packets.

Average connecting delay, which is defined as the average time interval between

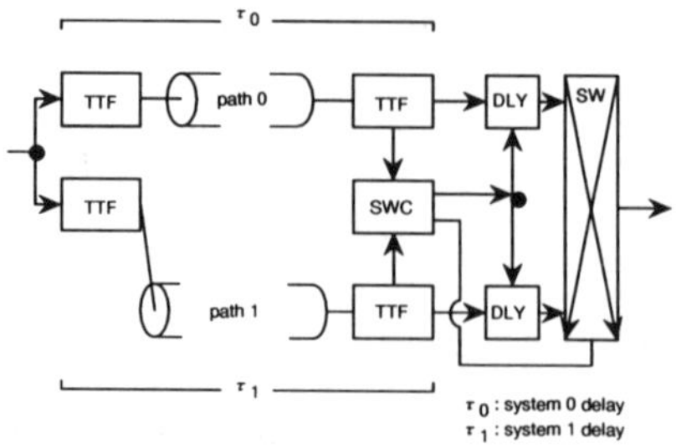

Fig. 5: Hitless switching circuitry for class A path.

the appearance of connection demand and connection establishment, depends upon the frequency of connection demand occurrence. If it is a purely random process, average connecting delay can be expressed as follows,

$$T = \left| \frac{1}{2(\rho - 1)} \left(\rho + 1 - \frac{\rho}{S} \right) + \frac{3}{2} \right| \tau(N) , \tag{1}$$

where,

$$\tau(N) = N \left(\tau_f \cdot \frac{L_h + (N+2)\log(nN+1)}{d} + tL + \tau_p \right). \tag{2}$$

τ_f is transfer time for SDH 1 frame (125 μs), d number of SOH bytes in 1 frame (12 bytes are assumed), t propagation delay (5 ns/m), L link distance (uniform 100 km), τ_p processing time for each packet (1 ms), L_h packet overhead bytes (8 bytes), and ρ connection demand frequency. Fig. 4 illustrates the relationship between average connection delay and connection demand frequency. The total number of nodes is 50 to mirror the Japanese long-distance network. The horizontal axis shows how many demands occur during every round trip of the Token in each ring. Average connection delay decreases with increasing ring number because the probability of the Token being held by a bridge node is higher. It should be noted that average connection delay is of the order of seconds, which is short enough for path-setup in trunk-networks.

4. Classified Path Restoration

This section clarifies the quality of each established path class in terms of survivability and transmission performance.

4-1. Class A

Class A paths are the most robust because hitless switching is employed. To adjust the protection delay to that of the working path, we detect the J1 byte in the Path Overhead (POH) of every frame. After storing both working and protection VCs in memory, both J1 bytes are read out at the same time. The system should add some fixed delay to the working VC ahead of providing services if the protection delay is long. By

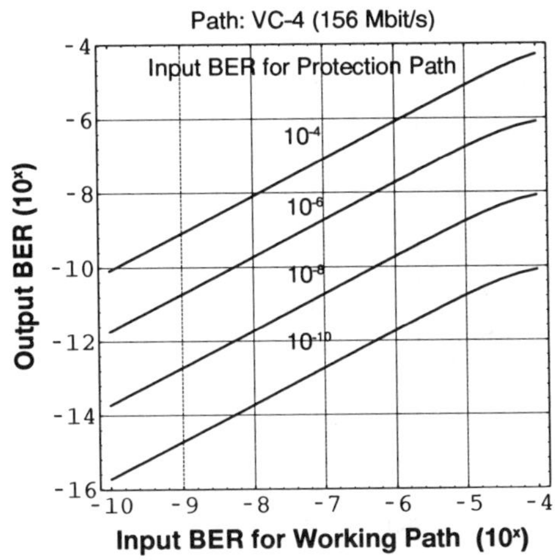

Fig. 6: Improvement of BER by hitless switching for class A paths.

detecting errors, we can protect against any possible troubles. VC errors can be detected by monitoring the B3 byte in POH, which means error is detected after receiving the next frame's B3 byte. In other words, this error monitoring scheme requires at least 1 frame period to reach the selector. If one or more errors are detected in the working VC but no error in the protection VC, the hitless switching circuit selects the protection VC. Fig. 5 shows a hitless switching system configuration, where TTF is a Transport Terminating Function executing section-termination and path-monitoring, DLY delay or memory, SWC is switching control which activates the switch by watching for error in the B3 bytes in VC path frame. It should be noted that error is detected frame-by - frame, and the switching circuit (SW) selects the path offering the best performance frame-by-frame. The scheme realizes not only switching without bit-loss against unpredictable failure, but transmission performance improvement (BER) in the normal operation state. Output BER ε^{out} of the switching circuit is described as follows,

$$Q(\varepsilon^{out}) = Q(\varepsilon^0) \times Q(\varepsilon^1), \tag{3}$$

where, ε^0 is the BER of the working path, and ε^1 is that of the protection path. $Q(\varepsilon)$ is the probability of detecting error in 1 frame by means of BIP-8 and can be represented as,

$$Q(\varepsilon) = 1 - \left|1 - P(\varepsilon)\right|^8. \tag{4}$$

$P(\varepsilon)$ is expressed as follows,

$$P(\varepsilon) = 1 - \frac{1 + \left(1 - 2\varepsilon\right)^{N_R}}{2}. \tag{5}$$

Fig. 6 depicts the relationship between output BER and input BER, where the path is assumed to be a VC-4 (156 Mbit/s). If input BER is 10^{-10} on both paths, working and protection, output BER can be improved to as low as 10^{-16}.

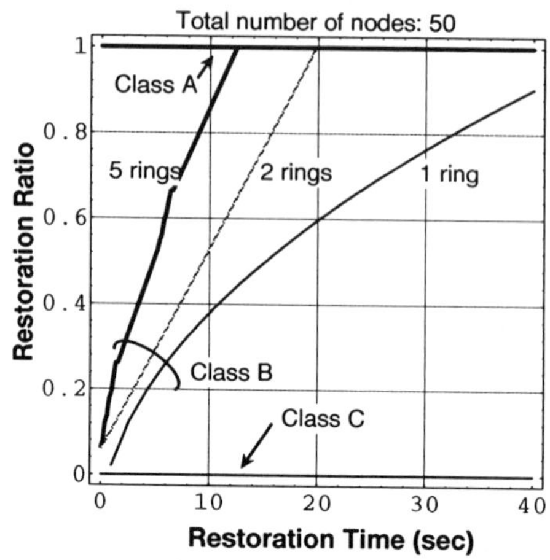

Fig. 7: Restoration performances of all classes of paths.

4-2. Class B

Class B paths require complete restoration with the acceptance of some bit-loss; they are cost-effective because they do not employ hitless switching and occupy only single routes. They suit services such as computer communications with Automatic Repeat reQuest (ARQ) or conventional telephone-calls. Shared switching is realized by using the same algorithm and equipment used in path-setup. When a failure occurs (here we assume a link failure), the two nodes adjacent to the failed link detect Loss of Signal (LOS) at the section layer and LOS information is embedded into the failed paths: the adjacent nodes generate the path-Alarm Indication Signals (AIS). Moreover, the adjacent nodes "loop-back" the Token in the firmware (or software) platform to enable communication around the failed ring: every node can get the Token in this single link failure state. Note that the Token is looped-back, while the main-signal is restored in the path layer. In the failure state, all class-B-paths, carried by the failed link, request path-connection. A node holding the Token re-connects a single class-B-path, and releases the Token. The next node seizing the Token also executes re-connection. Some nodes in the ring re-connect the paths, while the others only transfer packets. Here, restoration ratio R(T) is defined as "the ratio of the number of restored paths to that of failed paths, at time T". Calculated results are shown in Fig. 7, where the total number of nodes is 50, mesh connectivity is assumed, and the parameter is the number of rings (1, 2, and 5). The figure shows the restoration ratio of each path class. For class A, hitless switching realizes instantaneous restoration, at the expense of some transmission delay increase in the normal state due to buffer memory use. For class B paths, the perfect restoration time was beyond 40 seconds in a single ring, 20 seconds in double-rings, about 10 seconds in penta-rings. The perfect restoration time seems to be short enough for applications such as computer communications with ARQ, or conventional phone-calls: if one requires shorter restoration time, the class A path is recommended.

5. Discussion

The feasibility of deploying our multiple-ring architecture in full-scale trunk-transmission networks should be considered. It seems to be practical as seen by the following points. First, the fiber configuration of multiple-ring is the same as that employed in the ladder-based architecture which is conventionally employed in many countries. Accordingly, extremely expensive construction works such as brand-new fiber-deployment, are not necessary. Second, in terms of path-administration, classified path management is not as complex as existing path management because the number of paths will be drastically reduced through the introduction of large-capacity direct-connection paths.

Many operation functions already exist, and their relationship between the proposed scheme (Network management level) should be considered. The major issues will be the relationship with equipment OpS (Element management level) and that with network-planning OpS (Service management level). Our algorithm requires the automatic recognition of unused bandwidth of add/drop switches in a node; this information is administrated by the equipment OpS. This issue will be solved by appropriately structuring the interface to match the equipment OpS. The relationship betwee4n netwrok-planning OpS is more important because in our classified path operation scheme, link-capacity (number of paths in a link) strongly depends upon the ratio of the number of class A paths over the whole number of paths. This suggests that the new term of "survivability" is needed in network-planning OpS, for successful network design based on the prediction of customer needs and demand trends.

6. Conclusion

In conclusion, we have provided operation schemes for classified path restoration in a distributed control multiple-ring architecture, a candidate for the next generation of nation-wide trunk-transmission optical networks. We proposed a path-setup algorithm that uses the Token protocol in SOH bytes, by which network-nodes communicate with each other. This communication enables each node to reserve bandwidth and permit route determination. A classified path restoration algorithm was described that offers 3 classes of path in terms of restoration performance. Class A paths, the most reliable, never lose any bits even against unpredictable disaster. This is achieved by path-duplication at the source node, route diversity, and hitless switching at the destination node. Class B paths are restored by re-routing; the path-setup algorithm is reused which suppresses path-cost. Class C paths are the most economical because path failure is recovered only by maintenance action.

Acknowledgment: The authors thank Ikutaro Kobayashi, Tomonori Aoyama for their encouragement. The authors also thank Prof. Hidetoshi Kimura for his useful advice and comments.

References

[1]: K. Hagimoto, Opto-Electronics and Communications Conference (OECC) '96, 17A1-9, p.96, 1996
[2]: S. Matsuoka et al., GLOBECOM'95, p. 941, 1995
[3]: M. Tomizawa et al., IEICE Trans. Commun., vol. E79-B, No. 8, p. 1121,1996
[4]: N. Kawase et al., Electron. Lett., Vol. 30, No. 23, p.1962, 1994
[5]: ITU-T Recommendation, Synchronous Digital Hierarchy (SDH) Management, G.784, 1994

Core and ATM Networks
D.W. Faulkner and A.L. Harmer (Eds.)
IOS Press 1997 © AKM

High capacity SDH networks for new operators - Using the national electricity distribution network

Alistair M Henderson
Energis Communications, Carmelite House, 50 Victoria Embankment, London, UK

Abstract. This paper describes the stages of development of Energis as a national telecom operator, based on the infrastructure of the UK electricity distribution network. The development over the first 4 years is covered, and future strategies considered.

1) The strategic impacts of SDH

SDH equipment became available from an increasing range of suppliers during the early 90`s, promising increased reliability, greater flexibility, reduced cost, and improved management systems. Initially, many operators throughout Europe were reluctant to deploy SDH, as the established PDH infrastructure had been a heavy investment, and was providing a perfectly adequate service. New PDH products were still being developed by many manufacturers offering sophisticated features, and ATM was "only a couple of years away anyhow".

Within the UK, the major operators saw the opportunity and began the lengthy process of study, and the writing of detailed requirement specifications. Throughout Europe, operators unveiled plans to use SDH in critical parts of their networks, largely to complement or integrate existing PDH. Some worked from the core out, with 4/4 cross connects for fast restoration, and high capacity STM16 bit haulers to replace 565Mbit line systems. Others concentrated on the access networks, with STM1 ADMs, rings and 4/1 cross connects.

At about the same time within the UK, the deregulation of the Telecoms industry was creating exciting opportunities for new entrants. Cable TV companies, (mostly subsidiaries of the USA RBOC`s) commenced the deployment of dense regional networks in franchise areas. These new operators were licensed to carry both entertainment and telephony services within the area, and focused on the residential user.

It was also felt that there were opportunities for new national operators to compete with the established duopoly of BT and Mercury. The availability of the new technology SDH was seen as a key factor, offering the opportunity to construct a flexible highly reliable national network for the business community.

New technology was however not enough. The greatest problem in building a national network, is obtaining the wayleaves, and laying the fibre. During 1992/93, operators based on canals, railway lines, electricity pylons, water pipes and many other ingenious solutions were proposed, and business cases prepared.

2) The development of Energis

One such proposal for a new operator was set up by the National electricity distribution company (NGG). At the time, the NGG was owned jointly by the regional electricity distribution companies (RECs), and carried power between the generators and the RECs. The NGG has since been floated as an independent PLC. Initially named Telecom Electric, the licence to operate in the UK was first applied for in May 1992.

The proposal was exciting -

- The NGG network offered a fast and secure way of building a fibre core across the UK. Nearly 7000km of pylon routes mesh the UK, covering every major centre of population and industry.
- The technology was ready, and would allow the construction of the first 100% SDH national network.
- The National Grid had the financial resource and skilled staff (5000 employees) to make it happen.

Despite such advantages, projects can sometimes degenerate into endless planning loops. A major contract awarded by the BBC at the beginning however ensured that this project was for real. The contract involved a national network taking 2.5Gbit/sec SDH onto the customer site to provide video and radio contribution and distribution capabilities. The time scales were tight, and the demands of broadcast video technically exacting.

3) The first steps

The design and build of a new national telecoms network, by a company with no previous history as an operator, raises critical skill resource problems. The BBC contract also had critical time constraints, resulting in the need to compress activities into about 18 months. Substantial scarce skilled resources were required, firstly in network planning, then in build, installation, network integration, and network maintenance. Once each phase is complete however, the skills are no longer required in the initial volumes. A major problem for any new operator is the skill mix change in the early years, as described below.

Stage 1 The best SDH planners and designers are needed to ensure optimum network design, using the best products in the market. There is no time for extensive study.
Stage 2 The building begins. A large and experienced installation planning team, logistics managers, and an even larger field force must be marshalled. They must be experienced with SDH. There is no time for training.
Stage 3 The network must be fully integrated and tested, not as a series of individual products, but as a network. This skill is even scarcer to find!
Stage 4 When the network is built and operational, a small core team of designers, planners, logistics and integration remains for core network growth.
Stage 5 Operations, maintenance, and customer connection skills become the priority, as the new company fill the network with customers.
Stage 6 Product marketing and new services design skills are required to differentiate the customer offering and raise the margins.
Stage 7 Access network development skills are required to increase network density, and reduce dependence on other operators.

The Energis team had to rise rapidly to nearly 1500 skilled resources for 12 months, then fall back to a maintainable size, coupled with a skill mix change! Clearly this could only be done with assistance from someone in the business, and no existing UK (or European) operator was willing to make that offer.

Energis began the urgent search for a supplier partner, capable of providing access to the significant range and depth of SDH and management skills. Such a partner had to be able to work as member of a fully integrated team together with Energis staff, initially as leader, subsequently giving way to the permanent team as it built up. A further requirement was that the team had to provide access to a full range of SDH equipment, taking the best from the market as required by Energis.

4) Selecting a supplier

Selecting a network development partner did not necessarily mean we had selected a product supplier! Energis had to balance the need for competitive product supply with the problems of multiple source equipment integration. The chosen strategy initially was -

- Maintain competition by ensuring more than one supplier is always able to meet requirements.
- Minimise the complications by limiting the number of supplier interfaces. Use the network development partner.
- Ensure supplier has sufficient business prospect to make it worthwhile.
- Ensure relationships once begun are long term to maintain supplier interest in current job!

With SDH the problem is in some ways simplified, and in others greatly complicated. Of course SDH is designed to allow product interoperability, and indeed as long as the more esoteric functions of SDH are not required, most products do in practice inter work well at the traffic level. The problem comes when you try to pull it all together, and provide a common management system.

To meet the target of enabling competition, it was decided that Energis must maintain the ability to buy from two suppliers. We could split the network by geography, contract phase, product type, or by service. If any supplier had been able to offer a fully integrated SDH system managed from a single common platform system, we might have tried a Geographic split of suppliers. As we could find nobody offering this, the Geographic split meant doubling the trouble, for no perceptible advantage.

Supply by contract phase is all very well for fully interchangeable product, but SDH is not at that stage. (Will it ever be?). This does not meet the requirement of maintaining suppler opportunity either.

Supply by product type was eventually chosen, as it allows for two major supplier relationships, with both suppliers having a strong position in a well defined part of the network technology. It was essential though to maintain the competitive element by ensuring that both were able to supply in the others area of strength if we were being let down.

Nortel was chosen to supply PKI STM-16 line systems, and DSC 4/4 cross connects. Essentially, Nortel was to build and integrate the VC4 level of the network. GPT was chosen to supply the ADM155/ADM622 products, together with access multiplexers, effectively supplying Energis with the VC12 and below network.

The supplier strategy is fundamental, as it leads naturally to constraints in the future on network operations.. By splitting the product suppliers into the SDH levels of VC4 and

VC12, care must be taken to ensure that customer end to end path management over the VC4 and VC12 layers is maintained..

5) Planning the network

Having chosen partners and suppliers, the real task began. A core network of fibre was planned over the NGG network, taking us to the initial target population centres. An ambitious plan covering 4000 km of the NGG 7000 km power network was agreed, taking us within 8km of over 70% of the population of England and Wales. Access points were also planned for the 20 largest population centres in the first stage.

For initial construction, time was critical in order to maximise return on investment. The options available at the time included OPGW, where the fibre is buried inside a replacement earth wire, and 24 fibre wrap, where the fibre is wrapped around the existing earth wire. The NGG developed techniques for delivering both solutions on a live network, however wrapping techniques offered the fastest route to market. A wrapped network was planned for the majority of locations, with some OPGW, mainly in places where the earth wire was already due for replacement. Within London, extensive use was made of London underground tunnels to plan the routes.

The basic transmission interconnection between sites was planned at STM16 across the entire network. The Optical reach between sites is of vital importance to a new operator, as premises for repeaters are expensive, and have to be minimised. To get the reach necessary, a 28dB optical budget was essential. Indeed if Optical amplification had been more readily available in 1993, it would certainly have eased the planning task greatly.

Network resilience can be significantly improved with an SDH network. A number of options are available, each offering various advantages and penalties. For the high capacity high integrity services such as the BBC broadcast video service, two diverse routes were planned across the network, each taking a physically separate route from source to destination. Path protection mechanisms ensure the first level of resilience. The penalty here is cost, which cannot always be justified for all service types.

To enable shared protection, major node flexibility points are required. The Energis network was planned with a series of 4/4 cross connects, allowing VC4 service provisioning and restoration. A number of network configurations were planned to cover various time of day requirements, and disaster fallback scenarios. Current management capabilities restrict the speed at which such plans can be implemented in practice, so this remains a secondary protection mechanism for the near future.

One interesting feature of television broadcast is the unidirectional nature of traffic. SDH is entirely appropriate for carrying unidirectional traffic, however care must be taken with any bi-directional path level features! For example, an STM-1 terminating in Bristol may carry an incoming VC4 from London, with the opposite direction being routed to Glasgow, and so on, forming a complex thread of interlinking paths. You do not want far end receive fail (FERF) alarms in Glasgow, alerting a failure on the link from London to Bristol! As new path features are provided in SDH, recognition of the special nature of unidirectional traffic must be maintained.

The access network design was based on establishing equipment rooms in the major population centres, each linked back to a core network equipment housing (EAM) at the nearest point on the national grid network. These sites (commonly called POP's or points of presence) were called SNAPS (Synchronous network access points) in Energis, to emphasise the 100% SDH nature of the network. Each was linked to the main network with STM16 capacity, and equipped with ADM155, and access mux banks.

In practice, there are a number of parallel networks to be planned and built. The traffic network was one of the simplest! Others are,

- The sync network.
- The DCN network.
- The Management system network.
- The switch signalling network.
- The environmental alarm network.

Each network has its individual peculiarities, however each interrelates with the other in a complex manner. Care must be taken to avoid nepotism, e.g. routing of alarm traffic over paths reporting the alarm etc.

The sync network gives particular planning problems. For any given network failure, restoration of traffic may be trivial. The sync trail restoration however will require careful pre planning and implementation, perhaps requiring sync source reversal down an entire chain of line systems.

One of the "features" of SDH is the vast amount of information available. It sometimes seems that each network element has been designed to tell you everything it ever knew (except what you want to know) at the slightest excuse - and they also seem to do it in crowds. This requires a substantial data communication network, built for resilience and capacity. Alarm floods and capacity choke points need to be carefully calculated.

The planning of the management system can be the poor relation. In practice the integrity of the entire network is dependant on the ability of operators to rapidly pinpoint failures and take corrective action. Great care was taken in the planning of the Energis network management system, allowing each of the features of SDH to be used to maximum effect. The substantial complications of managing mixed SDH/PDH networks were avoided.

6) Build and test

Building of the network commenced before the design was compete. The truth is, the design was not complete until some time after it was all built and working! In practice, parallel design and build is becoming more common as the time scale pressures of our competitive environment increase.

A major aspect of the build programme was the fibre wrapping. This was carried out for us by the National Grid contracts division. Fibre digs supplemented the wrapping, being carried out in all conditions, often in remote locations throughout the UK. The fibre testing produced a mass of data for analyses, indicating fibre condition. A point to watch is to ensure all contractors use the same OTDR equipment, to enable analyses and comparison.

For a new operator, premises are a big issue. Nortel provided pre assembled units called EAM`s (Equipment accommodation modules) which could be built and tested at the factory before being shipped to site. Each EAM was positioned on pre-prepared sites, and hooked up to the fibre. This method enabled very rapid build and installation.

Unfortunately, build never occurs as planned. Unforeseen difficulties delay sections, whilst others prove easy. With a virtual workforce of 1500 people, working to very tight time scales, logistical control was a major problem. It was vital that the NGC, Nortel, Energis and other contractors worked as a seamless team, with a common strong point of control. (Our war room!). As sections were linked, network integration could begin.

The integration process for the network involved a series of steps -

- Fibre test. (loss and OTDR)
- EAM installation test. (Power, environment)
- Section tests. A series of tests carried out between each Network Element and its neighbour.
- Span tests. A series of tests between network node points.
- Ring tests. Full ring testing of each STM16 subnetwok.
- Inter-ring tests.
- Network wide feature testing.
- War Gaming!

Each test step logically built on the proceeding one, gradually bringing in testing of DCN, sync and environmental alarms at the appropriate stage. At one stage over 5000 separate exceptions and bugs of varying seriousness were being tracked. Many were exciting, requiring much lateral thinking to solve. Often problems in one part of the network caused strange effects in totally unrelated parts. One thing became clear - the various alarms and events currently defined in SDH are by no means comprehensive!

Having been through the process, there was much scope for making SDH roll out more efficient in the future. Analyses of failures picked up at each test stage indicates that much testing effort detected little or no failures, and a more elementary test sequence has been developed as we trend towards "plug and play". Much can be learnt from this process in terms of next generation equipment requirements!

7) Operations

The final stage of build and test was war gaming. This involved teams of field staff, complete with a collection of failed cards from integration test, attempting to beat the network operators! Of course, such fun can rarely be had with a live network. A serious competition developed, as more obscure faults were simulated, with multiple fault scenarios, and a final climax involving everything the field force could throw at them.. After going through that, the operators were somewhat disappointed when the network went live - and nothing much happened. As with all networks, the vast majority of faults and problems are caused by field activity, and finger trouble. Left alone, the network works with failures a rarity.

In October 1994, the Energis SDH network went live, initially for basic telephony traffic, and in January 1995 for broadcast video.

8) Subsequent developments

The Energis network has evolved rapidly, with growth rates of up to 20% per month at one point. There are currently 8 DMS100 switches operational, with 3 further on the way. Traffic has risen from 0 to nearly 4 million minutes per day, and is continuing to rise rapidly as more customers take up the service. Applications running on the network include VPN services, Frame relay, ATM, leased line, broadcast video etc. One interesting feature of the traffic profile is the high degree of traffic during the evening and weekend, due to the capture of over 50% of the UK internet inbound business! All this traffic has created an ever increasing demand for bandwidth, and a high degree of build churn. SDH with centralised management has proved its worth many times over as network growth has progressed.

During this period, the ranges of SDH equipment have evolved rapidly. Some features will be more critical to Energis than others. Key drivers will as always be, will it cost less to buy or will it cost less to run. or will it bring in additional revenue.

Substantial savings have already been made over the costs of running an equivalent PDH network, but much remains to be done in the field of network management. For example, currently, the routing of a private circuit involves data entry into numerous different systems, and substantial time delay. The interlinking of different suppliers management systems will be an important step in reducing operational costs. The next step will be in convincing customers to manage their own networks, (and charging them for the feature.)

SDH developments of interest include optical amplifiers, WDM, STM64 systems, optical switches, STM16 ADM, and a VC12 flexibility layer.

Clearly optical amplifiers are of great interest to new operators. Each equipment accommodation module is an expensive investment, and repeaters in a thin network contribute little. Optical amplification should allow greater reach between flexibility and SNAP points, obviating the need for many current EAM sites. (Which can then be re-deployed to expand the network in other areas.)

Capacity will always grow. Many will remember forecasts which predicted a total of eleven 565Mbit line systems as the total UK market! (Another operators forecast). The Energis network, with 24 fibres capable of carrying STM16 should have no problems for a long time! In practice however, there seems no limit for the UK appetite for raw bandwidth! Energis will require higher capacity systems as soon as they become available. The debate between WDM and STM64 is intense, with trials of both in advanced planning.

Network availability and restoration are key issues. The greatest source of path failures has proven to be the fibre, and not the equipment. (By a substantial margin). This applies for dig, wrap and OPGW. As capacities on a single fibre increase, and consequences of failure become intolerable, it becomes clear that rings are insufficient for the core network. A minimum of three diverse routes are required between any two places. Consequently, Energis is in the process of lighting third routes across the network. SDH is fundamentally designed on the concept of rings, and this has caused complex network design issues to arise.

In practice however, as failures occur mostly at the fibre level, it may be more cost effective and efficient, to carry out restoration at the optical layer of the network, rather than at the SDH layer in future. Clearly optical switches offer one alternative. No suitable high capacity high reliability device has yet been identified for this purpose however. An alternative, which may be preferable, would be the use of WDM for optical layer traffic restoration. Particularly on routes where optical amplification has been deployed between network restoration nodes, each lambda could be fed to a number of route alternatives if carefully planned. Fast traffic recovery would take place at the SDH layer, with WDM optical layer recovery increasing resilience again at a slightly slower pace.

Such a scheme would be more easily managed in a network which implemented a consistent interface between the SDH and optical layers, i.e. all at STM16 2.5Gbit/sec. The deployment of STM64 in such a network scenario could be a complicating factor better avoided.

STM16 ADM systems are another design factor. In practice, the bulk of traffic on a given route transits directly between flexibility nodes composed of 4/4 cross connects. Only limited traffic needs to drop off at intermediate SNAP sites. This implies that the majority of the market may remain as low cost, simple bulk bit haulers between nodes. That said, there is a smaller, but still significant requirement for STM16 ADM at SNAP sites, allowing protection, flexibility and increased capacity.

A further trend emerging from the Energis network is the degree of network interconnect between operators. As the UK operators SDH networks expand, it is becoming increasingly common for circuits to transit a number of operators. This has only been possible with the definition and agreement between UK operators of a common interconnect recommendation. The Public Network Operators transport interconnect group (PNO-TIG) is a working group managed by the public network operators group on behalf of the Network Interfaces Co-ordination committee. (NICC). This group has published recommendations for the technical, operations, and maintenance aspects of interconnecting VC12, VC3 and VC4 over STM-1/4/16 optical links between.

A series of recommendations have been published giving guidelines for the technical issues of interconnect, the practicalities of shared link installation and commissioning, and the process for joint operations and maintenance. Copies of these recommendations are available on request to the author.

One interesting problem to arise from the studies within PNO-TIG is the utilisation of path identification. Energis already delivers SDH directly to customers, for connection to SDH interfaces on their equipment. Such customers will often have multiple operators connected to their premises, and circuits may transit a number of different operators in the UK, and other countries. If path identification codes are to be implemented, who will administer the allocation of the unique codes? Customers will clearly own the codes for their own equipment, and will demand SDH number portability between operators. Operators will use identification codes which must be unique nationally and internationally, and must allow for route switching.

Perhaps the future is a single network in the UK, owned and managed by different operators, but all providing a common flexible platform for service providers. If that can be achieved, SDH will have more than justified its existence.

For Energis, it now seems inconceivable that anything other than SDH could be used to provide a service platform. The most particular of customers seem to be notice the improvement. Broadcast video engineers are not easily satisfied, and the rising tide of Internet service providers are aware of every bit error. As customer awareness of the advantages of SDH becomes more universal, and the volume of special and directly connected services increases, it will become increasingly difficult for operators to provide service on any medium other than SDH.

Core and ATM Networks
D.W. Faulkner and A.L. Harmer (Eds.)
IOS Press 1997 © AKM

Full-service access network architectures: an analysis of different options

M. D'Agrosa, D. Lanzoni, L. Zanetta
Sirti SpA, via E.Fermi 2, 20060 Cassina de'Pecchi (MI), Italy
Tel.:+39-2-66775129, Fax:+39-2-66775105, E-mail: l.zanetta@sirti.it

Abstract. In order to provide all the communication services that residential customers need and will need in the future, many challenges have to be faced in particular in the access segment of the network, where new architectures have to be defined and implemented. In this paper, an integrated HFC, an FTTx and an ADSL solution, some of the most promising full-service access networks, are presented and compared from a technical and economical point of view in several different scenarios. This analysis can provide some useful indications to a Network Operator that is planning the evolution of its access network.

1. Introduction

The telecommunication world is quickly evolving modifying the way of living and working of people. Traditional telephone services are not enough sufficient: the growing demand to communicate ask for the provision of new telecommunication services, like interactive multimedia services (broadband on-line services, VoD), and the implementation of new applications. Unfortunately, traditional networks are not able to support all this hunger for communication, so a Network Operator has to carry out a careful network planning taking into account present and future service requirements. All the parts of the network have to be considered in this process in order to harmonise the network evolution. The challenge is more critical in the access network for residential users due to the time and the amount of the investments required. Many architectures are proposed for a full-service access network. A lively debate has been going on among network planners in order to identify the cost-effective solution. However, the great variation of the parameters that characterise a geographic area and the uncertainty about the service penetration, allows to identify several scenarios. This paper presents some of the most promising full-service access network architectures for residential and small business customers and discusses advantages and drawbacks from a technical and an economical point of view, in order to help a Network Operator to plan the implementation/evolution of its own access network.

2. Full service network architectures

Among the several architectures that can be proposed for a full-service access network, three solutions, i.e. an integrated HFC, an FTTx and an ADSL network, characterised by the integration of all the services on a single network have been identified. This feature makes them particularly interesting due to the possible reduction of the investments required. A brief description of the selected solutions is given in the following.

2.1 Integrated HFC architecture

The structure of this architecture is similar to that of traditional CATV networks already installed in several countries (figure 1): the main differences are the bi-directionality of the transmission and the equipment that allows the provision of the services.

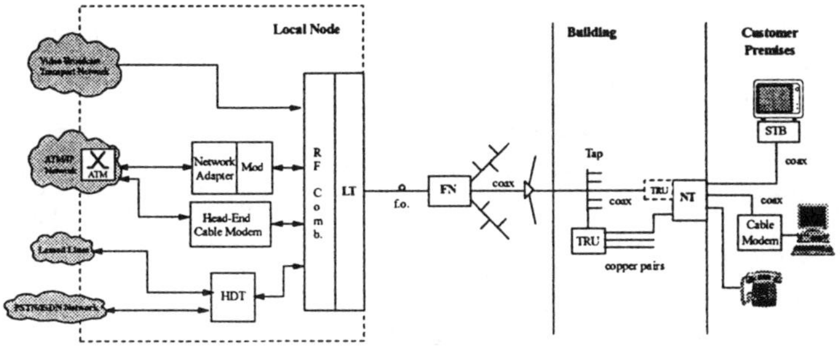

HDT: Host Digital Terminal
LT: Line Terminal
FN: Fiber Node
TRU: Telephone Remote Unit
NT: Network Termination
STB: Set Top Box

Figure 1- Integrated HFC architecture

In particular, in order to provide broadband on-line services cable modem systems are used, while to support VoD service a set of digital modulators have been deployed in the local exchange and some functionality has to be added in the Set Top Box. Moreover, for telephone services the use of cable telephony equipment is requested.

The main advantage of this kind of architecture is the high level of scalability. Using specific devices for each service allows to install the equipment according to the real service penetration. On the other side, the upstream bandwidth limitation due to the coaxial section of the HFC network could be a problem when high penetration of bi-directional services, especially telephony and leased lines, is considered or the number of living units covered by a single Fiber Node becomes too high: therefore these solutions are particularly interesting for a video oriented Network Operator.

2.2 *FTTx (x= Cab/B) architecture*

The considered solution (figure 2) is based on an ATM-PON that allows the integration of traditional telecommunication, multimedia and switched video broadcasting (SDVB) services. The information is transported by mean of an high speed cell-based protocol from/to the OLT into the local exchange to/from the ONUs located at the multi-flat block (FTTB) serving about some tens of living units or at the cabinet level (FTTCab) covering some hundreds of living units. The transmission capacity is shared among all the

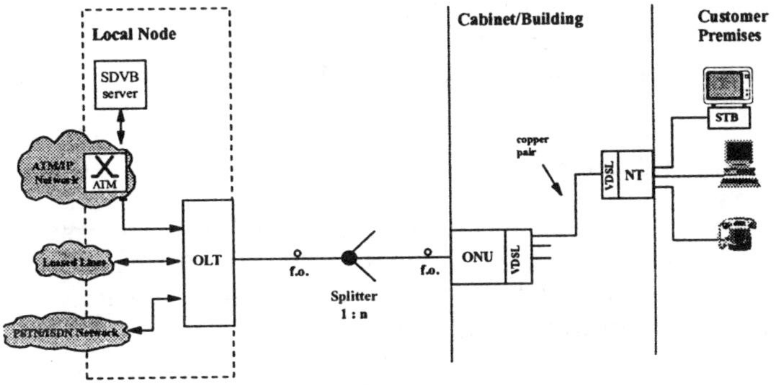

SDVB: Switch Digital Video Broadcast
OLT:Optical Line Termination
ONU: Optical Network Unit
NT: Network Termination
STB: Set Top Box

Figure 2- FTTx architecture

ONUs connected to the OLT. Typically the transmission capacity is different in the downstream and in the upstream direction due to the asymmetric nature of the services.
Customers are connected to the ONU through a single copper pair using a point-to-point VDSL transmission system able to provide from 52 Mbit/s to 13 Mbit/s.
FTTCab solution allows the reuse of the last section a pre-existent copper network while FTTB needs its the replacement by optical fiber cables till user buildings. The former nevertheless, presents some drawbacks due to the use of VDSL systems. In fact, transmission capacity decreases with the user distance from the ONU, thus bandwidth provided in FTTCab architecture is lower than in FTTB one. In addition, due to interference problems, the number of VDSL systems that can be active in the same cable is limited: this fact has to be taken in consideration in case of high penetration of SDVB and IMM services. This problem could be solved by decreasing the transmission rate over copper pairs and therefore the typology of the services provided or by moving to hybrid HFTTx architectures, in which a HFC network for video broadcast services overlays the FTTx network.

2.3 ADSL architecture

This solution (figure 3) allows the complete reuse of the existing copper network: ADSL equipment, able to transmit up to 8 Mbit/s downstream and 512 kbit/s upstream over a single twisted pair, are installed in the local exchange and at the network termination.

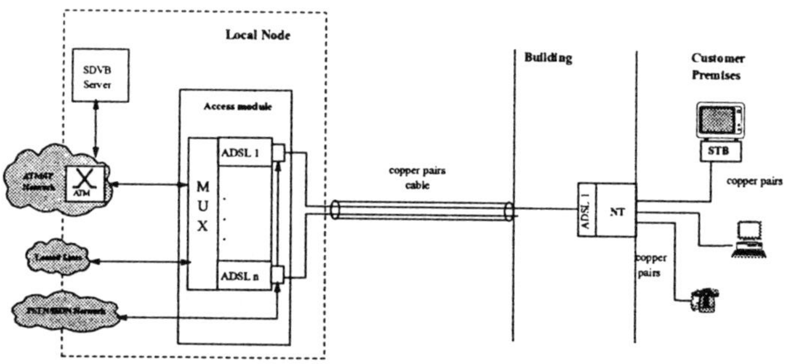

Figure 3- ADSL architecture

The main drawback of this architecture, as a real full-service solution, is the "little" bandwidth that can be provided to the user: 8 Mbit/s could not be sufficient if also SDVB service has to be supported. Moreover the bandwidth is in inverse relation to the distance.

3. Economic evaluation

In network planning, besides technical comparisons, an economic analysis has to be performed, too. Unfortunately, a unique scenario does not exist due to the different objectives of different Network Operators, the uncertainty about the service penetration and the variation of the parameters that characterise a geographic area. Therefore, in this paper, several economic evaluations are presented taking into account the cost of the equipment and the network infrastructures, from the local exchange to the network termination, to build up a full-service network. A 5 year study period, starting from 1999, is considered. The services provided and the relative final penetrations are indicated in table 1.

Services	Final penetration (%)
CATV/SDVB	30
POTS/ISDN	10-30-100
VoD	5
On line services	15
Symmetric Data	2

Table 1

The study area covers about 10.000 living units: two user densities have been selected, 1000 units/km^2 and 2500 units/km^2, depending on the number of users per building, 16 and 4 users per building respectively.

In the first scenario, the case of a Cable Operator with its own diffusive HFC network is analysed. Figure 4 shows the investment per living unit required to build up an integrated HFC network considering two user densities. The figure highlights the sensitivity of the investments to variations of POTS/ISDN penetration due to the increasing amount of required cable telephony equipment. On the other side the network infrastructure cost, that is the cost to make bi-directional the previous diffusive network, remains unchanged. Of course the infrastructure cost increases moving to the case of the lower user density (4 users per building) due to longer distance of the user from the local node.

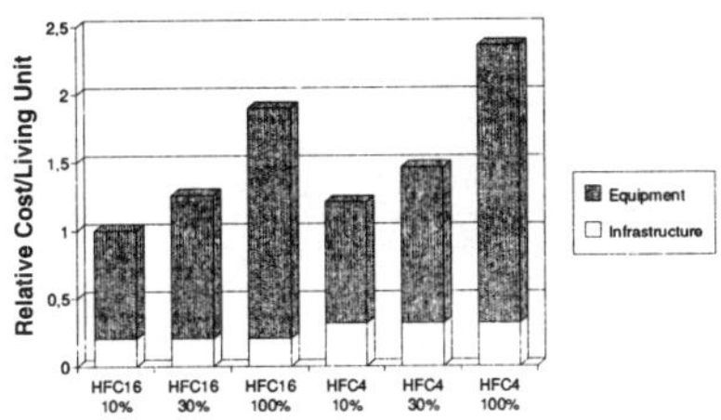

Figure 4

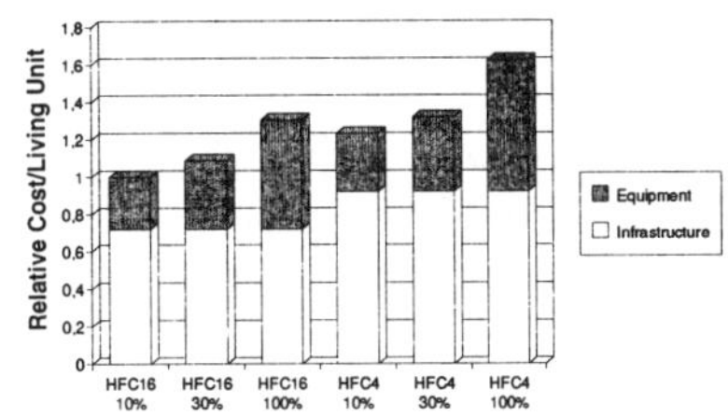

Figure 5

The investments per living unit in case of building from the scratch an integrated HFC network are shown in figure 5. The considerations are similar to those on figure 4, apart from the higher impact of infrastructure cost.

For a Telecommunication Operator, FTTcab/B and ADSL but also integrated HFC architectures are interesting. A distinction can be made supposing or not the reuse of a pre-existent copper network. In the first case FTTCab and ADSL solutions are possible while in the second one FTTB and integrated HFC can be considered.

Figure 6 compares FTTCab, supposing a different number of users per building, and ADSL solutions. ADSL looks to be lightly cheaper due to the absence of infrastructure costs, but the previously mentioned limitations as full-service solution have to be taken into account. As for the FTTCab, the cost impact of a variation of the users density is slight and limited to the infrastructure cost.

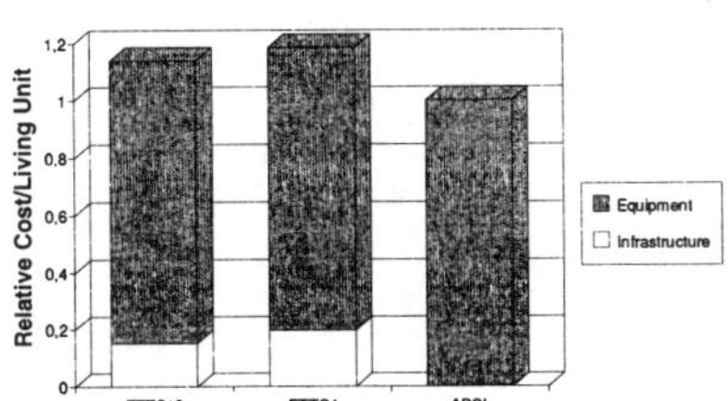

Figure 6

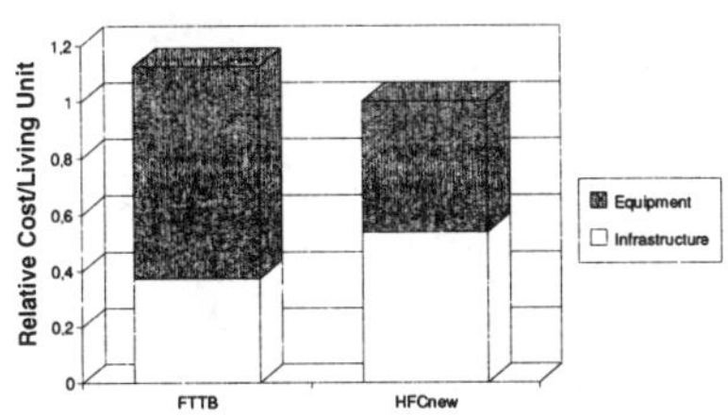

Figure 7

In figure 7, a comparison between FTTB and integrated HFC networks is performed supposing a telephone penetration of 100% and 16 users per building (1000 units/km^2). Both these solutions need a complete construction of the network from the local exchange to the users buildings. Moreover, considering the integrated HFC architecture, the network inside the building, from the basement to the network termination, has also to be built. Therefore the HFC network infrastructures cost is higher. On the other hand, due to the better scalability of the HFC solution and its little more consolidated technologies, the equipment cost is higher in the FTTB architecture.

4. Conclusions

Three full-service networks, i.e. an integrated HFC, an FTTx and an ADSL architecture, have been presented and compared from a technical and an economical point of view in different scenarios. "The best" solution is not universal but is tightly related to the scenario considered. Therefore a Network Operator has to carry out a careful network planning in order to identify its own solution on the basis of its objectives and constraints.

Core and ATM Networks
D.W. Faulkner and A.L. Harmer (Eds.)
IOS Press 1997 © AKM

Cell Broadcasting in ATM Photonic Switches*

Maurizio Casoni, Carla Raffaelli
D.E.I.S. - University of Bologna - Italy
email: *mcasoni{craffaelli}@deis.unibo.it*

Abstract. Two different photonic switches that implement output queueing by means of fibre delay lines are considered in order to support point-to-multi-point traffic. Different techniques are applied and their performance are evaluated in the presence of mixes of point-to-point and point-to-multi-point traffic by means of simulation. The main results are presented to show cell loss and delay behaviours and to outline the limitations that must be imposed to this heavier traffic in order to achieve ATM cell loss target.

1 Introduction

In the last few years many studies have been developed on the design of photonic switching fabrics to support the very high bandwidth demand of future broadband telecommunication networks [1]. The feasibility of architectures based on optical information forwarding and electronic control has been demonstrated within the constraints of present-day technology [2], [3].

Traffic evaluations of ATM switch architectures have been performed with reference to the ATM environment by mainly assuming point-to-point forwarding of cells and show how it is possible to meet ATM requirements expressed in terms of cell loss probability [4].

On the other hand, one of the main aims of future broadband networks is to support multi-service traffic so that further investigations are necessary for the design and the evaluation of techniques to support traffic with different characteristics.

In this paper two different photonic switches that implement output queueing by means of fibre delay lines [5], [6] are considered. In order to support point-to-multi-point traffic, different techniques are applied and their performance are evaluated in the presence of mixes of point-to-point and point-to-multi-point traffic by means of simulation. The main results are presented to show cell loss and delay behaviours and to outline the limitations that must be imposed to this heavier traffic in order to achieve ATM cell loss target.

2 Switch architectures for multi-point traffic

The switch architectures that are here considered for multi-point traffic are presented in figures 1 and 4. For more information about them see [5], [6]. They both perform cells forwarding on the basis of suitable wavelength cell encoding and provide for output buffering. Cell header is processed by electronic control.

OASIS Switch. In the architecture of figure 1 incoming cells are encoded with a wavelength related to their output address. A common set of fibre delay lines is provided to realise N output queues of length Q, where different queues are distinguished by different wavelength. So a cell is available in the switch for only one output with the delay corresponding to the delay line to which it has been sent on its arrival.

The single OASIS switching element cannot support multicast traffic. In fact once a cell has been encoded with a wavelength it can be read by one output only. However, different delays can be associated to the same cell. This can be exploited in order to support multi-point forwarding. To this end two cascading switching elements are considered as presented in figure 2. When a multicast cell arrives at the switch it is encoded with a wavelength associated to a dummy output and the necessary number M of copies are generated and sent to different delay lines. So M copies of the cell will be available on the same output at different cell times. The second stage switching element provides for the correct input/output linking for each cell. A freedom degree is represented by the choice of the dummy output and can be exploited to enhance performance. The following different alternatives are here considered:

*This work has been partially supported by the Commisson of the European Community, ACTS Project AC043 KEOPS (Keys to Optical Packet Switching).

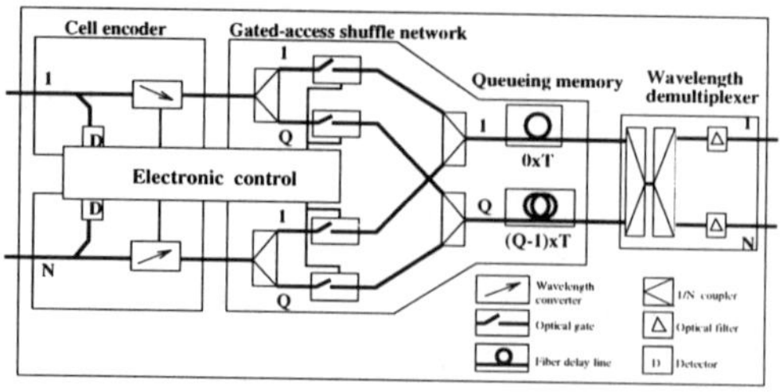

Figure 1: OASIS Switching Element

- *random choice*: a random choice of the output of the first stage is performed for multi-point cells;
- *shortest queue*: both point-to-point and point-to-multi-point cells are sent to the shortest queue buffer, thus achieving usage optimisation of first stage buffers (figure 2).
- *shortest queue with priority*: point-to-multi-point cells are assigned the highest priority thus achieving a better buffer draining at the first stage (figure 3).

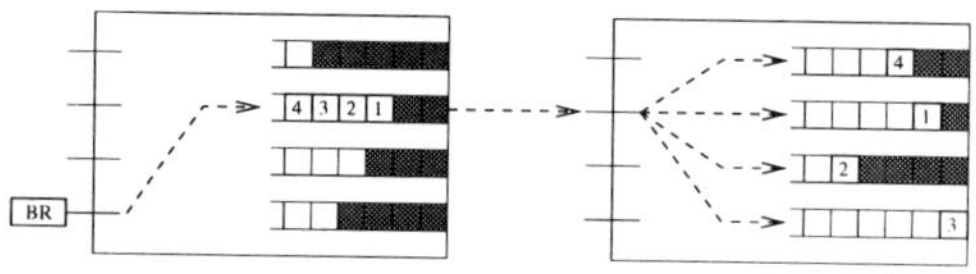

Figure 2: OASIS: Two-stage architecture for multi-point cell forwarding. BR is a broadcast cell.

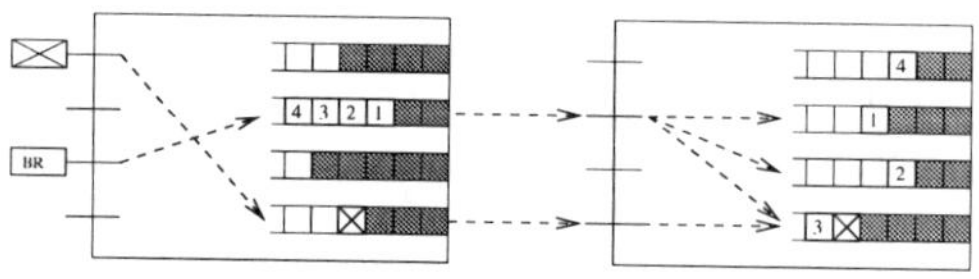

Figure 3: OASIS: Cell forwarding in the case of highest priority to multi-point cells. BR is a broadcast cell. Point-to-point cells are crossed.

Broadcast and Select Switch (BS). In the architecture of figure 4 each cell is encoded with a wavelength related to the arrival input. Then it is sent to all fibre delay lines where it is delayed of all possible quantities. The input/output linking is realised by the choice of delay and wavelength on the basis of the electronic control associated to destination output.

A single switching element of this kind can support generation of copies for multi-point traffic. In fact this is achieved by closing simultaneously the optical gates of the cell buffer that connects the delay line with the outputs involved in the multi-point transfer. In comparison to the OASIS mechanisms, half the hardware is necessary for multi-point connection. In any case it is possible to enhance performance by adopting a two stage organization of broadcast and select switches, thus also allowing comparisons between the OASIS and the broadcast-and-select solution at the same hardware cost. In this case copies can be generated by both stages. Different options are available to this end. In particular we have analysed the following ones:

- *static generation*: in this case each switching element of the two stage architecture produces half the copies necessary for the multi-point connection. At the first stage the generated copies are sent to the shortest queue buffer (figure 5);
- *dynamic generation*: the control logic dynamically decide how many copies are generated at the first or at the second stage depending on buffer state. In particular until buffer at the second stage have occupancy below a prefixed threshold, all copies are generated at the second stage. As soon as the threshold is reached generation of copies is shifted to the first stage

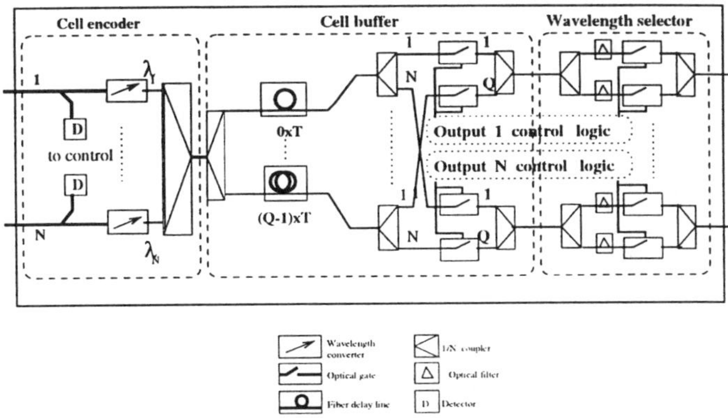

Figure 4: Broadcast and Select Switching Element

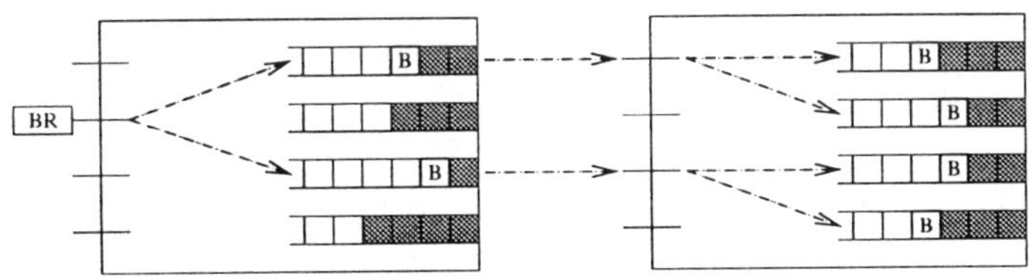

Figure 5: B&S: Cell forwarding in the case of generation of half the copies in the first and half the copies in the second stage. BR is a broadcast cell.

3 Simulation of multi-point traffic

Simulation of the presented architectures have been performed to evaluate cell loss probabilities up to 10^{-6}. This limitation is related to the practical simulation time and extrapolation is necessary to obtain values in the range of ATM applications. We consider as ATM target a cell loss less than 10^{-9}. The results obtained are very useful for comparisons between the two architectures and between the different techniques applied to get multi-point forwarding. Bernoulli input traffic is assumed as a mix of uniform and multicast traffic. Generation of multicast traffic is performed by randomly choosing for each multicast cell the number of copies and the related destination output. Broadcast traffic is a particular multicast traffic where cells are forwarded to all output. The mix is characterised by the percentage of multicast or broadcast traffic with respect to the total traffic amount. It is worthwhile pointing out that each multicast incoming cell increases the output load ρ_{out}. In the presence of broadcast traffic, output load for an $N \times N$ switch is expressed by

$$\rho_{out} = (N-1)\rho_{in}x_{br} + \rho_{in} \tag{1}$$

where ρ_{in} is the input load and x_{br} is the percentage of broadcast traffic on input. This equation gives the limitation on input load and on the percentage of broadcast traffic given the switch dimension N to maintain output load less than unit.

In figure 6 performance of the OASIS couple of 10×10 switching elements are presented for the different forwarding techniques in the presence of broadcast traffic. Limitation of the total load is necessary to obtain ATM target within the physical limitations of photonic technology, that, in our switch architectures, allow a maximum of 32 cell position buffers. However, as expected, a good exploitation of buffer capacity, achieved with the shortest buffer techniques, together with traffic load limitation can lead to a feasible dimensioning of the switch.

Figure 7 shows the delay distribution at the first and at the second stage for a 10×10 switch adopting the shortest buffer technique without priority and with broadcast traffic. It can be seen the effect of the generation of copies at the first stage that make the distribution of delay more "uniform" than at the second stage.

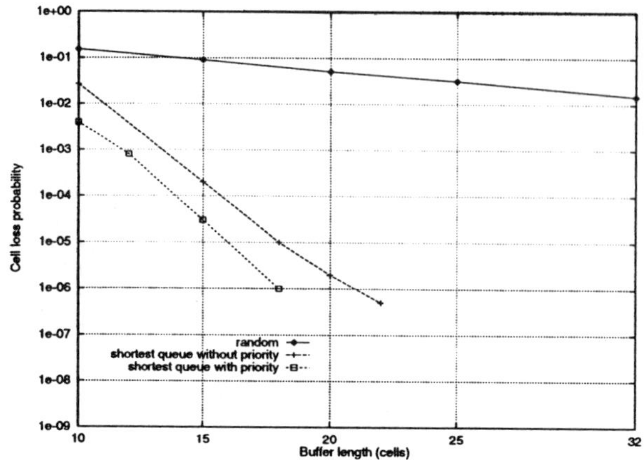

Figure 6: Cell loss probability for a mix of uniform point-to-point and broadcast traffic as a function of switching element buffer length. Switch size 10×10. Total load= 0.4. Broadcast traffic percentage 10%. Comparison with the case of point-to-point traffic only.

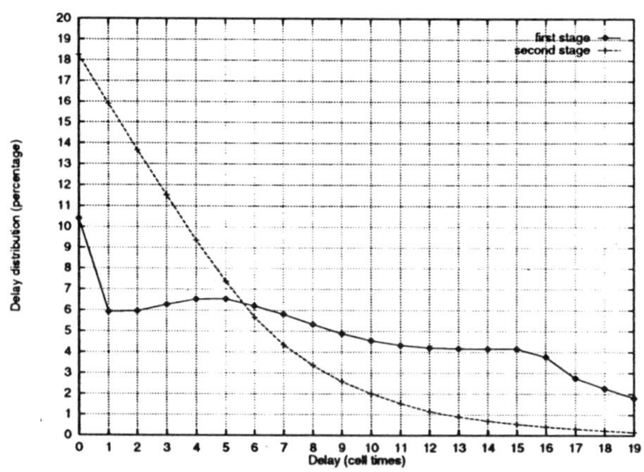

Figure 7: Delay distribution for a mix of uniform point-to-point and broadcast traffic in the case of shortest buffer techniques without priority. Switch size 10×10. Total load= 0.4. Buffer length= 20 cells. Percentage of broadcast cells= 10%.

The distribution of delays in the case of the shortest queue technique with priority is similar to figure 7 but more concentrated at low delays, as expected. This is also confirmed by the computation of the average delays which are shorter for the priority techniques.

Cell loss performance obtained with the broadcast-and-select switch are plotted in figure 8 together with the corresponding results obtained for OASIS with the shortest queue techniques. We can see that the availability of buffer in two stages make the OASIS performance better than those of broadcast-and-select, except for small buffer. In this case in fact the cause of loss is due to the serialization of copies that takes place at the first stage of the OASIS organization. Figure 8 also reports the analytical results obtained by the application of a classical model [7]. It can be seen a good agreement between analysis and simulation for a broadcast and select switch loaded with uniform point-to-point traffic with the average value obtained by equation (1), that is 0.76. It is important to note the slight difference due to correlation among copies generated for broadcast cells which, however, tend to reduce the output contention.

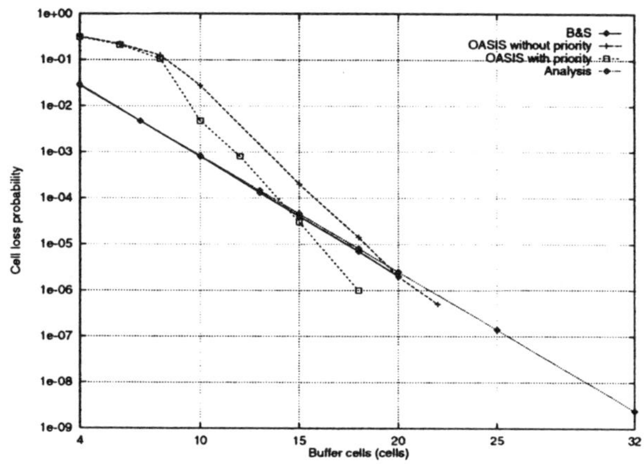

Figure 8: Performance of BS and OASIS 10×10 switches. Load=0.4. Percentage of broadcast cells= 10%. Comparison with analytic results obtained for pure output queueing.

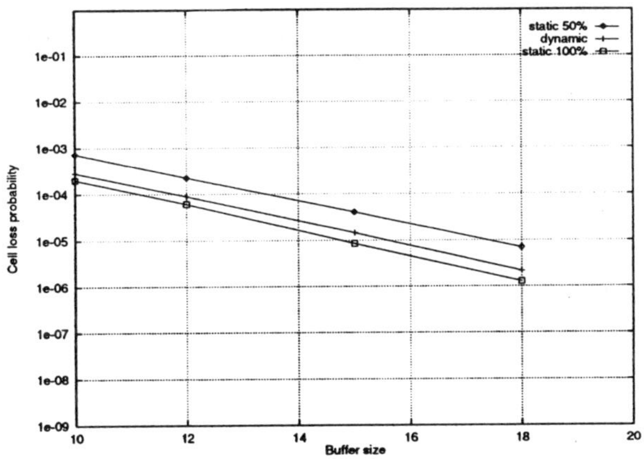

Figure 9: Performance of a couple of BS 10×10 switches. Load=0.4. Broadcast cells=10%.

Figure 9 shows results concerning the two stage organization of the broadcast-and-select switch. The static approach with half the broadcast cells generated by the first stage performs worse than the dynamic one which balances better the load between the two switches (the threshold has been set to 0.75 the buffer size). However, it is worth pointing out the even better performance of the static solution with all broadcast cell generated by the first switch: in this case all the buffers are exploited at the best, both the ones of the first switch, behaving as shared buffers, and the ones of the second switch. With this last approach the architecture is divided into the two logical parts of copying and routing networks.

4 Conclusions

In this paper two photonic switches have been analysed with point-to-multi-point traffic. Different techniques to generate copies are proposed and evaluated by simulation for mixes of point-to-point and point-to-multi-point traffic. The main results show that it is possible to match ATM requirements within the physical limitations imposed by technology. Also, a suitable organization of the switching matrix and of its operation may lead to better buffer utilisation and saving in their dimensioning.

Acknowledgements

The authors wish to thank Dr. Francesco Masetti for the many useful discussions.

References

[1] D. J. Blumental, P. Prucnal, J. Sauer, "Photonic Packet Switches: Architectures and Experimental Implementations", *Proceedings IEEE*, Vol. 82, N. 11, Novembre 1994.

[2] J.M. Gabriagues, J. Benoit, D. Chiaroni, D. De Bouard, P. Doussiere,= T. Durhuus, P. Gavignet-Morin, E. Grard, J.B. Jacob, J. Jacquet, C.G. Jorgensen, D. Leclerc, F. Masetti, P.A. Perrier, K.E. Stubkjaer: Design, Modelling and Implementation of the ATMOS Project Fibre Delay Line Photonic Switching Matrix. Optical and Quantum Electronics, Vol. 26, No. 5, May 1994, pp. 497-516.

[3] D. Chiaroni, P. Gavignet-Morin, P.A. Perrier, S. Ruggeri, S. Gauchard, D. De Bouard, J.C .Jacquinot, C. Chauzat, J. Jacquet, P. Doussiere, M. Monnot, E. Grard, D. Leclerc, M. Sotom, J.M. Gabriagues, J. Benoit, J.B. Jacob: Rack-Mounted 2.5 Gbit/s ATM Photonic Switch Demonstrator, Proc. ECOC '93, Sep. 1993, Montreux, Switzerland, Post-deadline paper, ThP 12.7.

[4] M.De Prycker, "Asynchronous Transfer Mode Solution for B-ISDN", *Ellis-Horwood*, II edition, 1993.

[5] J.M. Gabriagues, J.B. Jacob: OASIS: a High Speed Photonic ATM Switch. Results and Prespective. Proc. of ISS 1995, Berlin (D).

[6] F. Callegati, M. Casoni, G. Corazza, C. Raffaelli, F. Masetti, J.M. Gabriagues, M. Sotom, D. De Bouard, D. Chiaroni, P. Parmentier: Performance Evaluation of a New Photonic ATM Switching Architecture based on WDM. Proc. of ATNAC 1995, Sydney (Australia), December 1995.

[7] M.J. Karol, M.G. Hluchyj, S.P. Morgan, "Input Versus Output Queueing on a Space-Division Packet Switch", *IEEE Trans. on Communications*, Vol. Com-35, no.12, pp.1347-1356, Dicembre 1987.

Core and ATM Networks
D.W. Faulkner and A.L. Harmer (Eds.)
IOS Press 1997 © AKM

An Accelerated Simulation Technique for Evaluating Communications Systems utilising FEC

Simon Fragiacomo, Yi Bian *, Andrew Popplewell †, Chris Matrakidis, and John O'Reilly

Department of Electronic & Electrical Engineering, University College London, Torrington Place, London WC1E 7JE, UK
**Illgen Simulation Technologies Inc, USA*
†GEC Plessey Semiconductors, Microelectronic Centre, UK

Abstract

This paper presents a new simulation technique suitable for application to communication systems which utilise linear block codes for error correction. The proposed technique uses channel noise statistical information to reduce the overall number of words required for achieving an accurate result without sacrificing accuracy.

1 Introduction

Recent technological advances in optical telecommunications systems have made very low *residual bit error rates* (RBERs) possible. The use of *forward error correction* (FEC) has reduced this even further. However, the accurate evaluation of such a system in the region of interest is a very challenging task. This is because when a conventional Monte-Carlo simulation technique is used, the execution time required for obtaining a reasonably accurate result is excessive.

This paper will present a novel simulation technique which will significantly reduce the amount of time required to achieve accurate results, when a block code is used.

2 Present methods

Assume that the performance of an (n, k) BCH block code (which can correct upto t errors) must be evaluated.

A conventional simulation will perform three distinct operations i.e. encode, transmit and decode. Once all three are completed the decoding performance of the code for various SNRs will be assessed.

In order to measure decoding performance a given number of residual bit errors (usually 1000 errors are sufficient) are required to be present at the decoder's output. However, codewords with a number of errors less than or equal to t do not contribute anything to the RBER, since it is certain that they will be completely corrected.

Nevertheless they will still have to go through the whole process of simulation which can be time consuming. In order to enhance efficiency and reduce the total computational time required, a new technique was proposed.

This technique utilises knowledge of channel noise statistics and completely eliminates the words with less than t errors. If the word doesn't contribute to the RBER it is only calculated and not simulated.

To describe the proposed technique a white noise channel is considered. Let p be the bit error probability. Therefore p^u is the probability of having exactly u errors in u bits, while the probability of no errors in $(n-u)$ bits is $(1-p)^{(n-u)}$.

It now becomes obvious that if all possible combinations of having u errors in a total of n bits are taken, then the probability that a codeword of length n contains u transmission errors is

$$P_u = \frac{n!}{(n-u)!u!} p^u (1-p)^{(n-u)}$$

Probability p can be theoretically found for various SNRs. Therefore, using this formula the percentage of codewords from a user defined initial set which contain u errors can be statistically determined. If a high enough initial number of codewords is selected we can determine what percentage of those words will have more than t errors within them and simulate these only. The rest of the words (with less than t errors) are of no interest since they will always be correctly decoded. Thus they are not simulated but are only used in the final calculation of the RBER. Simulation of the codewords of interest is done using the same techniques as those described in [1].

So for example, if a $(127, 106)$ BCH code at a SNR of 5 is used with 28.593×10^6 transmitted words, only 208×10^3 will contain more than $t = 3$ errors and will need to be simulated, see figure 1. The words with fewer than t errors will be corrected by the HD decoder and are ignored.

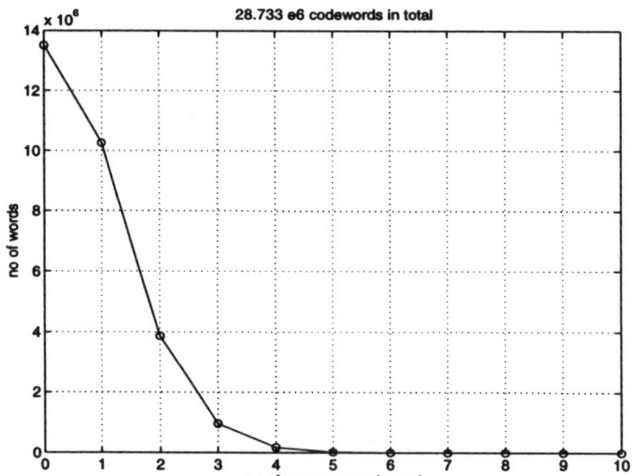

Figure 1: BCH (127,106,3) at SNR=5 simulated error distribution

Finally once the complete set of words has been decoded, the RBER needs to be calculated (and not simply measured) because the words with less than t errors were not included in the simulation. Since RBER is defined as

$$RBER = \frac{Number\ of\ information\ bits\ in\ error\ after\ EC}{Total\ number\ of\ information\ bits}$$

it is obvious that the non-simulated information must be added to the denominator if a correct result is to be obtained.

The proposed acceleration technique can thus be summarised as follows:

1. Select the total number of codewords to be transmitted which should be chosen to provide a sufficient number of residual errors.

2. Determine the number of codewords with more than t errors and find their distribution.

3. Generate a message and conventionally encode to create a codeword. Then modify a given number of bit positions so that they are in error, as determined from the distribution obtained in step 2.

4. Generate the remaining noise samples (as described in [1, 2]).

5. Conventionally decode the codeword.

6. Measure the number of residual errors and calculate the RBER.

The results of this technique were implemented in software and were validated [2]. Near identical results to a conventional simulation were achieved but at the same time RBERs of 10^{-9} or even lower could be simulated within a reasonable amount of time. Table 1 indicates the approximate times required to achieve 1000 residual errors for various SNRs.

(127, 106) BCH HDD code		
SNR	Time required using conventional simulation	Time required using accelerated simulation
1	55 sec	20 sec
2	59 sec	21 sec
3	1:31 min	29 sec
4	3:43 min	31 sec
5	25:50 min	50 sec
6	7:14:07 hours	1:05 min

Table 1: Simulator time required for obtaining 1000 residual errors at varying SNRs

The first column indicates the signal to noise ratio of the channel, while the second column indicates the corresponding simulation time required, if a conventional simulation technique is used. The third column indicates the time required if the proposed algorithm is used.

It should be noted that in the proposed algorithm the total number of words simulated does not increase exponentially with increasing SNR, which explains the very low simulation times.

Thus by using the proposed algorithm for simulating the performance of systems employing FEC significant reductions in the time required can be achieved, without affecting the accuracy of the results.

3 Conclusions

In this paper a novel simulation acceleration technique was presented. This has enabled the simulation of very low RBER floors within a reasonable amount of time and without any loss in accuracy. This is achieved by concentrating only on those codewords that do contribute to the RBER and not simulating any other words which will be corrected by the decoder.

References

[1] Y. Bian, JJ O'Reilly, A. Popplewell, "Novel Simulation Techniques for Assessing Coding System Performance", *Electronics Letters*, Vol 30, No. 23, pp. 1920-21, November 1994.

[2] Y. Bian, JJ O'Reilly, A. Popplewell, S. Fragiacomo, "New Simulation Techniques for Evaluation Telecommunications Transmission Systems with FEC", *IEE Brighton Conference in Telecommunications*, March 1995.

Core and ATM Networks
D.W. Faulkner and A.L. Harmer (Eds.)
IOS Press 1997 © AKM

Multimedia System Development using a CORBA based Platform

Pradeep Ray
Department of Computing, University of Western Sydney, Nepean, Kingswood, NSW 2747, Australia, Tel:+61 2 9685 9251, Fax:+61 2 9685 9245, email:pradeep@st.nepean.uws.edu.au

Abstract

Recent developments in multimedia communication have created the need for the interoperability of heterogeneous multimedia systems and services. The Common Object Request Broker Architecture (CORBA) from the Object Management Group (OMG) addresses the issue of distributed object modelling of heterogeneous computing applications. However, it does not address some important issues related to multimedia, such as streams and their quality of service. The Interactive Multimedia Association (IMA) has announced a recommended practice for the development of multimedia systems based on CORBA. This paper presents our experience in the development of a World Wide Web (WWW) based video-on-demand service, using IMA recommended practice, and a commercially available CORBA platform.

1. Introduction

Thanks to the recent developments in multimedia technology, it is now possible to have a number of interesting information services, such as video-on-demand, teleshopping, videoconferencing, cooperative computing etc. Most these applications require mechanisms to specify, implement and control multimedia protocols and formats. Presently, these mechanisms offered by vendors are proprietary in nature. Future growth of multimedia business is expected to depend on the standardisation of the design and delivery modes of heterogeneous services [2].

International standards bodies and vendor groups have been working on generic object-based distributed computing models, such as ISO Open Distributed Processing (ODP), TINA, and OMG Common Object Request Broker Architecture (CORBA) [6], [4], [7]. While CORBA is being adopted as the standard object model, it does not yet support some multimedia functions, such as streams. An industry group called Interactive Multimedia Association (IMA) has announced a Recommended Practice (RP) for the development of CORBA based multimedia services [5]. This paper presents our experience in the development of a video-on-demand service based on IMA-RP, and a commercially available CORBA-based platform, DIGITAL Object Broker.

2. CORBA based multimedia

The IMA recommended practice for Multimedia Systems Services (MSS) is based on the Common Object Request Broker Architecture (CORBA). To support interaction with remote objects, the MSS depends upon an underlying object infrastructure, such as that provided by the CORBA Common Object Support Services (COSS). Since the present version of CORBA (ver 2.0) does not support facilities for the definition and implementation of time-critical multimedia streams, the MSS defines a private Media Stream Protocol [8].

This paper presents the modelling of a multimedia service based on IMA recommended practice.

3. Multimedia service development

This section illustrates the use of IMA-RP in the development of our World Wide Web (WWW) based video-on-demand service.

3.1. Multimedia service workflow

The recent growth of World Wide Web (WWW) services has created a demand for for the browsing of live and stored multimedia information. It is necessary to consider two aspects in this multimedia service:

- **Capture and storage of video and audio information (Items)**
- **Selection, replay and delivery of video/audio items to remote clients**

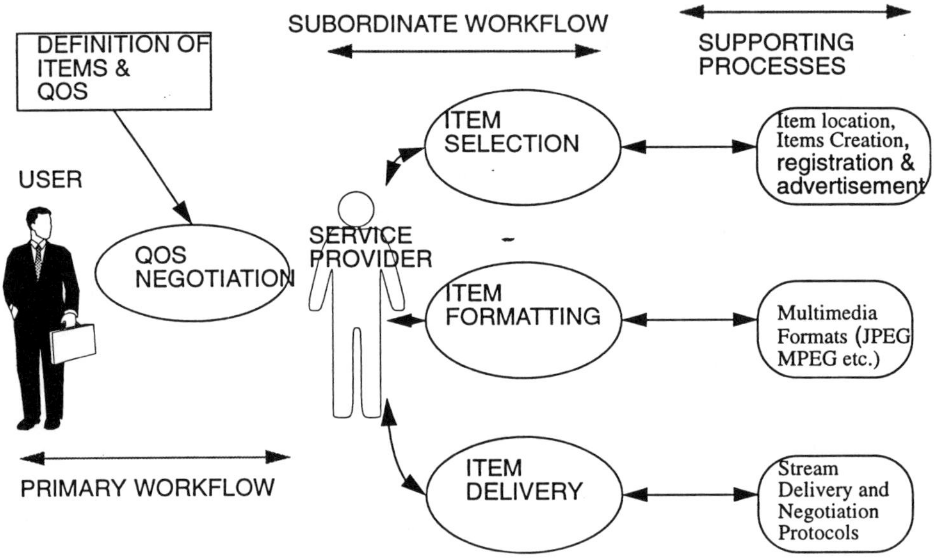

Fig. 1. Video-on-demand workflow

Fig.1 shows this video-on demand application using a workflow model where each loop shows an independent task with customer on the left and performer on the right of the loop. This representation based on Action Workflow, gives us a methodology for the description of various multimedia processes in the context of the business relationship between a customer and a service provider [1].. The service delivery task is divided into three sub-tasks:

- **Multimedia Item Selection**

  -This task is responsible for the browsing, selection and location of a multimedia item required by the client.

- **Multimedia Item Formatting**

  -This task represents the process of formatting and unformatting of the multimedia item selected

- **Multimedia Item Delivery and Negotiation**

  -This task represents the process of the negotiation for the required QoS and the delivery of the item as per that QoS.

3.2. Video-on-demand model based on IMA-RP

This Section illustrates the use of IMA MSS connection model in the description of our video-on-demand application, as shown in Fig. 2.

As shown in Fig. 3, the Video-on-Demand can be modelled as a CORBA application module that includes:

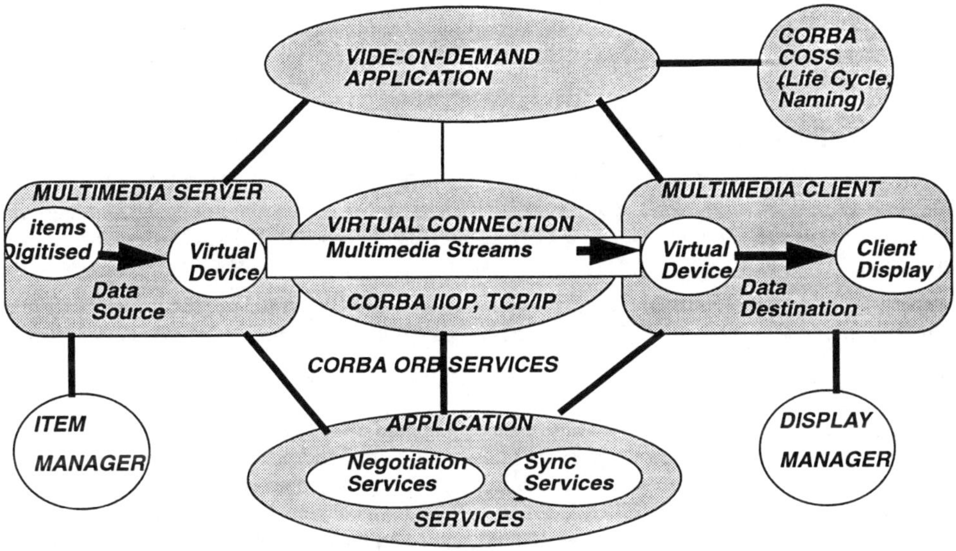

Fig. 2. CORBA/IMA based video-on-demand service

CORBA Common Object Support Services (COSS), such as COSLifeCycle, COSNaming, COSEventServices, COSPropertyServices. A Multimedia Server provides the service of browsing, selection and delivery of audio/video items that are distributed over the network. The Multimedia Client repre-

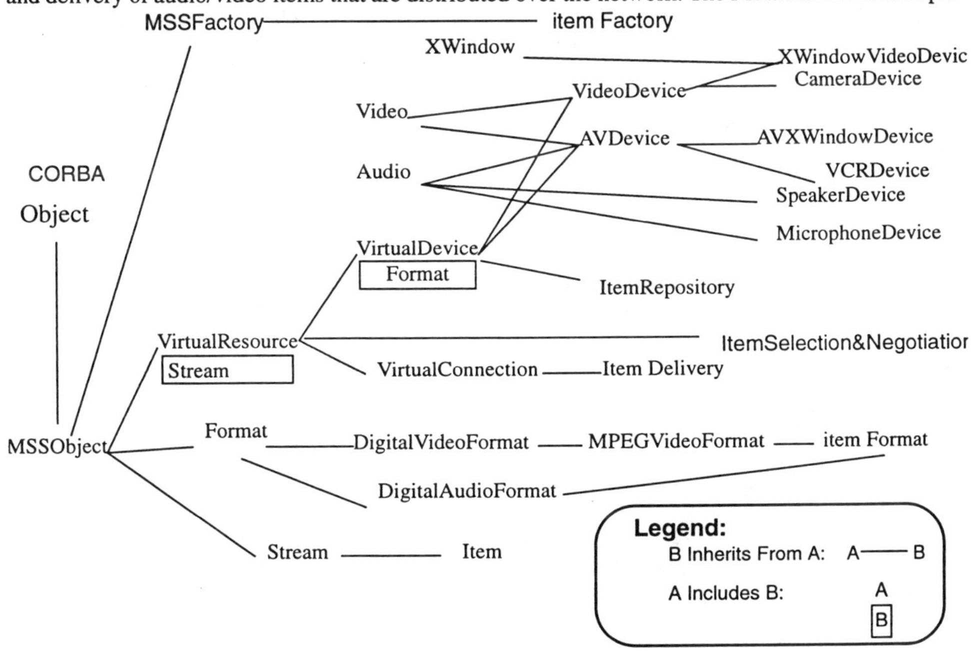

Fig. 3. Object relationship in the video-on-demand application

sents the client side equipment and software, such as a PC with multimedia pack. While various objects in this diagram communicate through CORBA Object Request Broker (ORB) services, the video-on-demand is delivered through a multimedia stream object outside the CORBA specification. Application services, such as negotiation and synchronisation services are based on MSS application services. The Item manager is responsible for the capture, identification and storage of audio/video Items. The Display Manager is responsible for the WWW based user interface of the Video-on-Demand service.

3.3. Video-on-demand service objects

Having defined the overall service model, the next step is to define the CORBA/IMA objects for the realisation of the video-on-demand application discussed above. As shown in Fig. 3, the relationship amongst objects defined in our video-on-demand application, based on IMA RP object inheritance diagram (Fig. 3). described in [5].

This system was designed for implementation on the Digital CORBA based product called Object-Broker [3]. The details of the implementation issues are not discussed here due to space constraints.

4. Conclusion

This paper has discussed our experience in the development of a CORBA based multimedia application, using IMA Recommended Practice for Multimedia System Services. This application uses a standard based CORBA object model for the interoperability of heterogeneous services. The object interface definitions, expressed in CORBA IDL, inherit from the objects defined in IMA-RP for multimedia services. Finally we have used the DIGITAL ObjectBroker ORB platform for the realisation of our experimental WWW Video-on-Demand service.

Although the CORBA based development had a steep learning curve, some features of ObjectBroker, such as OBBGUI and Quickstart were quite useful in the initial debugging of the application. Our difficulties were compounded by the evolving nature of today's commercial ORB platforms. While IMA-RP provides a good guideline for the development of CORBA based multimedia applications, more powerful object definitions are required for the design of a complex multimedia service.

Acknowledgements

The work reported in this paper has been funded in part by the Cooperative Research Centres Program through the Department of the Prime Minister and Cabinet of the Commonwealth Government of Australia.

References

[1].•Action Technologies Inc USA, "Action Workflow", http://www.actiontech.com/action/ 1995
[2].•J. F. K. Buford, "Multimedia Systems", Addison-Wesley Book, ISBN:0-201-53258-1, ACM Press 1994
[3].•Digital Equipment Corporation, "Object Broker: Designing and Building Applications", March 1996
[4].•Luis A. Fuente et al., "The TINA-C Management Architecture", 5th IFIP/IEEE International Workshop on Distributed Systems Operations and Management (DSOM'94), Toulouse, France, October 1994.
[5].•Interactive Multimedia Association, "Recommended Services for Multimedia System Services", May 1995; URL:http://ima.org
[6].•ISO 10746-X, "Reference Model for Open Distributed Processing (RM-ODP) Parts 1,2, 3 and 4
[7].•Object Management Group (OMG), "The Object Management Architecture Guide", 1992
[8].•Pradeep Ray et al., "Middleware for Multimedia Services", Proceedings of the European Conference on Networks and Optical Communications (NOC'96), Heidelberg Germany, June 1996

Core and ATM Networks
D.W. Faulkner and A.L. Harmer (Eds.)
IOS Press 1997 © AKM

A Hybrid Rate/Reservation MAC Protocol for HFC Networks

Mário S. Nunes, Mário M. Macedo
INESC, R. Alves Redol 9, 1017 Lisbon - Portugal - msn@inesc.pt

Abstract. A new MAC protocol is proposed for HFC (Hybrid Fiber Coax) based networks, named SARP (Simple ATM based Reservation Protocol). It is an hybrid reservation and rate based protocol, based on ATM, whose main characteristics are its simplicity, high efficiency and support of different classes of traffic. Two types of sources are considered for the SARP scheduling, requiring two different but coexisting and cooperating algorithms: one for continuous sources (CBR and VBR) the other for bursty sources (UBR and VBR).

1. Introduction

The SARP protocol is being developed in the scope of ATHOC [1], a project of the European program ACTS (Advanced Communications Technologies and Services) in Europe. The main objectives of ATHOC are to build a prototype of an ATM-based HFC system and to demonstrate on this system PC-oriented data services, namely Internet services

The SARP protocol and its associated traffic scheduler are located in the ANA (Head-End) subsystem of the ATHOC network. SARP relies on the existence of individual transmission grants (TG) in the downstream frame. Each TG allows the transmission in the upstream direction of a single cell of a particular ANT (Access Network Termination) identified by its TG address.

Two types of sources are considered for the SARP scheduling, which requires two different but coexisting and cooperating algorithms: one for continuous sources (corresponding to CBR and VBR audio and video sources), the other for bursty traffic sources (UBR and ABR, e.g. LAN type traffic). For the first kind of sources, the algorithm is based on a rate-based and dynamic scheduling protocol, where the adjustment of the allocated bandwidth to each terminal is based on implicit feedback of the terminal activity, either by empty cell detection or by piggybacking of queue occupation on the transmitted cells.

For the second type of sources, the algorithm is based on a request/credit method, using contention slots. In figure 1 a basic overview of the different messages used in the SARP are shown.

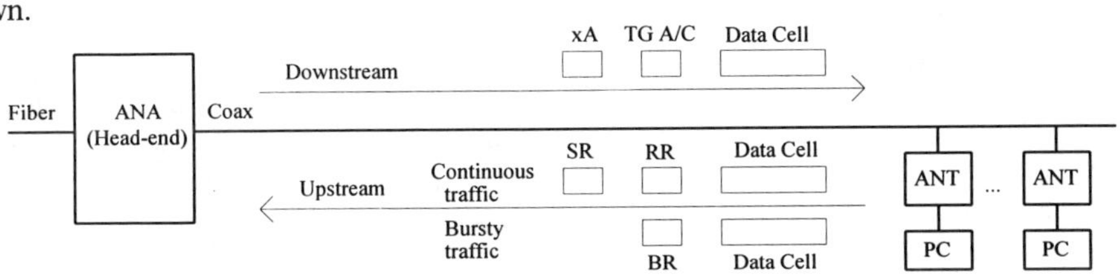

Figure 1- SARP messages

In the downstream channel there are three types of messages: ATM Data Cells, Transmission Grants (TG) control messages and Acknowledge (xA) control messages. Two types of TG are defined: TGA is used to indicate to a specific ANT, identified by its address, that it can transmit a data cell in the upstream channel, and TGC is used to allow the transmission of a control message in the upstream channel to all ANT, with contention.

In the upstream channel different logical messages are defined for data and for continuous or bursty traffic control, identified by specific ATM or MAC identifiers. SARP uses mini-slots for the transport of control messages in the upstream channel, in order to increase the bandwidth efficiency utilisation of the this channel.

For the continuous traffic two control messages are used, RR and SR cell. The RR (Rate Request) message is used to request a minimum guaranteed bit-rate to the Head-end. The SR (Synchronisation Request) message is used to synchronise the low bit-rate CBR sources with the grants transmitted by the Head-end. While the Data cells are sent always without contention, using the TGA, the control messages can be used with contention.

For the bursty traffic only one control message is used, the BR (Burst Request). When the ANT receives a burst of cells from the terminal equipment, it generates a BR message with a parameter indicating the number of cells in the buffer. The bandwidth manager at the Head-end generates as many TGA grants as the number of cells requested, scheduled accordingly the algorithms defined in the SARP, taking into account the traffic of all the sources active at the moment.

All requests are acknowledged with a specific downstream message (xA in the figure). Three acknowledged messages were defined: RA (Rate Acknowledge), BA (Burst Acknowledge), and SA (Synchronisation Acknowledge).

The basic algorithm of the SARP protocol is described in the following rules for the two types of sources.

2. Algorithm for continuous sources

1) In SARP, each continuous source (CBR/VBR) has an assigned instantaneous grant rate Gi, sharing the available bandwidth with the other sources. A pool of free bandwidth is available for this purpose. An overload monitor acts upon all sources, establishing limits on bandwidth use, specially when the bandwidth use approaches the upstream channel capacity.

2) The grant bit-rate Gi is continuously updated, based on the activity of the source, starting with a fraction of the bandwidth available. The controller is continuously checking if the sources are using the grants for sending cells in the upstream direction, checking empty cells and the terminal queue occupation.

3) When grants are satisfied or the queue occupation increases to a level higher than a specified threshold, the controller increases the grant rate given to the terminal. When one grant is not satisfied, a rate computation is done to estimate the present terminal transmission rate. The new grant rate complies with the minimum and maximum rate specifications, defined in the RR control message.

4) If a source stops transmission, all cells in the grant window become empty and Gi falls down quickly to Gmin and is considered idle for the controller. The communication restarts only when the first not empty cell arrives, or with an explicit request message.

5) The overload monitor watches for overload conditions, giving criteria for source rate increase and decrease. It bases its decisions on the free rate monitored over a sliding window. If there is overload, sources start to decrease its grant rate slowly and then faster if overload persists. Sources which have higher ratios of grant rate and guaranteed minimum rate, are the first to decrease its rate. This way we achieve fairness and maximal bandwidth use.

3. Algorithm for Bursty traffic sources

Bursty sources are managed together, serviced with a "round-robin" discipline. A global burst grant rate Gb is calculated to service all active bursty sources.

To accommodate the bursty sources and control messages, not all the channel bandwidth capacity is allocated to the continuous sources, with some bandwidth reserved for contention slots and bursty traffic sources. Optimal balancing between these types of traffics is yet a matter of research.

When a request for burst transmission (BR) arrives to the controller in a contention slot, some of the bandwidth Gb is reserved for transmission of the indicated number of grant cells to that source.

Each time a new burst request arrives to the controller, it is added to the present burst list and the Gb burst grant rate is recalculated.

4. SARP Control messages

The Control messages and associated parameters are summarised in Table 1. The parameter Message Type (1 octet) is common to all messages and identifies the control message. The parameter Terminal Identifier (2 octet) identifies the ANT.

Table 1 - *SARP control messages and parameters*

Name	**Description**	**Direction**	**Parameters**			
TGC	Transmission Grant with Contention	Downstream	Message Type	Probabi-lity		
TGA	Transmission Grant with Address	Downstream	Message Type	Terminal Identifier		
RR	Rate Request	Upstream	Message Type	Terminal Identifier	Min_Rate	Max_Rate
RA	Rate Acknowledge	Downstream	Message Type	Terminal Identifier		
SR	Synchronisation Request	Upstream	Message Type	Terminal Identifier	Cell_Waiting Time	
SA	Synchronisation Acknowledge	Downstream	Message Type	Terminal Identifier		
BR	Burst Request	Upstream	Message Type	Terminal Identifier	Num_Cells	
BA	Burst Acknowledge	Downstream	Message Type	Terminal Identifier		

Transmission Grant with Contention (TGC): This downstream control message is used by the Head-end to indicate to the ANTs on the bus that they can transmit requests on the contention mini-slots. Probability (1 octet) indicates the probability of ANT transmission requests on the mini-slots, with value 255 corresponding to probability 1. The parameter Probability is used to implement a P-Persistent slotted aloha protocol, in order to reduce the bus collisions on high load conditions.

Transmission Grant with Address (TGA): The TGA is a downstream control message used by the Head-end to indicate to a specific ANT on the bus that it can transmit a data cell on the upstream slot. This message is used both for continuous and bursty sources.

Rate Request (RR): The RR is an upstream control message used by the ANTs to request to the Head-end a continuous channel, with or without minimum and maximum rate requirements. Min_Rate (2 octet) is used by the ANT to require a minimum rate for the source, in units of 4 Kbit/s. If no minimum rate is needed, a zero value is assumed. Max_Rate (2 octet) is used by the ANT to request a maximum rate for the source, in units of 4 Kbit/s.

Rate Acknowledge (RA): This downstream control message is used by the Head-end to notify an ANT that the previous Rate Request has been received, and also that the ANT can transmit a data cell in the upstream slot. This last function is similar to the Transmission Grant with Address, and is the first message of a series of TGA, after a RR message.

Synchronisation Request (SR): This upstream control message is used by an ANT to notify the Head-end of cell delays on CBR links that exceeds maximum value requirements. It is used to synchronise the Head-end TGAs with the cells generated by the ANT. The ANTs issue as many SR as needed to decrease cell delays to acceptable values. Usually the SR will be needed at the beginning of the transmission, or when jitter or drift in the source makes it necessary. Cell_Waiting_Time (2 octet) specifies the actual delay of cells in 10 μs units.

Synchronisation Acknowledge (SA): This downstream control message is used by the Head-end to notify an ANT that a Synchronisation Request has been received and processed, and also that the ANT can transmit a data cell in the upstream slot. This last function, is similar to the Transmission Grant with Address, because it is used both as an acknowledge and as a transmission grant. SA is transmitted instead of TGA, when a SR has been issued by the ANT, but serves also the purpose of the transmission of a data cell.

Burst Request (BR): This upstream control message is used by an ANT to request to the Head-end the transmission of a group of cells, specified in the parameter Num_Cells (1 octet). After reception of the request, the Head-end will issue a BA and as many TGA with the ANT address as requested.

Burst Acknowledge (BA): This downstream control message is used by the Head-end to notify an ANT that a Burst Request had been received and processed.

5. Conclusions

The SARP protocol presents interesting properties of fairness, low implementation complexity, low delay and flexibility for adaptation to very different traffic categories (CBR, VBR, UBR, ABR). Different traffic categories have been simulated and in all cases the algorithms behaves well, with the available bandwidth being efficiently used. When the bit-rate of all the sources exceeds the available bandwidth, the CBR sources keep a guaranteed bit-rate, while the UBR sources compete for the remaining available bandwidth.

Part of the work reported in this paper has been carried out within the ACTS project AC037 ATHOC. The authors would like to thank the European Commission for supporting this project.

References

[1] C. Sierens, K. Krijntjes, M.S. Nunes, et. al., "A first description of the Hybrid Fiber Coax concept in the framework of the ACTS project ATHOC", IEE Meeting on Optical and Hybrid Access, BT Laboratories, Martlesham Heath, Ipswich, UK, March 1996.

[2] Mário S. Nunes, Mário M. Macedo, "A Simple Reservation MAC Protocol for HFC Networks", IEEE ATM'96 Workshop, San Francisco, USA, August 1996.

Core and ATM Networks
D.W. Faulkner and A.L. Harmer (Eds.)
IOS Press 1997 © AKM

Implementing Dynamic Unicast VCC Routing in ATM Networks

P. Aranzulla, J. Mellor
School of Engineering, University Science Labs., South Road, Durham City. DH1 3LJ (UK). E-mail: P.J.Aranzulla@Durham.ac.uk

Abstract. *A hop by hop algorithm is proposed which is shown to perform well under symmetrical and non-symmetrical loading, for both highly and sparsely connected VPC topologies. The signalling overhead required for updates of the network state to nodes is then limited by combining various strategies for obtaining global network state information via local means, the aim being to provide performance approaching that obtained by a full network-wide flood of information.*

1. Introduction

The history of routing shows algorithms have always been closely aligned with the capabilities of technology. As technology has developed, so routing algorithms have progressed to be dynamic and non-hierarchical in nature. ATM networks pose new routing problems, such as a dynamic VPC topology and QoS requirements, requiring new routing strategies.

The possibility of a reconfigurable logical or VPC network, either by modification of the allocated bandwidths of VPCs or by the reconfiguring of the physical paths taken by VPCs, results with a network that is no longer fixed for connection routing purposes. Therefore there is a strong requirement for routing algorithms that are dynamic and not limited to certain kinds of logical network topology. This is re-enforced by having to guarantee a QoS for a connection, for sub-optimal routes that cannot support the required QoS will cause increased blocking of connections attempts, unlike previously where the connection was accepted with all connections' QoS being degraded.

By guaranteeing QoS for a connection, a new factor is added to the routing problem. The guarantee is for an end-to-end QoS, this being assured by reserving an effective bandwidth for the source type and characteristics [1]. If the route is of one logical hop in length, meaning that it traverses only one VPC from its source to its destination, then that end-to-end QoS can be guaranteed by assigning an effective bandwidth for the end-to-end QoS on the VPC. If the route traverses two VPCs however, then that end-to-end QoS can only be guaranteed by assigning a higher effective bandwidth on each VPC than previously, equal to QoS / 2, so that in a worst case scenario the end-to-end QoS is still guaranteed. Therefore the longer the path chosen on the logical network, the higher the effective bandwidth required on each VPC to guarantee the end-to-end QoS. This strengthens the requirement for minimum hop routes still further when compared to previous network technologies where a longer hop route did indeed use more network resources, as it traversed a greater number of links, but the reserved capacity on each link was fixed however long the route chosen.

Some previously suggested algorithms for unicast VCC routing have not taken into account the higher effective bandwidth required for longer routes. For example Minimum

Sum of Loads (MSL) [2] routed over least loaded routes, so sometimes choosing longer paths over shorter ones that might still have sufficient free bandwidth to accept the connection. Other algorithms have assumed a highly connected logical network topology, such as attempting to adapt currently used circuit switched network routing algorithms [3]. Adaptive Minimum Hop (AMH) was designed to overcome these deficiencies, and was shown to out-perform MSL in most cases [4]. This algorithm assumes a more sparsely connected network, choosing the least loaded route out of the set of possible minimum hop routes to the destination node.

Alternate Adaptive Minimum Hop (AAMH) was proposed to free this topological restriction [5], initially choosing the least loaded route from the set of minimum hop routes, and if all these are saturated obtaining the set of routes closest in length to the minimum hop routes, and routing over the least loaded of these. Improvements of using AAMH over AMH were most noticeable under non-symmetrical loading, which is the loading type generally found in real networks. Under symmetrical loading AAMH performance was found to be equal to or slightly better than AMH performance, when trunk reservation was used in conjunction with AAMH. Under non-symmetrical loading for both a four node fully connected and a seven node sparsely connected network topology a lower blocking probability of up to 40% was achieved [6]. This paper continues by examining implementation issues of AAMH.

2. Implementation Issues of AAMH and Simulation Results

All the simulation results were obtained with the OPNET simulation environment, using an ATM model library which was constructed in-house due to the standard ATM model library being too restrictive for use in this and future research. The full ATM Forum UNI signalling protocol [7] has been implemented for VCC set-up and tear-down.

The above results are obtained when each node has full and instantaneous network-wide loading knowledge. This is not possible in a real network, as the signalling and processing overhead of the network loading packets would be prohibitably high. Therefore this next section is concerned with obtaining good performance from an implementable version of the AAMH algorithm.

Previous studies have shown that these continuous measures of dynamic load are not needed to improve routing decisions over a purely random or fixed alternate path method. Instead, the dynamic load measures can be mapped into discrete categories, with load changes from one of these categories to another then being updated to other nodes [8]. The following load bands were chosen to be used in this study: 50%, 80%, 90% and 95%. As the concern is to avoid saturation of VPCs as much as possible, so the finer grained information is only required at loadings close to saturation.

Even by using load bands, the signalling overhead is still prohibitably high for updating other nodes of a changed load band when using full flooding in large networks. Therefore further methods are required to reduce the signalling when updating other nodes about a change in load band category of a VPC. These might include limiting the number of hops permissible when flooding the new VPC loading state information, so that only nodes within a certain radius from the source node of that VPC receive the update signal packet. Such methods would always provide more local information, never provide any information on VPCs which are further away. Other methods would circumvent this by using the VCC set-up signalling protocol packets to store loading information. This would require no extra signalling, and would return limited knowledge on VPC loadings near and far to the source node for that VCC. The load information for the end VPC can be included in the returning 'connect' message, each intermediate node and the source node then receiving that information. This would give a mixture of near VPC and far VPC information to nodes. A further degree of local information is gained by updating the loading at a node for each outgoing VPC from that node. Finally further information for loading of near VPCs is obtained by storing the loading on VPCs that are saturated, including this information with the returning 'release' packet which indicates that the connection attempt has been blocked at that node.

For evaluating the efficacy of these various techniques for obtaining network loading information via reduced or no signalling, a noticeable difference in blocking probability was required between alternate routing using no loading information (AR), and alternate routing using full network information (AAMH). For the previous four and seven node networks, this difference was found to be no greater than a 1% higher blocking probability for AR. A larger topology network was therefore constructed, which is a twenty-eight node simplified version of the US Internet backbone. This returned a difference of about 2% blocking probability as is also shown in the first graph. The following results were obtained under symmetrical network loading conditions.

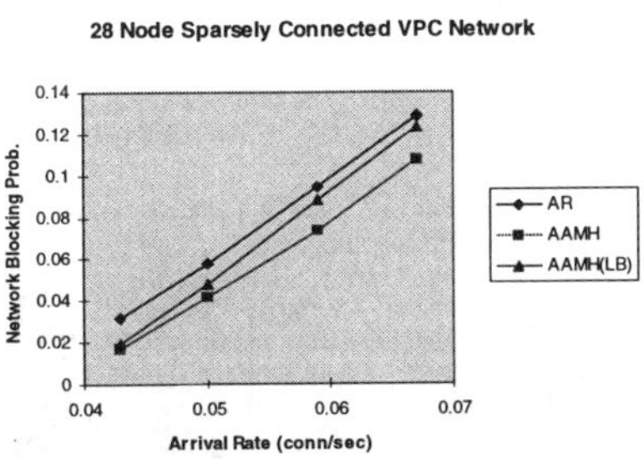

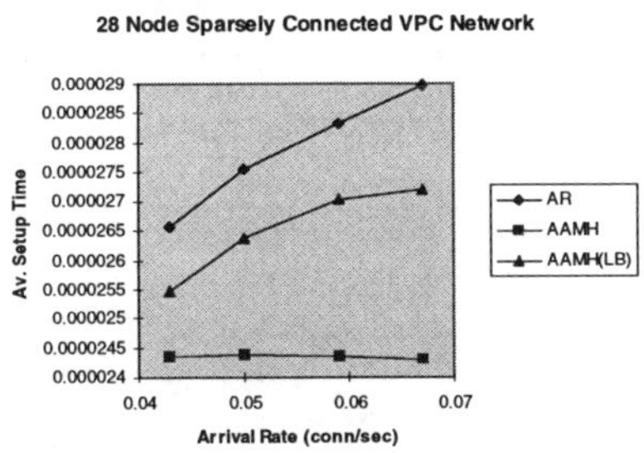

The effect of having load bands and therefore increased granularity in the loading information, is shown by the middle line of this graph. This indicates that there is a negligible difference at arrival rates causing low blocking probability, but as the loading rates increase so the network blocking probability for AAMH when using load bands increases to be almost that of AR. However the second graph, which plots average set-up time against the arrival rate, indicates that the benefits of using loading information continue as the set-up time is still noticeably improved on alternate routing using no loading information (AR).

These results were obtained using full network-wide flooding of VPC load band changes. The following include results for various implementation schemes at an arrival rate of 0.05 connections per second. The first chart gives the network blocking probability returned by AR first and then AAMH using full flooding to obtain the network loading information. It indicates that there is a 1% blocking probability difference between the two. The network topology has a diameter of 7 hops, and the lack of available load information from most of the network when using 2 hop limited flooding is shown by the higher blocking probability. What is of most interest however, is the fact that the 'connect' and 'release' packet loading information (indicated by the Route Acknowledgements and Blocking, or RA+B) is almost equal to the performance obtained by full flooding of the loading information at this arrival rate. The average set-up times are again indicative that AR returns a higher average set-up time than AAMH which uses any method for obtaining some network loading information.

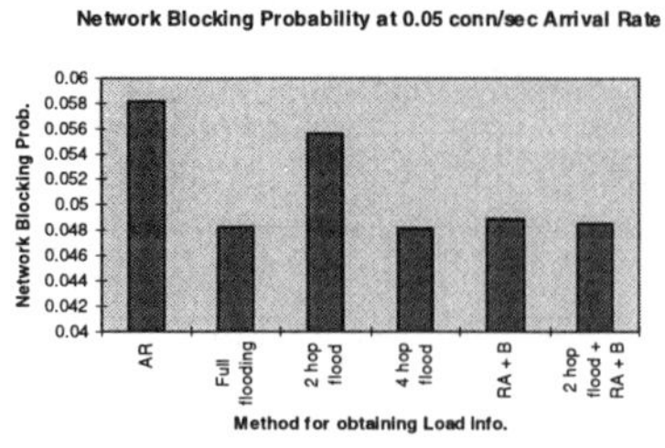

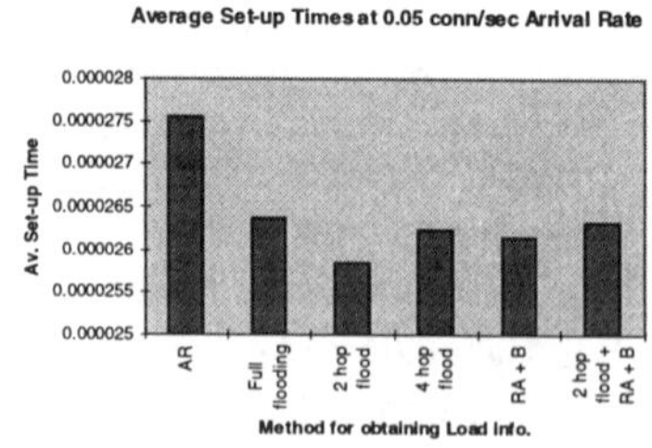

The results for non-symmetrical network loading are shown below and indicate that AR out-performs AAMH when using load bands. This was due to AR exhaustively attempting all possible routes, whilst AAMH(LB) would attempt two; first a minimum hop route, then an alternate route. Using load bands means that a number of routes might be possible, all with the same indicated spare capacity, but the one continuously attempted might actually be saturated. To avoid this, a modification to AAMH might be to attempt all routes which are equal in greatest free capacity. Finally, the RA+B results are shown with the FF results, indicating that again very similar performance is achieved under non-symmetrical network loading.

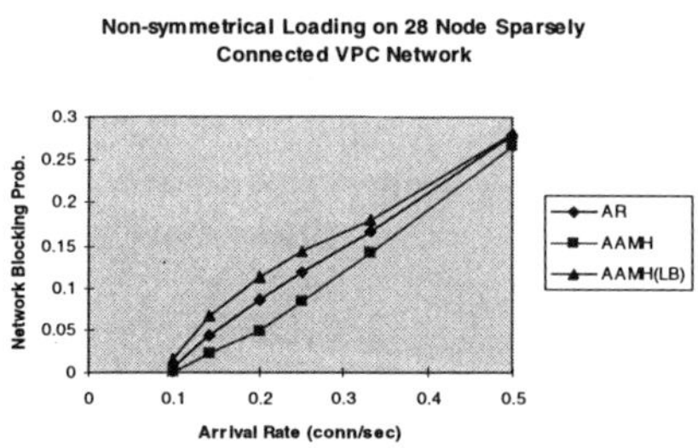

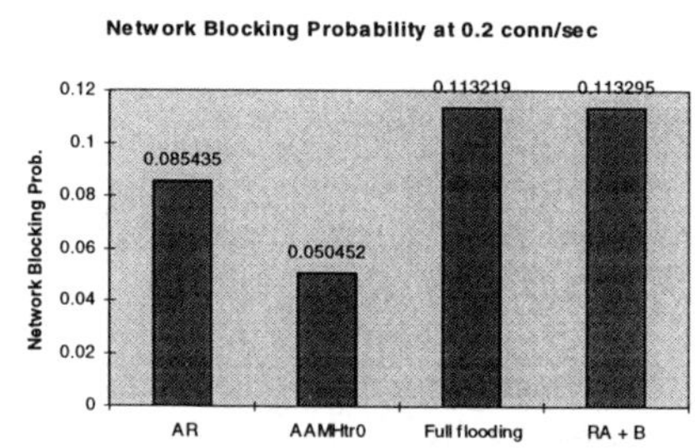

References

[1] Guerin R., Ahmadi H., Naghshineh M., "Equivalent capacity and its application to bandwidth allocation in high-speed networks", *IEEE Journal on Selected Areas in Communications,* 1991, Vol. 9, pp. 968-981.

[2] Chugo A., Iida I., "Dynamic path assignment for broadband networks based on neural computation", *IEICE Transactions on Communications,* 1992, Vol. E75-B, pp. 634-641.

[3] Gupta S., Ross K.W., El Zarki M., "On Routing in ATM Networks", *Routing in Communications Networks,* Prentice-Hall, 1995, pp. 49-72.

[4] Jordan T.P., Morse M.J., "Routing in Realistic ATM Networks", *Proceedings of IEE Thirteenth UK Teletraffic Symposium,* 1996.

[5] Aranzulla P., Mellor J., Mars P., "Dynamic Routing in ATM Networks", *Third Communication Networks Symposium,* 1996, pp. 159-162.

[6] Aranzulla P., Mellor J., "Comparing Two Routing Algorithms Requiring Reduced Signalling when applied to ATM Networks", *Proceedings of the Fourteenth UK Teletraffic Symposium,* 1997, pp. 6/1-6/7.

[7] *ATM User-Network Interface Specification, Version 3.0,* ATM Forum, 1993.

[8] Mitra D., Gibbens R.J., Huang B.D., "Analysis and optimal design of aggregate-least-busy-alternative routing on symmetric loss networks with trunk reservation", *Proceedings of the 13th International Teletraffic Congress,* June 1991.

Core and ATM Networks
D.W. Faulkner and A.L. Harmer (Eds.)
IOS Press 1997 © AKM

PMD Measurements Below 0.05 ps by the Fixed Analyser Technique

F. Ravet, Y. Choquet, B. Heens, X. Wang, L. Meuleman, M. Blondel
Service d'Electromagnétisme et de Télécommunications - MULTITEL Telecom
Faculté Polytechnique de Mons, Boulevard Dolez, 31, B-7000 Mons, Belgium

Abstract. We developed a set-up based on the fixed analyser technique using coupled 1.31 and 1.55 µm LED's, an optical spectrum analyser and an automated datas processing. So we measured quickly PMD values lower than 0.05 ps. We successfully characterised fibre lengths from 0.5 to 25 km. Five standard single-mode fibres and one dispersion shifted fibre have been measured. The PMD measurement uncertainty is also examined.

1. Introduction

Since the beginning of the eighties Polarisation Mode Dispersion (PMD) appears as an important limiting factor in optical fibre systems. It affects the bit-rate of digital links and contributes to distortions in analogue communications. In order to limit the impact of this effect, fibre and cable manufacturers have developed new products with very small PMD, typically lower than 0.1 ps over 1 km. Mainly four characterisation techniques have been discussed in the standardisation committees[1,2] and studied within the context of international round-robins[3,4]. Good agreements have been obtained between the different methods. These last years, important efforts have been made to improve the PMD measurement techniques in order to be able to characterise PMD values[5,6] smaller than 0.1 ps. This is important for cable manufacturers because they should guarantee PMD values as low as possible and PMD could be considered as a quality control parameter. These are the reasons why works carried out in such a field should interest the standardisation bodies as well as the optical fibre industries.

In this work we use an alternative way[7] with regard to small PMD characterisation techniques discussed in references 5 and 6. Our method is based on well-known wavelength scanning techniques[1,2,3]. The basic idea lies in the observation that the wavelength bandwidth of the source is the main limitation in low PMD measurements[2]. This operation has to be carefully managed because broadening the wavelength window increases the measurement duration. We also studied the uncertainty of the PMD measurement results we got. We used a experimentator oriented relation that we derived from a recently established formula[8].

2. Experimental set-up

When a short polarised pulse is launched in a classical telecommunication fibre, it undergoes spreading due to the combined effects of local birefringences and polarisation mode coupling. The shorter the pulse is, the smaller the measurable PMD is. A very short pulse corresponds to a very low coherence source, i.e. a broadband source in the spectral domain. For instance, when measurements are done with a source scanning from 1470 to 1570 nm, the lower PMD value are about 0.16 ps. Increasing the wavelength window up to 550 nm reduces this limit below 0.03 ps. Our idea is to couple two LED's respectively centred at 1.31 and 1.55 µm into the Fibre Under Test (FUT).

The measurement is carried out in two steps. In the first step, the polarised light is launched into a polariser and then into the FUT. The output power is then detected by means of a wavelength scanning device. In the second step, the launching operation is identical but an analyser is added between the scanning device and the FUT output in order to measure the output power. The spectral power density ratio, $R(\lambda)$, is then calculated. Then we derive then the PMD by computing the Fourier Transform (FT) of $R(\lambda)$. So we get the autocorrelation function, $E(t)$, of the detector output photocurrent intensity. The measured PMD value is then given by the standard deviation of the $E(t)$[1,2,3]. The measured PMD, $\Delta\tau_{mes}$, is then the standard deviation of the $E(t)$. $E(t)$ can be directly measured by an interferometer[1,2,3] analysing the FUT output intensity in the time domain.

In reference 8, it has been theoretically demonstrated that the measurement uncertainty is reduced when the wavelength window width is increased. The authors derived the following relation

$$-\frac{0.9}{\sqrt{\overline{\Delta\tau}\Delta\omega}} \le \Delta\tau_{mes} - \overline{\Delta\tau} \le \frac{0.9}{\sqrt{\overline{\Delta\tau}\Delta\omega}} \qquad (1)$$

where $\overline{\Delta\tau}$ is the theoretical mean PMD delay and $\Delta\omega$ is the spectral window width. Relation (1) gives the PMD range centred on the theoretical mean PMD delay where it is probable to find $\Delta\tau_{mes}$. This important relation is very useful when we have informations on the theoretical PMD mean value. We propose to derive an experimentator-oriented relation from this expression. We express $\overline{\Delta\tau}$ as a function of $\Delta\tau_{mes}$. So relation (1) becomes

$$\frac{0.2}{\Delta\omega}\left(\sqrt{1+5\Delta\tau_{mes}\Delta\omega}-1\right)^2 \le \overline{\Delta\tau} \le \frac{0.2}{\Delta\omega}\left(\sqrt{1+5\Delta\tau_{mes}\Delta\omega}+1\right)^2 \qquad (2)$$

In order to validate this relation, let us consider the following example. When $\Delta\omega=518\ ps^{-1}$ (it means a scanning from 1150 to 1700 nm), $\Delta\tau_{mes}=1\ ps$ has an uncertainty of about 4 %. The same uncertainty[8] is found when using relation (2) for $\overline{\Delta\tau}=1\ ps$.

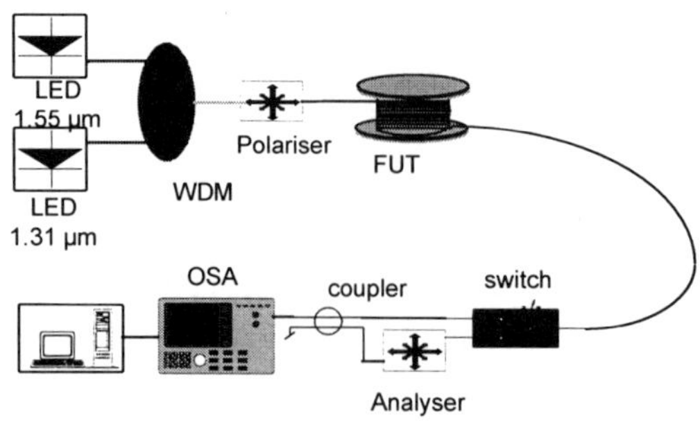

FIGURE 1. Set-up used in the experiments.

The experimental set-up we implemented is shown in figure 1. The broadband source is constituted of two coupled LED's. One is centred in the 1.31 µm window while the other one is centred in the 1.55 µm. So we have an useful spectrum from 1150 nm to 1700 nm, permitting the measurement of PMD below 0.05 ps. This spectrum is coupled into a broadband polariser and then into the FUT by means of a WDM coupler. The transmitted spectrum analysis stage is constituted of an optical switch, a polariser, a coupler and an OSA. All these components have broadband characteristics. Data processing, Fourier analysis and PMD computation are done by a home-made software running on a classical computer (PC 386, 33 MHz, RAM 4 MB). It integrates the last standardised recommendations and systematical errors analysis[2,9]. The switch reduces the handling duration between the two spectrum acquisition. As the OSA is a spectrum fast detection equipment, it contributes to the measurement speed improvement.

3. Results and comments

We characterised five Standard Single-mode Fibres (SSF) and one Dispersion Shifted Fibre (DSF). We carried out at least 10 measurements on each fibre. The table joined here presents the measured PMD mean values, the PMD RMS and the uncertainties. The measured $R(\lambda)$ of different fibres are shown in figure 2. The corresponding autocorrelation functions can be examined in figure 3. In the case of the SSF, the measured PMD values vary from 0.024 ps to 0.131 ps, respectively for fibres lengths of 500 m and 25 km. 0.024 ps is the lowest value we are able to characterise due to the finite spectrum of our source. As appearing in the table, our measurements have a good reproducibility. The uncertainty varies from 11 % to 24 % and thus seems high. Meanwhile our set-up significantly improves the measurement technique. When the source has a narrower spectrum, for example $\Delta\omega=100\ ps^{-1}$ (it means a scanning from 1470 to 1600 nm), $\Delta\tau_{mes}=0.04\ ps$ has an uncertainty of about 42 % . In our broadband case we get an uncertainty of about 19 %. No trials have been performed on fibres longer than 25 km. The absorption peak in the 1.4 μm window should constitute a limitation for long haul links.

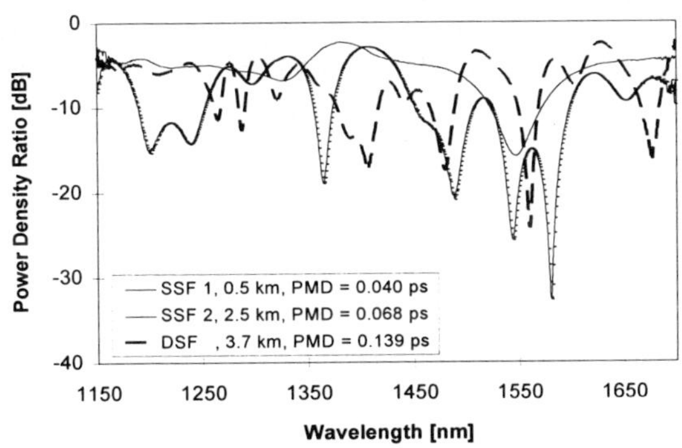

FIGURE 2. Power density ratio of two SSF's and one DSF.

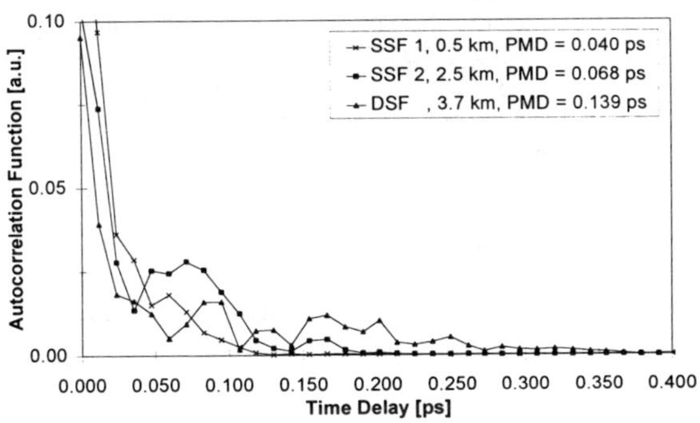

FIGURE 3. Autocorrelation function of two SSF's and the DSF derived from the curves presented in figure 2.

Our set-up can be considered as a fast technique when taking into account the measurement spectral range needed for so small PMD. Both the OSA scanning and switching operation never exceeded 10 seconds in all the tests we performed. The only time consuming factors are the computer data processing steps. The total duration of the operation is less than two and half minutes involving all the steps and the computations.

TABLE. Measurements results

	SSF 1 500 m	SSF 2 500 m	SSF 3 2.5 km	SSF 4 25 km	SSF 5 2.5 km	DSF 3.7 km
Mean PMD [ps]	0.040	0.027	0.068	0.124	0.080	0.129
RMS PMD [ps]	0.004	0.003	0.004	0.006	0.004	0.012
Uncertainty	19 %	24 %	15 %	11 %	14 %	11 %

4. Conclusions

The set-up has been successfully tested. SSF's as well as DSF, from 500 m to 25 km, have been characterised. The smallest PMD value we measured is lower than 0.030 ps and can be considered as the inner set-up limit. The broadband source we use improves significantly the PMD measurement uncertainty. If we take the very large spectrum into account, the full measurement operation can be considered as fast. By optimising the software it is possible to increase the computer operation speed. The dynamic range does not allow measurement over 25 km. These last comments as well as the comparison with other techniques are the possible developments of this work.

Acknowledgements

F. Ravet, Y. Choquet, B. Heens, X. Wang, L. Meuleman are working in the frame of an R&D contract commonly supported by *Ministère de la Région Wallonne* and the *European Commission*. F. Ravet would like to warmly thank Prof. Nicolas Gisin from University of Geneva for the helpful discussions he had with him and Mr Laurent Hubaut from MUTITEL TCTS for his technical support.

References

1. ITU-T SG15-WP4, Draft Recommendations, November 1995.
2. IEC SC 86A/WG1, Draft Recommendations, June 1996.
3. N. Gisin et al., "Definition of Polarization Mode Dispersion and First Results of the COST 241 Round-robbin Measurements", *Pure Appl. Opt.*, N°4, 1995, pp. 511-522.
4. B. Perny et al., "Polarisation Mode Dispersion: Large Scale Comparison of Jones Matrix Eigenanalysis Versus Interferometric Measurement Techniques", to be published in *Electronic Letters*; 1996.
5. A. J. Barlow et al., "Measurements of Ultra-low Polarisation Mode Dispersion in Single Mode Optical Fibers by a Technique of PMD Biaising", *Proceedings OFMC'95*, Liège, September 1995, vol. 12, pp. II.4.
6. Ph. Oberson et al., "Interferometric Polarization Mode Dispersion Measurements with Femtoseconds Sensitivity", *Proceedings SOFM'96*, Boulder, October 1996, pp. 143-146.
7. F. Ravet et al., "Low PMD Measurements on Middle Distance Optical Fibres: a Fast Wavelength Scanning Technique", paper presented at *Annual Symposium of IEEE/LEOS*, Enschede, 28 November 1996.
8. N. Gisin et al., "How Accurately Can one Measure a Statistical Quantity Like Polarization Mode Dispersion", *Proceedings SOFM'96*, Boulder, October 1996, pp.131-134.
9. P.A. Williams, "Accuracy Issues in Comparisons of Time-and Frequency Domain Polarization Mode Dispersion Measurements", *Proceedings SOFM'96*, Boulder, October 1996, pp.125-129.

Core and ATM Networks
D.W. Faulkner and A.L. Harmer (Eds.)
IOS Press 1997 © AKM

MO-16, a Modular Optical 16-Channel Transmission System - a First Step towards a Photonic Network

Authors: U. Kohn, E. Kremers, G. Müller, E. - U. Scheuing, F. - C. Tischer
Bosch Telecom GmbH, D-71522 Backnang/Germany

1. Summary

Wavelength Division Multiplexing (WDM) is a very potential means to enlarge the capacity of existing fibre plants and to provide greater flexibility in network planning. With the modular approach of the MO-16 equipment a long distance carrier is able to migrate from a low capacity point-to-point link towards an optical network including add-/drop functionality of optical carriers. The network is expandable just when more capacity or a higher degree of flexibility is required.

2. General Features

The equipment is a fully engineered optical WDM transmission system for the 1550 nm band which is able to transmit up to 16 channels with a channel spacing of 200 GHz, in conformance with recent ITU proposals [1]. Transmission is, dependent on the laser sources used, possible on links of up to 600 km total length with repeater (optical amplifier) span "target distances" of either 80 or 120 km, using optical amplifiers (OFA) whenever necessary. It is primarily meant for standard G.652 singlemode fibre applications but may, with appropiate channel spacing, also be used for other types of fibres. All channels are bitrate transparent for digital data streams up to 2.5 Gbit/s. Different versions of equipment allow adaptation to individual user needs.

The equipment consists of Line Terminal equipment and of Optical Inline Amplifiers, the latter either without or with add-drop facilities. In line terminals, up to 16 channels are multiplexed and arranged in groups of four contiguous channels, with multiplexing resp. demultiplexing done in two steps to achieve greater flexibility, modularity and simplified mechanical design. Transparent transponders usable up to 2.5 Gbit/s provide stable transmitter sources with distinct wavelengths.

Table 1

Output power per channel (8-channel-system)	6 dBm
Output power per channel (16-channel-system)	3 dBm
Channel spacing	200 GHz
Target distance per repeater section	80km; 120 km
Total link length (direct intensity modulation)	150 km
Total link length (electro-absorption modulator)	600 km
Bitrate	100 Mbit/s ... 2.5 Gbit/s

Operation and maintenance is carried out according to the standards set by SDH equipment, i.e with local and remote access, and connection to a Telecommunication Management

Network (TMN) via a Q-type interface. Supervision and management facilities are an integral part of the equipment. An Optical Supervisory Channel (OSC) located outside the OFA band cares for transportation of supervision and management signals within a link and besides provides an engineering order wire for servicing and repairing purposes. Basic transmission parameters are contained in **Table 1**.

3. Line Terminal Units

Terminal equipment cares for optical multiplexing resp. demultiplexing of the 16 channels, alone or in conjunction with transmitting and receiving optical amplifiers, dependent on the actual repeater spans. The principal circuit diagram is shown in **Fig.1**.

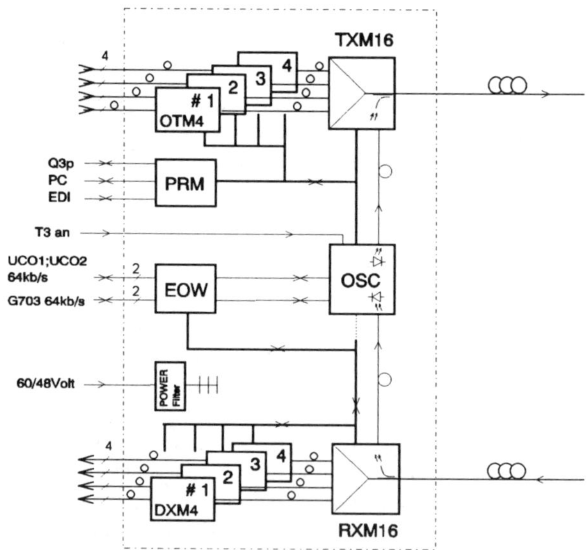

OTM4 : Optical Transponder-Module for 4 Channels
TXM16 : Optical Transmit - Module for 16 Channels
RXM16 : Optical Receive - Module for 16 Channels
DXM4 : Optical Demultiplexer - Module 1/4
OSC : Optical Service - Channel
PRM : Processor - Module
Q3p : Management - Interface

Figure 1: FlexPlex MO-16/T Optical Terminal Multiplexer

Supervision and management information of all relevant parameters is handled by a processor module and accessible via a local maintenance PC interface as well as via a Q-type TMN interface. All remote line alarms are broadcast along the line by means of the OSC, in order to make all information is accessible at every piece of equipment.

4. Transponders

Transponders of the O/E/O type (i.e. optoelectrical and electrooptical conversion with intermediate electrical processing) provide frequency adaptation for incoming signals with a random frequency either in the 1300 or the 1500 nm window, transferring the signals to optical carriers with a dedicated frequency within the 1550 nm OFA window grid. Different versions using either directly-modulated lasers or lasers with integral EA modulators may be chosen according to the total link length. The transponders are suitable for a wide range of input power levels and provide amplitude stabilization by means of the amplitude-limiting transfer characteristic of the electrical processing part. The circuitry consists of a custom-designed ASIC pair which is implemented on a ceramic substrate in order to achieve a good high-frequency response. Several versions of transponders are available in groups of up to four on one plug-in module. The physical layout of one transponder substrate is shown in **Fig. 2**.

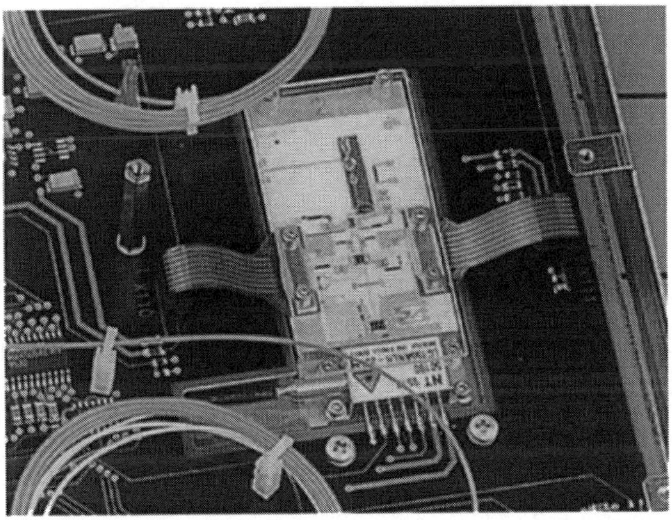

Figure 2: Optical Transponder Substrate

5. Inline Amplifiers

Optical amplifier equipment cares for common optical amplification of the 16-channel aggregate signal, with different types of OFAs dependent on the actual repeater spans. The principal circuit diagram is shown in **Fig. 3**. OFAs are of the Erbium-doped (EDFA) type, with 980 nm pumping sources [2].

The supervision and management scheme of the line amplifiers corresponds with that of the terminal units, i.e., all information is accessible along the link as well as via a local maintenance PC or a Q-type TMN interface.

Adding and dropping of one or more optical channels may be accomplished by using inline amplifiers with additional add-drop-functions.

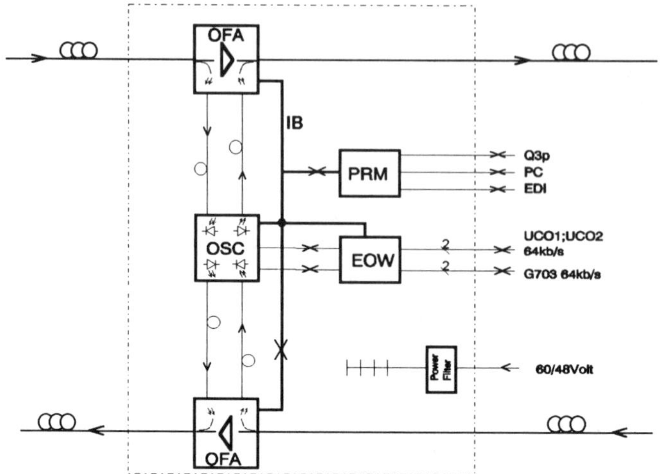

OFA	: Optical Fiber Amplifier
OSC	: Optical Service - Channel
PRM	: Processor - Module
Q3p	: Management - Interface

Figure 3: FlexPlex MO-16/L Optical Line Amplifier

6. Mechanical Design

The equipment is designed to be as modular as feasible, in order to allow maximal flexibility for the supplier and for the user. A single type of shelf of the dimensions 600 mm * 450 mm * 260 mm accommodates either one terminal for up to 16 channels, one add-drop amplifier, or up to two inline amplifiers without add-drop-facilities. All plug-in modules are of the dimensions 370 mm * 230 mm.

References

[1] ITU Draft Recommendation G.mcs

[2] Edited by Michael J.F.Digonnet, *Rare Earth Doped Fiber Lasers and Amplifiers*, Stanford University (1993)

Core and ATM Networks
D.W. Faulkner and A.L. Harmer (Eds.)
IOS Press 1997 © AKM

Simulation and Modeling of an Interconnected Network using Time and Wavelength Division Multiplexing Scheme

Bénédicte BOULET[•], François TILLEROT[*], Jacques SALAÜN[•] and Marc LE LIGNÉ[*]
[•]*Pouyet Réseaux Publics et Nationaux*
Le sémaphore - 6/8 rue du vieux chemin - 94207 Ivry sur Seine Cedex, France
[*]*France Télécom CNET, LAB/RIO/ARO*
Technopole Anticipa - 2 avenue Pierre Marzin - 22307 LANNION Cedex, France

Abstract: We consider a two-level lightwave network using Time and Wavelength Division Multiplexing (T/WDM). Stations are connected together by star couplers into clusters according to geographical considerations and clusters are interconnected to a star coupler through nodes. Stations and nodes are equipped with fixed transmitters and tunable filters. Simulations are carried out on 130 stations and 10 clusters. Various curves are drawn for different parameters variation. We can show that the network has better performance when optimal values of the dimensioning features are chosen. A model of a star topology is also proposed, which gives results close to the simulation.

1. Introduction

The optical bandwidth of a fiber is very large (approximately 25 to 30 THz) and not very efficiently used. To reduce the electronic bottleneck, several experiments on WDM (Wavelength Division Multiplexing) [1-4] networks were carried out. Single-hop networks (no intermediate station) [2] need one tunable or N fixed components (transmitter or receiver) while multihop networks (one or more wavelengths are assigned to a node) [2, 3] can be fitted with fixed transmitters and receivers.

WDM networks are limited by the WDM component technology and particularly by the number of wavelengths and the wavelength spacing. A lot of hybrid access schemes like T/WDM have been studied by a number of authors [4 - 6]. Time multiplexing may be carried out by the Dynamic Allocation Scheme (DAS). In this case, the slots in a data frame are dynamically assigned to the transmitters.

In this paper, we consider T/WDM with a dynamic allocation scheme. In the following part, we describe the T/WDM scheme from Kovacevic *et al.* [5], the network architecture considered to interconnect star networks (clusters). Then, we evaluate the network performance when clusters are linked by a star coupler. The third part discusses the simulation results and the fourth part deals with the simulation result modeling.

2. Principle

The objective is to interconnect several star networks where each uses a T/WDM scheme that allows a time and frequential transmission media division. Various configurations may be offered by the near-future optical component technology. According to the recent advances in tunable filters performance, the first level of flexibility could be obtained with tunable receivers switching on the reception wavelength and fixed sources.

2.1. The T/WDMA scheme

This section refers to previous studies [5], particularly in the network topology, the time frame structure, the slot allocation protocol and in the transmission and reception components. The following notations will be used : N_S the number of stations, N_λ the number of wavelengths, $N_{S/F}$ the number of subframes per frame and $N_{S/S}$ the number of slots per subframe.

2.1.1. Slot allocation protocol

The time frame is subdivided into subframes and each subframe into slots. Tuning occurs between subframes and not between slots as in [5]. The great advantage of this subframe subdivision is that a station receives from all stations transmitting on the same wavelength within a subframe. The minimal subframe size depends on the receiver tuning (i.e. no free slot is available for this wavelength in the frame or the destination station already receives on another wavelength), the call is blocked and rejected.

2.1.2. Simulation results on star topology [5]

In this study, the frame size is set to 125 μs. The simulation results show a minimal blocking probability for an optimal number of wavelenghts when the number of slots per subframe is greater than 1. From [5], we can point out three kinds of blocking. The transmission blocking occurs when the slot allocation matrix is full, the wavelength conflict appears when the receiver is already tuned on another wavelength than that of the call request. The third blocking is an intermediate one generated by a combination of both effects. The network performance increases with the number of slots per subframe and with the number of subframes per frame. Thus, there is a trade-off between the number of wavelengths, the number of slots per subframe and the number of subframes per frame.

2.2. Network architecture

The single star topology considered above is not cost effective when many stations are located in a same geographical zone, which are all distant from the central star coupler. To interconnect all stations, it is better to group the nearer stations into clusters and to link all clusters together through a star topology as described in figure 1.

We propose a distributed control system for management. Each station (or node) knows the slot assignment in the current frame, computes the slot that should be taken up at the next frame and sends a frame update. In the physical layer, control information is conveyed through a time multiplexed dedicated channel. The control channel is also divided into subframes and slots. A guard time between information in the control channel and data transmission in the data communication channel is added. Each station is provided with a dedicated pair of transmitters and receivers at a specific wavelength apart from the wavelengths used for data transmission.

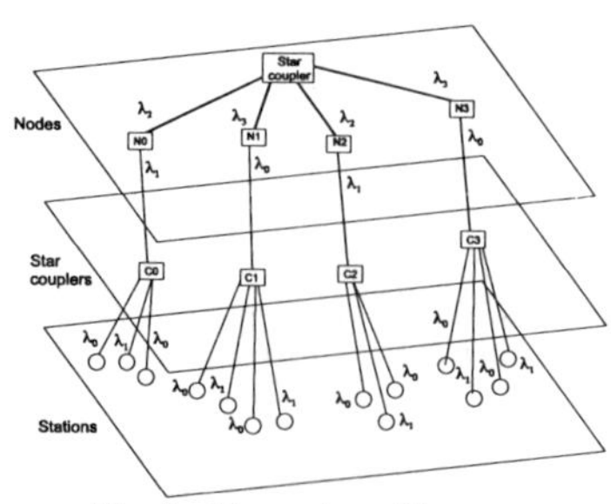

Fig. 1 : Network architecture.

2.3. Slot allocation protocol

In an interconnected network, two frames are used : the node frame flows from a node to another node and the cluster frame flows between the stations (with one cluster frame per cluster). With the network topology described above, local and distant communications may be transmitted. Local communications do not pass through a node (the call will only pass through the cluster star coupler) and generate one single slot allocation in the cluster frame with the T/WDM slot allocation algorithm [5]. When a station communicates with a station located in another cluster (distant communication), the call is conveyed in several hops. In this scheme, one slot is allocated in the source station cluster frame, one in the node frame and one in the destination station cluster frame. Each of these three allocation schemes uses the same slot allocation algorithm as for local communication.

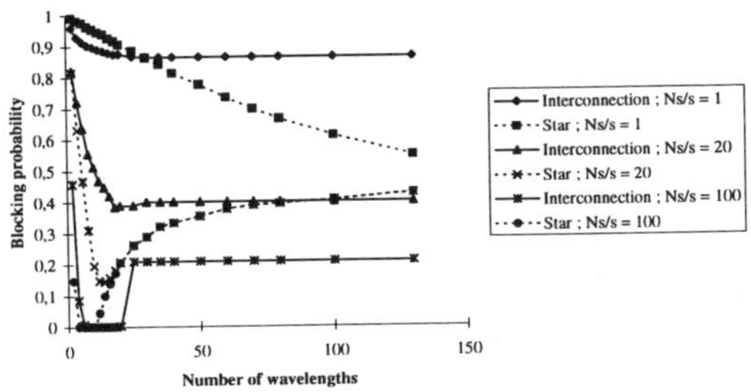

Fig. 2 : Blocking probability versus N_λ for $N_{S/F} = 10$.

2.4. Evaluation of network performance

In the performance study of an interconnected network, we use the following assumptions. We consider circuit switching communication. Each station generates a call request according to a Poisson process [7] with an exponentially distributed connection duration. We do not use any queuing system (if a station does not find any slot for its request, the call is blocked and discarded). The traffic is assumed to be uniform and the data rate is the same for all stations (one slot per frame per request). Since we choose a dedicated channel for the control traffic, we focus on data traffic only. The number of available wavelengths is equally split between stations and nodes (Fig. 1). In the node layer, the wavelength comb differs from that of the station layer with optical/electrical/optical conversion in the nodes. The allocation proceeds independently for all clusters. A same wavelength allocation scheme may be used for several clusters and the scalability issue as regards clusters (i.e. adding a new sub-network) is limited to the upper layer. In each cluster, the assignment of transmitters wavelengths is determined in a balanced way, so that each wavelength is used by a same number of stations in a cluster.

3. Simulation results

In our simulation, we consider 10 nodes and 130 stations with a fixed transmitter and two tunable filters. The 130 stations generate a traffic of 2500 calls/s, with an average duration of one second. We assume a frame duration of 125 μs (100 μs for the information subframes and a 25 μs guard time). We compare the simulation results for a single star and for interconnected stars. In the interconnected network, distant calls can be blocked either in the transmission cluster (receiving blocking for the node access), in the nodes layer or in the destination cluster (transmission blocking from the node).

Wavelengths are equally split between clusters and nodes ($N_\lambda/2$) up to $N_\lambda/2 = 10$. Then, N_λ in nodes is kept equal to 10. The existence of an optimum value is also pointed out for interconnected networks.

• When $N_{S/S} = 100$ and if $N_\lambda > 5$, the interconnected network performance is better than for a single star network. For an interconnected network, the blocking occurs mostly :

- at the nodes level (transmission blocking) when $N_\lambda < 20$. The nodes frame has not enough available slots to flow all the traffic load.
- in the transmission cluster (receiving blocking) when $N_\lambda > 20$. The transmission cluster cannot flow the traffic towards the cluster related to the destination node.

The main blocking for single star network is a transmission and mixed blocking combination when $N_\lambda < 3$ and a receiving blocking when $N_\lambda > 10$.

• If $N_{S/S} < 20$, the single star network is more performant. In this case, we get a mixed blocking for both networks which combine to other blockings in the transmission cluster level and in the node layer.

Figure 3 shows the results for an irregular wavelength distribution (up to 10 wavelengths per node). This figure shows that up to 50% of node wavelengths, increasing the number of wavelengths for nodes improves the performance for $N_\lambda < 20$. In this case, the main blocking is the node blocking (transmission and mixed blocking). Most calls are distant communications. No geographical considerations were taken into account in traffic

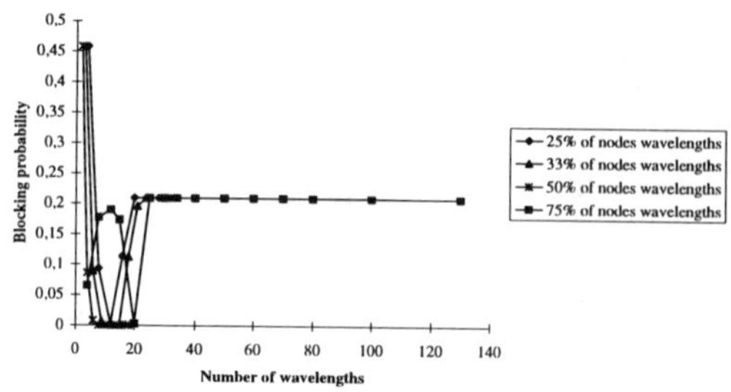

Fig. 3 : Blocking probability versus N_λ for $N_{S/F}$ = 10 and $N_{S/S}$ = 100.

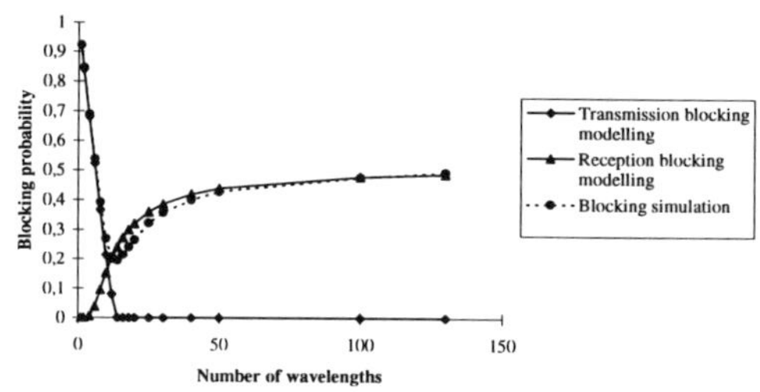

Fig. 4 : Blocking probability versus the number of wavelengths for $N_{S/F}$ = 10 and $N_{S/S}$ = 20.

assumptions. When the number of wavelengths used for nodes increases, the number of available slots in the node frame increases and the transmission blocking decreases. On the other hand, assigning 75% of wavelengths for nodes does not improve the network performance. Some blockings which were negligible previously (local communication, transmission cluster and receiving cluster blocking) become important because the number of available slots at the cluster level is too low for the number of requests.

4. Modeling

4.1. Transmission blocking

We consider calls arrival as a Poisson process [7, 8] in a server with an arriving call rate of λ_T, and with a service duration μ exponentially distributed. For a number of server of N_{Serv}, the blocking probability is given by the B Erlang formula with losses as follows :

$$P_{Trans} = \frac{\dfrac{(\rho_T)^{N_{serv}}}{N_{serv}!}}{\displaystyle\sum_{i=0}^{N_{serv}} \frac{(\rho_T)^i}{i!}} \qquad (1)$$

Where $\rho_T = \frac{\lambda_T}{\mu}$. The arriving call rate is $\lambda_T = \frac{T}{N_\lambda} * \frac{1}{\mu}$ with T the traffic for all stations in calls/s. Since we have to find a slot in a total number of slots = $N_{S/F} * N_{S/S}$, we can consider that the number of server is : $N_{serv} = N_{S/F} * N_{S/S}$.

4.2. Receiving blocking

We also consider the B Erlang formula with losses described in (1). Now, the aim is to find a subframe where the destination station does not receive on another wavelength. Thus, the number of server N_{serv} is equal to the number of subframe per frame $N_{S/F}$ and the load

ρ_R is equal to the average number of the used wavelengths : $\rho_R = \sum_{i=0}^{N_\lambda} i * P(i)$

where *P(i)* is the probability to use *i* wavelengths of the N_λ. Thus, $\rho_R = \sum_{i=0}^{N_\lambda}\left[i*\left(\frac{\mathcal{N}(N_\lambda, N_\lambda - i, N_{C/S})}{N_\lambda^{N_{C/S}}}\right)\right]$

where - $N_{C/S}$ is the number of calls per station given by : $N_{C/S} = \frac{T}{N_S} * \frac{1}{\mu}$.

- $\mathcal{N}(N_\lambda, N_\lambda - i, N_{C/S})$ is the total number of possibility to assign i wavelengths on the N_λ for $N_{C/S}$ calls. To calculate this number of wavelengths, we use the recurrent formula given by :

$$\mathcal{N}(N_\lambda, N_\lambda - i, N_{C/S}) = (i) * \mathcal{N}(N_\lambda, N_\lambda - i, N_{C/S} - 1) + (N_\lambda - i + 1) * \mathcal{N}(N_\lambda, N_\lambda - i + 1, N_{C/S} - 1)$$

In the end, the receiving blocking probability is :

$$P_{Rec} = \frac{\frac{\rho_R^{N_{S/F}}}{N_{S/F}!}}{\sum_{i=0}^{N_{S/F}}\left(\frac{\rho_R^i}{i!}\right)}$$

4.3. Modeling results

Both blocking modeling are compared to simulation results. Figure 4 represents the blocking probability versus the number of wavelengths for $N_{S/F} = 10$ and $N_{S/S} = 20$. On this figure, we can notice that the model results are close to the simulation curve.

5. Conclusion

In this paper, we describe an interconnected stars network using a T/WDM scheme. T/WDM is introduced at cluster and node levels. Simulations show that an interconnected network is more performant than a single star network when an optimal trade-off between N_λ, N_λ in nodes, $N_{S/S}$ and $N_{S/F}$ is chosen.

References

[1] C. A. Brackett, "Dense Wavelength Division Multiplexing Networks : Principles and Applications", IEEE J. Select. Areas Commun., Vol. 8, No. 6, pp. 948-964, August 1990.

[2] B. Mukherjee, "WDM-Based Local Lightwave Networks. Part I : Single-Hop Systems", IEEE Network, pp. 12-27, May 1992.

[3] B. Mukherjee, "WDM-Based Local Lightwave Networks. Part II : Multihop Systems", IEEE Network, pp. 20-32, July 1992.

[4] A. Ganz et al., "Time-Wavelength Assignment Algorithms for High Performance WDM Star Based Systems", IEEE Trans. Commun., Vol. 42, No. 2/3/4, pp. 1827-1836, February/March/April 1994.

[5] M. Kovacevic et al., "Time and Wavelength Division Multiaccess with Acoustooptic Tunable Filters", J. Fiber and Integrated Optics, Vol. 12, pp. 113-132, August 1993.

[6] M. S. Borella et al., "Efficient Scheduling of Nonuniform Packet Traffic in a WDM/TDM Local Lightwave Network with Arbitrary Transceiver Tuning Latencies", IEEE Infocom 95, pp. 129-137, Boston MA, June 1995.

[7] G. Fishman, "Principles of Discrete Event Simulation", Wiley J., New York, 1978.

[8] M. Schwartz, "Telecommunication Networks", Addisson-Wesley, 1987.

Core and ATM Networks
D.W. Faulkner and A.L. Harmer (Eds.)
IOS Press 1997 © AKM

Gaussian pulse transmission considered with higher order optical effects

Kaoru KINJYOU, Masanori HANAWA and Mikio TAKAHARA
Faculty of Engneering, Yamanashi University, Kofu 400, Japan

Abstract. We have reported the conditions that soliton-like pulses have been automatically generated in a long length fiber even if gaussian pulses have been launched into an optical fiber. Then, we have confirmed that there has been the region which long distance transmission over 3,000km has been achieved without additional systems such as dispersion compensator even if higher order optical effects have been considered.

1 Introduction

Wavelength division multiplexing (WDM) optical communication systems [1] have been expected as a candidate for next generation high capacity, long-haul optical communication systems. To increase the channel capacity of one channel is useful to increase the capacity of WDM systems, too [2]. While we have to use very narrow width optical pulses to realize very high bit-rate systems. However, when the pulse width is narrower than several pico-seconds, serious pulse distortion is caused by the higher order nonlinear optical effects or strong pulse broadening in fibers.

Many experimental and simulated results for high bit-rate and long haul transmission systems have been reported, however, almost all of them were used some compensation technologies such as dispersion compensation fibers (DCFs), optical phase conjugators (OPCs) [3] and soliton transmission. While, automatic soliton pulse generation from gaussian pulse launched has been reported [4]. However, previous reports were not shown any conditions to generate soliton pulses.

In this paper precise conditions to generate soliton-like pulses anutomatically will be shown in a long haul optical fiber system containing optical amplifiers to compensate the fiber loss and considering higher order nonlinear optical effects in fibers. Moreover, we show the region that long haul transmission could be achieved over 3,000km without any additional compensation technology except for optical amplifiers.

2 Evaluation method

Fig.1 shows the system model used in this paper. The chromatic dispersion of DSFs is exactly assumed constant along the fiber. Fiber loss is perfectly compensated the EDFAs set every 30 km, and the EDFA in front of the receiver is used as a pre-amplifier.

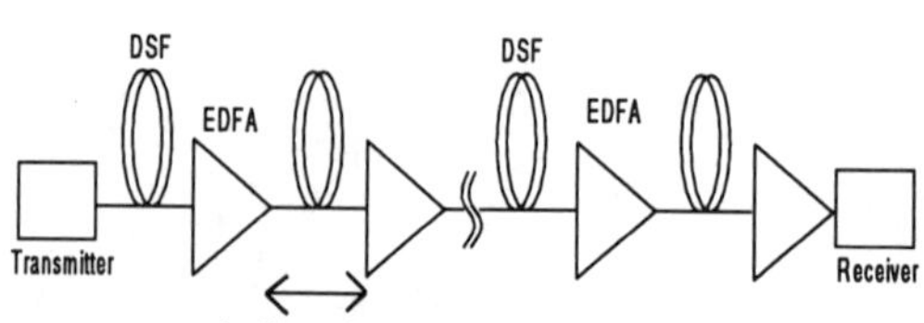

Figure 1 System model

The optical pulse propagation in the single mode optical fiber is usually expressed by the following nonlinear Schrödinger equation (1) [5].

$$\frac{\partial \phi}{\partial z} = -\frac{i}{2}\beta_2\frac{\partial^2 \phi}{\partial \tau^2} - \frac{\alpha}{2}\phi + i\gamma \mid \phi \mid^2 \phi \tag{1}$$

However, when the pulse width is less than single pico-seconds, we have to add higher order terms to Eq.(1), then the pulse propagation equation is changed into Eq.(2) [5].

$$\begin{aligned}\frac{\partial \phi}{\partial z} &= -\frac{i}{2}\beta_2\frac{\partial^2 \phi}{\partial \tau^2} - \frac{\alpha}{2}\phi + i\gamma \mid \phi \mid^2 \phi + \frac{1}{6}\beta_3\frac{\partial^3 \phi}{\partial \tau^3} \\ &- \frac{2\gamma}{\omega_0}\frac{\partial}{\partial z}(\mid \phi \mid^2 \phi) - i\gamma T_R \phi \frac{\partial \mid \phi \mid^2}{\partial z}\end{aligned} \tag{2}$$

In Eq.(2), from the fourth to sixth terms in the right hand side are additional higher order terms. They are the third order dispersion term, the time differential term for Kerr effects and the self frequency shift term, respectively. The third order dispersion term causes pulses with very narrow pulse width to give serious distortion. β_3 is the third order dispersion coefficient and is shown by the following equation.

$$\beta_3 = (\lambda dD/d\lambda + 2D)\lambda^3/(2\pi c)^2 \tag{3}$$

In this paper, transmitted pulses shapes are calculated based on Eqs.(1) and (2) using Split-Step-Fourier Transform method. The results are evaluated by the eye penalty, and the transmission distance limit is the point of the 1.0 dB eye penalty.

3 Automatic soliton-like pulse generation from gaussian pulse

The used parameters in our simulations are shown in Table 1. The noticeable point is the simulations are carried out in the anomalous chromatic dispersion region of DSFs.

Table.1 simulation condition

amplifier spacing	30 km
fiber loss	0.2 dB/km
fiber refractive index	1.45
fiber core diametter	10 μm
Kerr coeffcient	3.17×10^{-20}m^2/W
zero dispersion wavelength	1.55 μm
$dD/d\lambda$	0.07 ps/km/nm^2
T_R	5.0 fs
launching pulse sequence	...0100111...

Firstly, the generating conditions of soliton-like pulses are investigated by using Eq.(1). Fig.2 shows the characteristics of transmission distance versus chromatic dispersion in the case of pulse width 5ps, bit rate 20Gbit/s and launching peak power is 10 or 15dBms. Each curve in the figure has a peak point of the transmission distance because the larger chromatic dispersion than that of the peak point makes the transmission distance shorten due to pulse broadening and the less dispersion case makes that due to SPM effects. Therefore the peak point is the balanced situation between pulse broadening and SPM effects in the system. This fact shows that the soliton-like pulse generation can be achieved by gaussian pulse launching. As the result, very long haul transmission over 30,000km for 10dBm or about 17,000km for 15dBm, has been achived. If the larger peak power is used, the chromatic dispersion of the longest distance point (the peak point) has to becomes larger because of cancelling the pulse broadening in the system.

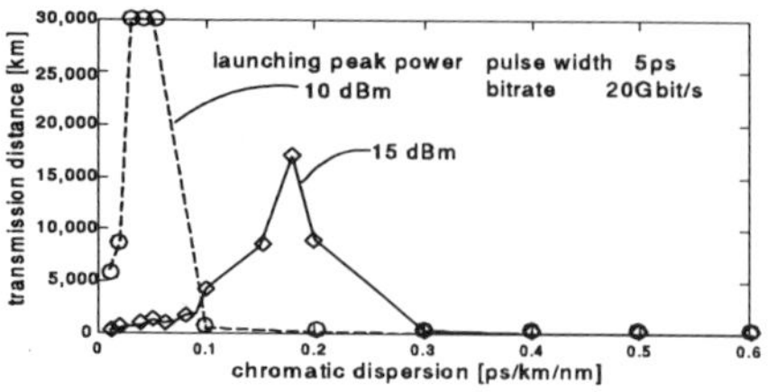

Figure 2 transmission distance for chromatic dispersion

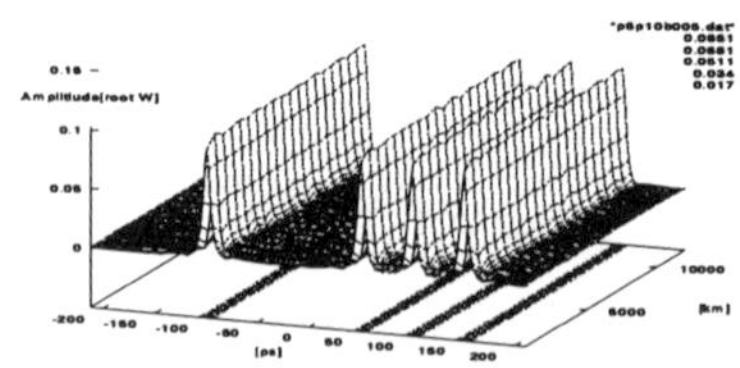

Figure 3 propagation waveform (5ps, 10dBm, 0.05 ps/km/nm)

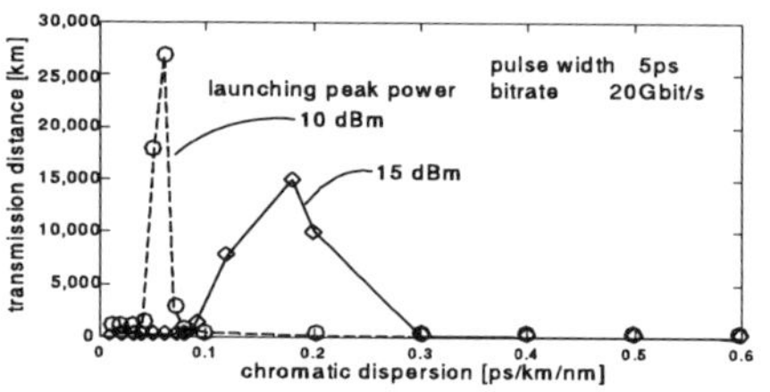

Figure 4 transmission distance for chromatic dispersion

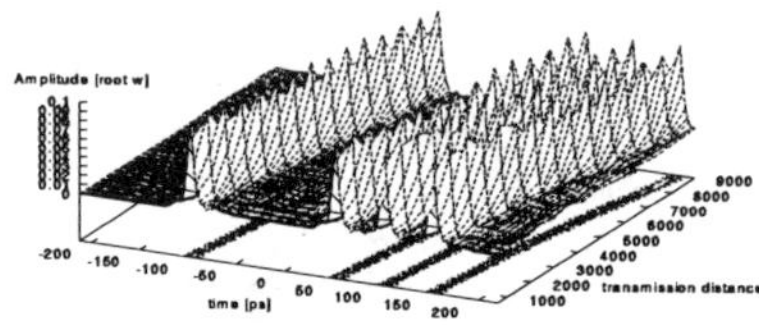

Figure 5 propagation waveform (5ps, 10dBm, 0.05 ps/km/nm)

Fig.3 shows propagation waveforms at the peak point. Then, pulse width is 5ps, the launching peak power is 10dBm and the chromatic dispersion is 0.05 ps/km/nm. In this figure, it is found that long haul transmission is achieved without distortion as if they are soliton pulses propagation. From the results mentioned above, the generation of soliton-like pulses in a long distance fiber with EDFAs has been confirmed by selecting system parameters suitable even if gaussian pulses are launched.

Secondly, the conditions of automatic soliton-like pulse generation have been considered by using Eq.(2) with the higher order nonlinear terms in fibers. The results are shown in Figs.4 and 5. In Fig.4, we can confirm that there is the peak point of transmission distance even if the higher order nonlinear terms have been considered. By comparing Fig.4 with Fig.2, the transmission distance becomes shorter due to higher order effects (mainly due to the third order dispersion). However, very long distance transmission about 26,000km for 10dBm or about 15,000km for 15dBm, has been also achieved.

Fig.5 shows propagation waveform for a parameter set (5ps, 10dBm, 0.05 ps/km/nm) of a peak point in Fig.4. Though the self frequency shift term causes propagation delay, the pulses have kept high peak power and almost same waveform over long distance. Based on these results, it can be concluded that long distance transmisson of a pulse train can be achieved by using automatic soliton-like pulse generation even if higher order optical effects in fibers have to be considered.

Lastly, we investigated parameter sets (pulse width, launching peak power and chromatic dispersion) by which long haul transmission could be achieved. Fig.6 shows contour plots of the transmission distance. Results for two cases (pulse width is 5ps and 10ps) are shown in Fig.6. In these figures, a contour line shows the transmission distance achieved even if any chromatic dispersion on the line is used. For example, a thick dashed-and-doted line shows that 3,000km transmission is available if a parameters set on the line is used. Longer distance transmission, could be achived in the inside of the line.

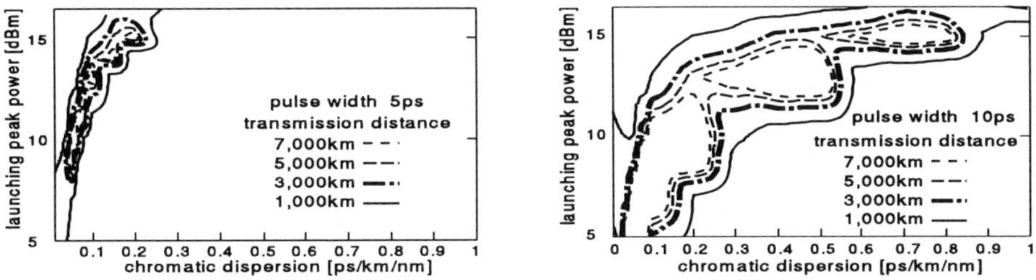

(a) pulse width 5 ps,bit rate 20 Gbit/s

(b) pulse width 10 ps, bit rate 10 Gbit/s

Figure 6 Capable regions of long distance transmission

The following conclusions are gotten from these results. The longer distance transmission region shifts into the longer chromatic dispersion with increasing the launching peak power, and the region is gradually changed into the narrow chromatic dispersion region according to narrower pulse width. This can be explained as follows. When the launching peak power is high, the SPM effects are strong. Then strong pulse broadening is needed to cancel it. On the other hand, when chromatic dispersion is small, pulse broadening is weak. Therefore the launching peak power needed is low. On the hand, when pulse width becomes narrow, the effects of pulse broadening is strong and the pulse power is low. Therefore, to balance the SPM effect and the pulse broadening in the narrow pulse width case, higher launching peak power is needed in the case of equal chromatic dispersion and if the launching peak power is equal, the smaller chromatic dispersion is needed. From these results, it is known that long haul transmission can be achieved by using parameters set inside of the region even if gaussian pulses have been launched, though the capable region of long haul transmission is narrow in the case of using narrow pulse width.

4 Conclusion

We have shown the conditions that soliton-like pulses are automatically generated from gaussian pulses. One can easily design long haul transmission systems over several thousand km by selecting suitable parameters set that soliton-like pulses are automatically generated. Evaluations of systems considering chromatic dispersion fluctuations in fibers and the amplified spontaneous emission noise in EDFAs are further works.

References

[1] N.Bergano,"40Gbit/s WDM transmission of eight 5Gbit/s data channels over transoceanic distance using the conventional NRZ modulation format",OFC795,PD-19,1995

[2] T.Morioka,S.Kawanishi,H.Takara,,O.Kamatani, "Penalty-free,100Gbit/s optical transmission of 2ps supercontinuum transform-limited pulses over 40km",Electron.Lett.,1995,31,pp124-125

[3] Bing Yao, Masanori Hanawa, Mikio Takahara, "High bit rate optical pulse transmission using OPC and DSCF", Proc. of 1996' IEICE Electron society conference, B-1117

[4] Hiyohuki Toda,"Dispersion managed optical soliton transmission", study of lazer, vol24, number 6, pp.27-28(1996)

[5] Govind P.Agrawal, "Nonlinear Fiber Optics", Academic Press,INC

Core and ATM Networks
D.W. Faulkner and A.L. Harmer (Eds.)
IOS Press 1997 © AKM

Nonlinear Jitter of Optical WDM-Systems due to Cross-Phase-Modulation

N. Hanik, Deutsche Telekom AG, Ringbahnstr. 130, 12103 Berlin, Germany

Abstract. Comparing the performance of a single channel 2.5 Gbit/s transmission system and a multichannel system, it is demonstrated that Cross-Phase-Modulation leads to an increasing jitter of the propagating data signals. A simplified method to determine the strength of this nonlinear perturbation along the transmission line is derived. It is shown that the maximum fiber distance and the input power level of a WDM system is restricted due to CPM-induced jitter. A method to increase system performance would be the use of dispersion compensation techniques already for 2.5Gbit/s systems.

1. Introduction

The use of a 2.5 Gbit/s data rate is common standard in optical signal transmission. Applying Erbium Doped Fiber Amplifiers, transmission distances of 1000 km are possible without signal regeneration or dispersion compensation. The evolution towards Wavelength Division Multiplexing using an increasing number of these 2.5Gbit/s channels is an efficient and simple method to handle the rapidly growing data rates expected in the future. Nevertheless, it is shown in this contribution that nonlinear interaction of the WDM-channels is not negligible. Using a 100 Ghz frequency grid for WDM systems, the main perturbations of the data signals are due to the interaction of Self-Phase-Modulation (SPM) and Cross-Phase-Modulation (CPM) with Chromatic Fiber Dispersion. The origin of these nonlinear effects is the nonlinear refractive index of glass fibers [1]. They result in a frequency modulation of the signal carrier, induced by the intensity modulation of the propagating data signals. SPM is caused by the particular intensity modulation of the considered channel, CPM results in an interaction of the adjacent channels of a WDM-system.

2. Theory

The mathematical formalism for the evaluation of the propagating channels of a WDM system is as follows:
Having modulated signal carriers of the form:

$$E_i(z,t) = \hat{E}_i \cdot s_i(z,t) \cdot e^{-\frac{\alpha_i}{2} \cdot z} \cdot e^{-j\beta(\omega_i)\cdot z} \cdot e^{j\omega_i t}, \qquad (1)$$

where s_i is the modulated complex signal amplitude of the i-th channel with carrier angular frequency ω_i, the nonlinear propagation of the signal amplitudes of the WDM system is governed by a coupled differential equation system:

$$\begin{aligned}&\left[\frac{\partial}{\partial z} + \beta'(\omega_i)\frac{\partial}{\partial t} - j\cdot\frac{1}{2}\beta''(\omega_i)\cdot\frac{\partial^2}{\partial t^2} - \frac{1}{6}\beta'''(\omega_i)\cdot\frac{\partial^3}{\partial t^3}\right] s_i(z,t) = \\ &= -j\cdot\frac{\mu_0\cdot\omega_i^2}{2\beta(\omega_i)}\cdot 3X_{eff}^{[3]}\cdot\left\{e^{-\alpha_i z}\cdot\left|\hat{E}_i\right|^2\cdot\left|s_i(z,t)\right|^2 + 2\cdot\sum_{\nu,\nu\neq i} e^{-\alpha_\nu z}\cdot\left|\hat{E}_\nu\right|^2\cdot\left|s_\nu(z,t)\right|^2\right\}\cdot s_i(z,t).\end{aligned} \qquad (2)$$

The summation on the right of eqn. 2 is over the remaining N-1 channels of the WDM-System giving rise to CPM in channel #i. β' denotes the group velocity, the higher order derivatives with respect to frequency describe the chromatic and cubic fiber dispersion. Finally, the factor $X_{eff}^{[3]}$ includes the effective nonlinear susceptibility of the glass fiber. A value of $X_{eff}^{[3]}=6.7*10^{-34}$

Asm/V^3 is assumed, corresponding with a nonlinear refractive index of $n_{nl}=2.3*10^{-22}m^2/V^2$ in [2]. Analog equations hold for the remaining channels of the WDM-system. This coupled equation system can be solved numerically.

3. Numerical Simulation of the System performance

The following figure shows the simulated transmission system. The data rate is 2.5Gbit/s per channel. An Erbium amplifier regenerates the signal power every 70 km.

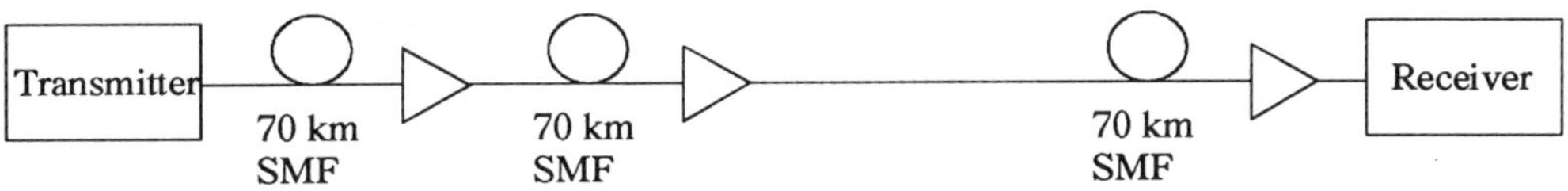

Fig. 1. Investigated Transmission system.

For simplicity of the representation, the transmitter includes the signal multiplexing unit of the WDM system, the channel selection is installed in the receiver unit.

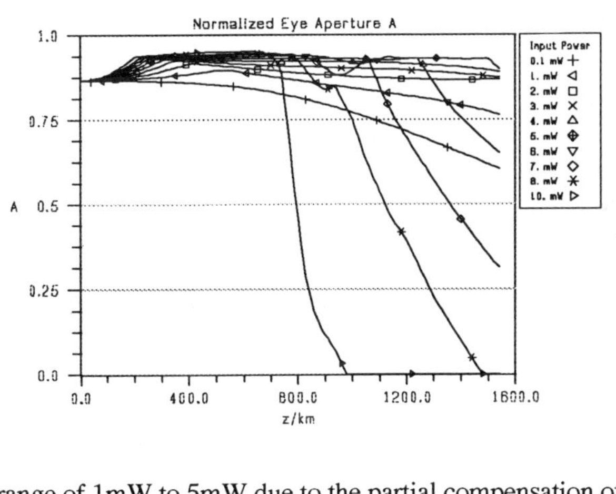

Fig 2. Normalized Vertical eye aperture A of a 2.5Gbit/s single channel-system.

The Figs. 2 and 3 show the vertical eye apertures of the received data signal of a single-channel system and of the central channel of a 5-channel WDM-system depending on the transmission length and the peak-power level (the power of a ..11111..-signal). A channel distance of 100 Ghz was chosen for the WDM system. Fig. 2 shows that, in case of a single channel system, there exists an extensive system optimum with respect to the eye aperture in the range of 1mW to 5mW due to the partial compensation of dispersion and fiber nonlinearity.

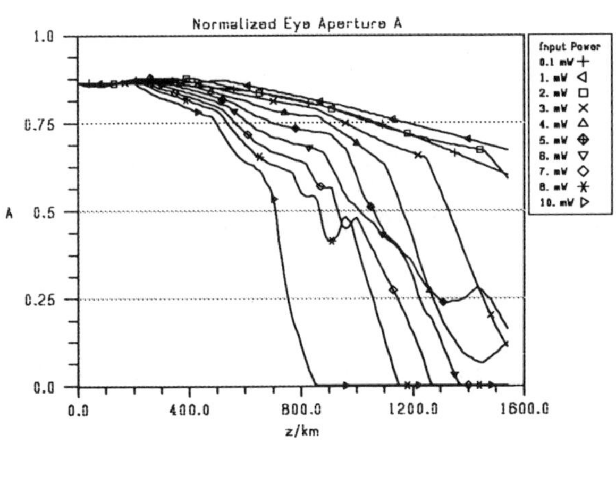

Fig 3. Vertical eye aperture A of the central channel of a 5-channel WDM-system as a function of the input peak power per channel.

In the WDM-case, this optimum range is restricted to a small region around 1mW. As Four Wave Mixing is negligible here because of the high dispersion of the conventional Single Mode Fiber, this effect arises only due to the interaction of Cross Phase Modulation and fiber dispersion. CPM induces a quasistatistical, time dependent frequency chirp along the propagating signals of the WDM system. This frequency chirp is transformed into a statistical jitter of the data signal by the chromatic fiber dispersion.

Fig. 4 shows the received eye pattern of the central channel after 750 km of fiber. Comparison with the single-channel case shows that the system degradation is caused by an increasing jitter due to Cross-Phase-Modulation.

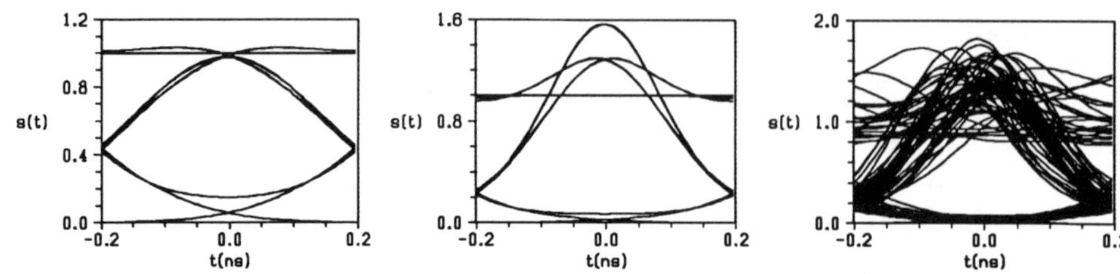

Fig 4. Eye pattern at the end of a 750 km transmission distance. Left: single channel system with 0.1mW peak input power. Center: single channel system with 5mW input power. Right: the central channel of a 5-channel WDM-system with 5mW input power per channel.

This result is a motivation for a further investigation of the jitter accumulation due to the nonlinear CPM-process. In the next section, a simplified calculation method is derived that avoids time consuming numerical simulations.

4. Estimation of the nonlinear Jitter

As an approximation, we assume that the perturbing channels with index ν in eqn.2 propagate without nonlinear degradation along the fiber. The equation for channel #i under consideration may then be solved seperately. We normalize the time coordinate with respect to the group delay along the fiber, i. e. $\tau_i = t - z/v_{gr}(\omega_i)$, v_{gr} being the group velocity of channel #i. Introducing the signal peak-power $\hat{P}_\nu$ of the respective channels, eqn. 2 may be integrated to give a result of the form:

$$s_i(z,\tau_i) = s_i(0,\tau_i)\cdot e^{j\Delta\varphi_i(z,\tau_i)}. \tag{3}$$

The nonlinear phase term due to CPM is given by:

$$\Delta\varphi_i(z,\tau_i) = -\frac{\mu_0\cdot\omega_i}{\varepsilon_0\cdot n_{co}^2}\cdot 6\mathrm{X}_{eff}^{[3]}\cdot\int_0^z\sum_{\nu,\nu\neq i}\frac{e^{-\alpha_\nu z'}}{A_{i\nu,eff}}\cdot\hat{P}_\nu\cdot s_\nu\left(\tau_i+z'\cdot\left[\frac{1}{v_{gr,}(\omega_i)}-\frac{1}{v_{gr,}(\omega_\nu)}\right]\right)dz'. \tag{4}$$

n_{co} denotes the refractive index of the fiber core. The time derivative of the nonlinear phase gives the nonlinear frequency shift:

$$\Delta f_i(z,\tau_i) = \frac{1}{2\pi}\cdot\frac{d}{d\tau_i}\Delta\varphi_i(z,\tau_i). \tag{5}$$

Finally, the group delay deviation $\Delta\tau_{gr,i}(z,\tau_i)$ along the data stream of channel #i at the fiber length z may be calculated as:

$$\Delta\tau_{gr,i}(z,\tau_i) = D\cdot\frac{\lambda_i^2}{c}\cdot\int_0^z\Delta f_i(z',\tau_i)\,dz'. \tag{6}$$

The standard deviation σ (z) of $\Delta\tau_{gr,i}$ is calculated using a time integration of the considered data stream according to:

$$\sigma^2(\Delta\tau_{gr,i}(z)) = \frac{1}{T} \cdot \int_0^T \Delta\tau^2_{gr,i}(z,\tau_i)d\tau_i, \qquad \ldots(7)$$

and taking the square root. As $\Delta\tau_{gr,i}$ takes positive and negative values with equal probability, the average value of $\Delta\tau_{gr,i}$ is zero. In Fig. 5, the emerging jitter of a 144 bit data stream is evaluated. The central channel of a 5-channel system according to Fig. 1 is considered. A comparison of this results with the vertical eye aperture of the exact simulation according to Fig. 3 shows that a penalty of the eye aperture of 1 dB is correlated with a nonlinear jitter of about 5%. If we take this value as an upper limit, the limitation of the total fiber length can be evaluated using Fig. 4a for the central channel of a 5-channel system, and Fig. 4b for the central channel of an 11-channel system at various input peak power levels. With this relatively simple and time effective evaluation method, the total transmission distance for a specific system configuration with respect to system degradation due to nonlinear jitter may be estimated. The accuracy in the range of small perturbations is sufficient.

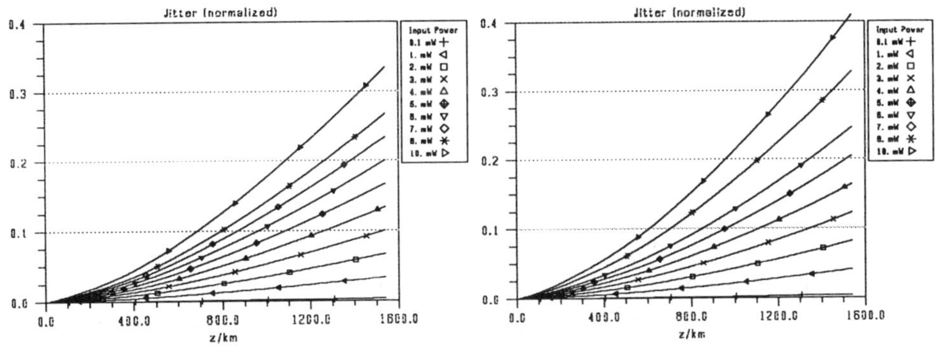

Fig. 5. Standard deviation of the nonlinear jitter (normalized by the bit-duration) according to eqn. (7) for the central channel of a 5-channel-system (left) and the central channel of an 11-channel system (right).

5. Conclusion

Using 2.5 Gbit/s data signals as a standard for the evolution of WDM-Systems, the CPM-induced jitter leads to severe system degradations at high channel numbers and long transmission distances. Methods to reduce this perturbation are the choice of a higher channel distances or low input power levels, both means that reduce system capacity. Introducing a 10 Gbit/s data signal, the implementation of appropriate dispersion compensation mechanisms would be obligate to achieve comparable transmission distances. In this case, investigations of the same kind as presented here have shown that the jitter induced by CPM, which is a product of nonlinearity *and* total dispersion, is likewise strongly reduced. Thus, a far-sighted implementation of dispersion compensation already for 2.5 Gbit WDM-systems would assure an easy upgrade to 10Gbit/s systems, and a significant reduction of the system perturbation by additive jitter.

References

[1] G. P. Agrawal, Nonlinear Fiber Optics. Academic Press, Boston, 1995.

[2] K. J. Blow, N. J. Doran, Nonlinear effects in optical fibres and fibre devices. IEE Proc. 134 (1987) 3, 138-144.

Core and ATM Networks
D.W. Faulkner and A.L. Harmer (Eds.)
IOS Press 1997 © AKM

A Wavelength Division Multiaccess Protocol for Optical Fibre Ring Networks

Ahmed E. Kamal
Dept. of Electrical & Computer Eng.
Kuwait University
P.O.Box 5969 Safat
Kuwait 13060
E-mail: kamal@eng.kuniv.edu.kw

Gerrit K. Janssens
Dept. of Computer Science
University of Antwerp
Middelheimlaan, Antwerp
Belgium 2020
E-mail: gerritj@ruca.ua.ac.be

Abstract

We present a multiaccess strategy for use over optical fibre ring networks employing two counter rotating rings. Wavelength division multiplexing is used in which case the spectrum is divided into groups of channels. The groups are skewed in time. Transmitters employ tunable lasers, while receivers employ a number of fixed wavelength filters, where a receiver is assigned at most one wavelength from each group. Therefore, a transmitter can tune to any of the wavelengths assigned to a receiver. In case the wavelength is already in use, the transmitter may try another wavelength in another group, and the time skew between groups facilitates this tuning operation. In addition, the strategy employs two approaches that results in an added network capacity, namely, slot reuse and wavelength reuse. The paper presents an approach to carry out wavelength assignment such that the interference between nodes accessing the same wavelength is minimized.

1 Introduction

In this paper we consider ring networks using WDM, and implemented using two counter rotating rings. Very few studies of such networks have appeared in the literature e.g., [9, 1, 2, 4, 5]. One of the disadvantages of such topologies is that the power loss due to the linear tapping pattern severely limits the number of nodes which can be connected to such a network [7]. However, with the recent advances in optical amplifier technology, interest in such topologies has been revived. The objective of this paper is to introduce a strategy for multiple access over slotted ring networks using WDM. The strategy should enable the resolution of conflicts between transmitters and receivers without introducing any added node latencies, or be a source of bandwidth wastage due to node coordination or device tuning. The approach that we take in this paper is to use tunable transmitters and fixed receivers. However, each receiver device may have multiple receivers and may receive on multiple wavelengths. The choice of such devices is based on the observation that due to the serial nature of packet arrivals from the node's host, usually a node is transmitting a single packet. However, a node ma be receiving multiple packets transmitted by severa nodes. This approach has the advantage of reducin the probability of packet loss due to busy receivers However, it may lead to bandwidth wastage due t the packet header processing time, and the transmit ter tuning time. We overcome this problem by split ting the packet header into two parts. The first part i the busy bit, which is transmitted on a common con trol channel, while the second part is the source an destination identities which are transmitted with th rest of the packet. The network uses a time-skewe arrangement of the control channel with respect t the data channels. It also uses slot reuse, as well a bandwidth reuse.

In section 2 we describe the network architecture and this will be followed by the protocol description i section 3. Section 4 considers the problem of wave length assignment to receivers. Section 5 conclude with some remarks.

2 Network Architecture

The network consists of two counter rotating rings Each of the two rings can be implemented using a sin gle mode fibre. Each node has two optical interfaces one to each ring. It has been shown in [3, 6] tha with this approach and under the uniform source des tination distribution, a system capacity of 8 can b achieved if slots are erased by their destinations.

On each ring, wavelength division multiplexing i used. Therefore, the low loss region on each ring which is between 1.2 to 1.6 μm, is optically divide into a number of channels, where the separation be tween adjacent channels is on the order of 1 nm. Th single channel bandwidth can be calculated to be o the order of several giga bits per second. One of th channels on each ring is designated as the control chan nel, while the rest of the channels are divided into G groups of channels. Each group consists of B channels and each channel is divided in time into slots such tha each slot can hold exactly one packet. The total num ber of channels on each ring is therefore equal to GB+ channels. For example, a total of 13 channels can b divided into 3 groups, with each group consisting o

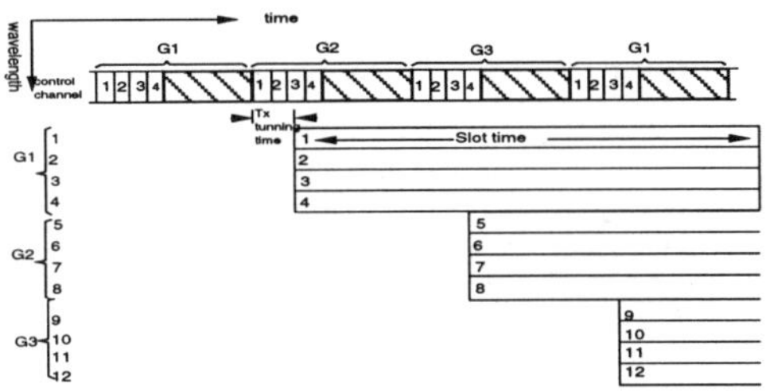

Figure 1: Channel grouping

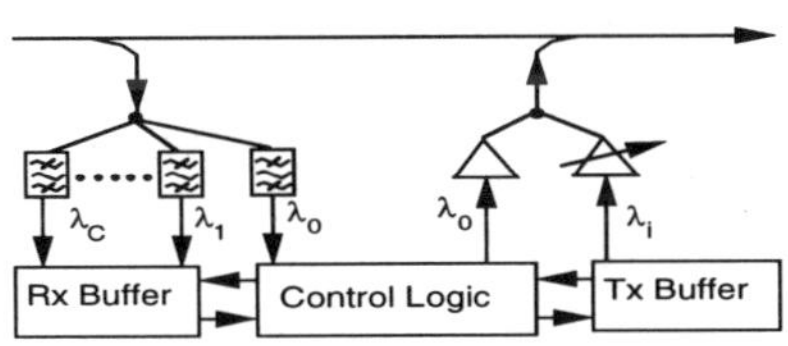

Figure 2: Node architecture

four channels, in addition to one control channel (see Figure 1). Within each group, the start of slots on all B channels is synchronized. Additionally, the groups of channels are skewed in time, as shown in Figure 1, and the amount of time skew is equal to the maximum of the $\frac{\text{slot time}}{\text{G}}$, and the worst case transmitter tuning time. The purpose of skewing is to allow the transmitters ample opportunity to use slots, as will be explained in the next section. The control channel is also divided into slots which is divided in turn into GB fields, with each field pertaining to one data channel. A light pulse in this field indicates that the slot is occupied, while the absence of such a pulse means that the slot is available for use.

The node architecture is shown in Figure 2. The transmission channel wavelength is λ_i, and the node is assumed to be able to receive on wavelengths $\lambda_1, \lambda_2, \ldots, \lambda_C$. The control channel is λ_0. To read from and write to the control channel each node is equipped with a fixed filter and a fixed laser tuned to the control channel wavelength. The output of this filter is converted to an electrical signal which is used by the control logic. The operation of the control channel laser is also controlled by the same logic. The transmitter section of each station is equipped with one tunable laser. The tuning range is assumed to be over the entire wavelength range. The receiver section of each station has C fixed receivers, where $1 \leq C \leq G$. This is necessary since each destination station is assumed to be reachable on C wavelengths, such that the destination is reachable on at most one wavelength per group. The outputs of these filters are decoded and fed to a common receiver queue if the slots contain any packets destined to the station. Since a node completely receives no more than one packet at any point in time (because of the single channel per group condition and the group skew arrangement), then it is guaranteed that no conflicts in packet queueing at the receiver will take place.

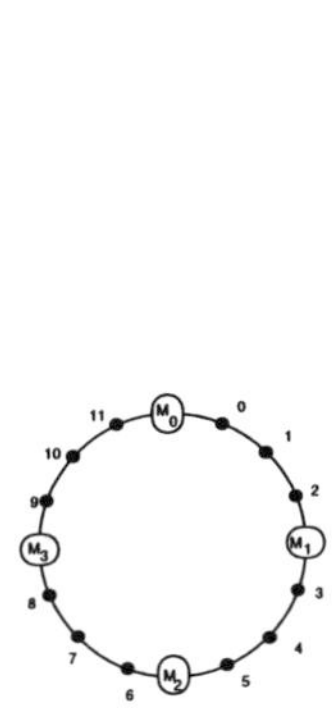

Figure 3: Example of a ring with four headends

The purpose of the headends, or the erasure nodes, is to empty slots which have already been read by their respective destinations. This allows reuse of both slots and wavelength assignments to receivers. For example, in Figure 3 in which there are four such headends, M_0, M_1, M_2 and M_3, and twelve normal nodes, nodes 9, 10, 11, 0 and 1 may use a certain wavelength to send to node 2. Since such a slot will be erased by M_1, then it can be used b nodes 3, 4, 5, 6, 7 and 8. This is an example of slot reuse. At the same time, a wavelength used by station 0 to transmit to stations 1, 2, ..., 5, can also be used by station 1 to transmit to stations 2, 3, ..., 6. this is an example of wavelength reuse.

Headends must be able to receive on all channels, and erase packets which have already been read. The implementation of those erasure nodes can be either done in a pure optical manner, using wavelength selective dividers and combiners [11], or can be implemented electronically[1]. The latter requires the use of optical-to-electrical and electrical-to-optical converters. In either case, if the packet contained in a slot has been already read, then it is replaced by an empty slot. Otherwise, the contents of the slot are left unchanged. (The setting of the switches which implement this replacement can be determined in advance through reading the control channel, as will be described in the next section.) The headend may also act as an amplifier, which is a much needed function if such linear topologies are used [5].

3 Protocol Description

The network protocol is based on the empty slot approach employed, for example, in the ATM Ring [6]. That is, a node that wants to transmit in a slot must find out if the slot is empty or not. If so, the node marks the slot as full and starts writing its information into the slot. If not, it looks for another slot.

Marking the slot as either empty or full is done over the control channel. Therefore, at the input to the node hardware, the control channel data is filtered

[1]The utilization of signal attenuation proposed in [8] can be used to implement the function of the erasure nodes, but this requires that the wavelength not be reused before a certain number of taps.

and read by the control logic, just prior to being written with a 1 if the node wanted to use it. Based on the status of the bit corresponding to the channel under consideration, if the bit was already 1, then the channel is in use, while if it was a 0, then the channel is available for transmission. In the latter case, the station tunes its laser to the channel's corresponding wavelength and starts transmitting. The transmitter starts by writing the identity of the destination station followed by its own (source) identity and then the rest of the packet. Since each receiver station is assigned a number of channels, such that there is at most one channel per group, a source that finds one such channel in a certain group to be already occupied, tries to find another channel in the following group. It therefore checks the control channel for the status of that next channel, and keeps on repeating this process until the packet is finally transmitted.

An example of channel assignment is shown in the column labeled *Source 0* in Figure 5 in which G=3, B=2, C=2 and N=24. The channel assignments on the clockwise ring, as seen by station 0, are shown in the figure. In the next section we will explain why these assignments will change from station to station. According to this example, if station 1 would like to transmit to station 6, it uses the clockwise ring since it provides the shortest distance. Noting that station 6 receives on two channels, namely channels 1 and 3, station 1 hunts for an empty slot on these two channels. If station 1 inspects channel 1 first, and finds the slot on that channel occupied, it tries the slot on channel 3. If channel 3 is also found to be occupied, station 0 keeps on inspecting channels 1 and 3 alternately, until a slot is found empty on either of these two channels. Station 0 then uses this slot for transmission to station 6.

The receiver protocol is very simple and follows directly from the description of the node architecture (see Figure 2). Receivers are equipped with C fixed wavelength filters, which are tuned to the wavelengths of their assigned channels. The received slots are decoded into electronic form, and if the slot is found to contain a packet destined to the station, then it is relayed to a queue for further processing by the station. Otherwise, i.e., if the slot did not contain a packet destined to the receiver, the receiver logic ignores the packet. Notice that since a slot can be used for transmission to multiple receivers, then detecting the status of the packet in a slot requires the receiver to inspect the destination address.

4 Wavelength Assignment

In describing the system we have made the assumption that channels are assigned to receivers according to certain criteria. In this section we discuss these criteria.

1. First, it should be noted that in order to allow receivers to queue multiple received packets without intervention, the packet queueing times should be distinct. Therefore, restricting the receiver to at most one channel per group guarantees that multiple packet receptions will not occur at the same instant of time.

2. Second, since the total number of channel assignments is equal to NC, where N is the number of stations in the system, which is usually greater than the total number of channels in the system, GB, then at least on channel will be assigned to more than one destination station. Notice that since a channel can be used to carry packets to several receivers, then the number of destination stations assigned to the same channel should be minimized. For this purpose, each channel should be assigned a number of destinations that is between $\lfloor \frac{NC}{GB} \rfloor$ and $\lceil \frac{NC}{GB} \rceil$. It is to be noted that this assignment also balances the load on all channels.

3. In addition to the above, we would also like to minimize the number of times two destination stations are assigned to the same channel. This minimization criteren is different from that in point 2 above.

   A binary integer programming (BIP) characterization of the channel assignment according to this interference minimization criterion has been developed, and can be solved using a method such as the branch-and-bound [10]. However, it is well-known that such a solution is NP-complete. In special cases, linear time algorithms can be found for the channel assignment problem, such that the intersection is minimal, such as the case in which $C = 2$, $G \geq 2$, and $B \geq 1$. In this case, the maximum number of destination nodes that can be assigned such that the number of times two destinations are assigned to the same channel is equal to $\binom{G}{2} \cdot B^2$. The number of destination nodes per channel, n, in this case is equal to $(G-1)B$.

   Based on the above, we introduce an algorithm in Figure 4 that performs the channel assignment for the source station 0 on the clockwise direction in exactly $2N$ steps. The channel assignment in the example in Figure 5 has been produced using this algorithm, and for sources 0 through 2. The algorithm is based on considering the destination nodes of groups of $\frac{(G-1)GB^2}{2}$ and assigning them to channels such that they exhaust all the $\binom{G}{2} \cdot B^2$ permutations. The last group of nodes, if any, is then

4. The fourth factor that influences channel assignment is wavelength reusability. We therefore introduce the concept of *station pairing*. A *station pair* with respect to a certain ring are two stations that are separated by $\lfloor N/2 \rfloor$ stations on the direction of flow of the considered ring. For such a station pair, a wavelength which is used for transmission by stations $(i - \lfloor N/2 \rfloor) \bmod N$ through station $(i-1) \bmod N$ to station i in the clockwise direction can also be used to carry transmissions from stations i through $(i + \lfloor N/2 \rfloor - 1) \bmod N$

RoundLength $:= \binom{G}{2} \cdot B^2$;
NoRounds $:= \lfloor \frac{\text{NoNodes}}{\text{RoundLength}} \rfloor$;

$S_{i,j} = \phi \quad \forall i,j$

for Round := 1 to NoRounds
    NodeOffset $:= (i-1) \cdot \binom{G}{2} \cdot B^2$;
    for i:=1 to $G-1$
        for j :=1 to B
            for k := 1 to (G-i)B
                Node := NodeOffset + $(i-1)(G-1)B^2 + (j-1)(G-i)B + k$
                $S_{i-1,j-1} = S_{i-1,j-1} \bigcup \text{Node}$;
                $S_{i+\lfloor \frac{k-1}{B} \rfloor,(k-1) \bmod B} = S_{i+\lfloor \frac{k-1}{B} \rfloor,(k-1) \bmod B} \bigcup \text{Node}$;

Figure 4: Assignment algorithm used for $C = 2$.

Group	Channel	Destinations Source 0	Source 1	Source 2
0	0	1,2,3,4	2,3,4,13	3,4,13,14
	1	5,6,7,8	5,6,7,8	5,6,7,8
1	2	1,5,9,10	5,9,10,13	5,9,10,13
	3	2,6,11,12	2,6,11,12	6,11,12,14
2	4	3,7,9,11	3,7,9,11	3,7,9,11
	5	4,8,10,12	4,8,10,12	4,8,10,12

Figure 5: Example of channel assignment for the transmissions of nodes 0 through 2 on the clockwise ring in a network with N=24, G=3, B=2 and C=2.

to station $(i + \lfloor N/2 \rfloor) \bmod N$ in the same direction. It should be noted that the assignment produced by this algorithm is optimal provided that $C \cdot \lceil \frac{N}{2} \rceil \leq (G \cdot B)^2$.

Finally, it should be noted that if the farthest station reachable by station i on a certain direction, e.g., the clockwise direction, is $(i + \lceil \frac{N}{2} \rceil) \bmod N$, and if station i uses wavelength λ_j for transmission to station $(i+1) \bmod N$ in the same direction, then station $(i+1) \bmod N$ may use wavelength λ_j to transmit to station $(i + \lceil \frac{N}{2} \rceil + 1) \bmod N$ in the same direction. This is the basis for constructing the channel assignment for other stations once the channel assignment for the source station 0 has been determined. The complete channel assignment can be obtained by employing the concept of channel pairing introduced above.

5 Conclusions

The paper has presented a multiaccess protocol for optical fibre ring networks. The protocol has the following merits: there are no channel collisions; there are no receiver collision; distinct receiver packet queueing instants, since a receiver can only receive one packet per group, and groups are skewed. The possibility of having an all-optical implementation of this network exists. Any fairness algorithm used in slotted rings such as the Orwell, the ATM Ring, the FECCA or the Metaring algorithms, can be used in this network.

References

[1] S. Banerjee and B. Mukherjee, "Near-Optimal Photonic Ring Configurations", Fifth LAN/MAN Workshop, Taormina, Italy, May 1992.

[2] I. Chlamtac and A. Fumagalli, "An Optical Data Double Ring Network", Jour. High-Speed Networks, Vol. 2, 1993, pp. 355–371.

[3] I. Cidon and Y. Ofek, "MetaRing—A Full Duplex Ring with Fairness and Spatial Reuse", IEEE Trans. on Comm., Jan. 1993, Vol. 41, No. 1, pp. 110-120.

[4] K. Falcone and O. Tonguz, "Access Methods for Fiber-Optic Interconnection of LAN's", IEEE Jour. of Lightwave Tech., Vol. 11, No. 5, May 1993, pp. 1113–1124.

[5] A. Grah and T. D. Todd, "Simplified Photonic Bus and Ring Networks Using a Headend Controller", in proc. IEEE Infocom '94, June 1994, Toronto, Canada, pp. 263–271.

[6] K. Imai et al., "ATMR: Asynchronous Transfer Mode Ring Protocol", Comp. Net. & ISDN Sys., Vol. 26, No. 6-8, March 1994, pp. 785-798.

[7] I. P. Kaminow, "Photonic Multiple-Access Networks: Topologies", AT&T Tech. Jour., Vol. 68, No. 2, March 1989, pp. 61–71.

[8] M. J. Karol, "Exploiting the Attenuation of Fiber-Optic Passive Taps to Create Large High-Capacity LAN's and MAN's", IEEE Jour. of Lightwave Technology, Vol. 9, No. 3, March 1991, pp. 400–408.

[9] M. J. Karol and R. D. Gitlin, "High-Performance Optical Local and Metropolitan Area Networks: Enhancements of FDDI and IEEE 802.6 DQDB", IEEE Jour. on Selected Areas in Comm., Vol. 8, No. 8, Oct. 1990, pp. 1439–1448.

[10] C. H. Papadimitriou and K. Steiglitz, *Combinatorial Optimization: Algorithms and Complexity.* Prentice-Hall.

[11] R. Ramaswami, "Multiwavelength Lightwave Networks for Computer Communication", IEEE Communications Mag., Feb. 1993, pp. 78–88.

Core and ATM Networks
D.W. Faulkner and A.L. Harmer (Eds.)
IOS Press 1997 © AKM

Internet, a future information superhighway

Dirk DE MAEYER
University of Antwerp (RUCA)
didem@ruca.ua.ac.be
Faculty of Applied Economics, Middelheimlaan 1, 2020 Antwerp, Belgium

Abstract. This paper gives an idea on what Internet's potential is for becoming an information highway. First, the information highway and the Internet are defined. Then the characteristics of an information highway are described. The concept of an information highway puts some requirements on the infrastructure. The users of an information highway, who are located in the residential and business environment, have their share of requirements as well, but they focus on the requirements put forward by the services or applications they will use on an information highway. At the same time we will check whether the Internet has implemented these properties and how, or if work is going on to develop them. The framework for this discussion is the TCP/IP reference model. Some emphasis will be put on the IP next generation protocol, IP version 6 (IPv6). It will become clear that the Internet certainly has potential for becoming an information highway.

1. Introduction

Some years ago the American vice-president Al Gore launched the term information highway. Since then many people use this buzz word to indicate some kind of data communications network. Very often Internet is mentioned as the information highway. After defining both the information highway and the Internet we will provide the distinctive properties that make any network into a highway. At the same time we will discuss how these properties are implemented on the Internet and evaluate whether Internet is already an information highway or it is merely a predecessor of a highway.

2. Definitions

Although most people talk about information highways when considering the modern networks, almost none provide an exact definition. The following definition tries to include as much of the capabilities inherent to a highway as possible: an information highway is a high-capacity, interactive electronic pipeline to the office or to the home, capable of simultaneously supporting a large number of electronic applications and providing interactive phonelike connectivity between users and services and between users and other users. [1] Already a lot of properties are included in this definition, yet others can be derived from it, such as mobility, universal access, universal addressing, and the like.

Before describing these properties, we will first provide a definition for the Internet and introduce the TCP/IP reference model.

The Internet is defined as a global network of networks based on the TCP/IP protocols, enabling computers of all kinds to directly and transparently communicate and share services throughout much of the world. [2] [3]

The term Internet is short for 'interconnected networks'. Key to the Internet are the TCP/IP (Transport Control Protocol/Internet Protocol) protocols. Basicly, these protocols can be divided into 3 layers: the internet, the transport and the application layers. However, in Internet's TCP/IP reference model we can distinguish a fourth layer, the host-to-network layer, which only specifies the protocols required for using the available network media, whereas the TCP/IP protocols are independent of the underlying media. The internet layer contains the IP, the ICMP, the IGMP and the routing protocols. The transport layer incorporates the TCP and the UDP protocols. At the application layer we distinguish amongst others the FTP, the Telnet, the HTTP, the Gopher, the SMTP, the NNTP, the DNS, and the SNMP protocols. [1] [4]

Let us now turn to the discussion of the highway properties and the evaluation of Internet's protocols and mechanisms.

3. Highway versus Internet

In this section we will discuss the properties of an information highway, namely the infrastructure, digital transmission, capacity, connectivity, simultaneity, applications' requirements, interactivity, openness, mobility, universal access, affordable access, network management and security. [1] [5] We will first describe every property in terms of a highway, stipulating the necessary requirements. Then we will see how it is implemented on the Internet and evaluate whether it meets the necessary requirements.

Infrastructure

The infrastructure of an information highway consists of three components: the network access equipment, this is the equipment used by the end-user to access the network and includes hardware (routers, switches, computers, set-top boxes) and software (browsers, operating systems), the access network, also called the local on-ramps, responsible for delivering or sending the information to or from the users, and the transport networks, responsible for transporting the information between the access networks. [1]

The same topology can be found on the Internet. As transport networks there are backbone networks, such as NSFNET, now called vBNS (very high Bandwidth Network Services) network, and regional networks, such as EuropaNET, Eunet and EBone. Access networks, often called stub networks on the Internet, provide two types of access: access for individual computers (terminal and host access) and access for autonomous networks (network and gateway access). [1] [3] [4] On the Internet the provider's networks often have the functionality of both access and transport networks. Currently the network access equipment for the Internet consists of some type of computer (PC, workstation, laptop, ...) equipped with network cards and/or modems, using TCP/IP stacks designed for various platforms, and Internet applications, like a World-Wide Web (WWW) browser. Thus the Internet has a highway-like infrastructure.

Electronic pipeline

What is meant here is that the information must be transmitted digitally end-to-end without digital-to-analog and analog-to-digital conversions, thus using digital bit streams.

In the case of the Internet all of the TCP/IP communications are digital, whereas at the host-to-network layer, in fact at the OSI physical layer, some modulation and demodulation, e.g. in the case of transmission over PSTN lines, is possible. But the TCP/IP protocols and especially the Internet protocol is independent of the hardware media, which allows the exploitation of any existing and new digital transmission technology of any kind. [3]

Capacity

Inherently to the information highway is that it is a high-speed broadband communication network. [6] In theory broadband means that the bandwidth must be at least 2 Mbps. In practice however much higher bandwidths will be used. Tens of Mbps for access networks and several hundreds of Mbps or even Gbps in the case of transportation networks.

On the Internet the bandwidths are very diverse: the access networks have bandwidths from 14.4 Kbps for dial-in connectivity using the normal PSTN up to 10+ Mbps for LAN connectivity or, and these are not yet common practice, for dial-in connectivity using ADSL and even up to 30 Mbps for cable TV access. The transport networks are even more diffuse: in the U.S. the vBNS network is said to have 655 Mbps in 1996 up to 960 Gbps in 1998, but for the rest of the world 2 Mbps is common. There are however initiatives for increasing the bandwidth in the near future up to 34 Mbps. [1] [4] [7] [8]

Here it can be clearly seen that the current bandwidth in general for the Internet is lacking. A serious increase in available bandwidth both for access as well as transportation networks is highly needed. The remark we immediatley pose here is whether the expected increase in bandwidth in the near future will be sufficient for the ever growing number of users and the rising number of bandwidth hungry (multimedia) applications.

Connectivity

Typically for a highway there is a need for three types of connectivity: point-to-point, point-to-multipoint and multipoint-to-multipoint. The counterparts on the Internet are unicast and multicast. Although available in IPv4, multicast mechanisms are much better implemented in IPv6. [9] Then we also have a series of new protocols to support multicasting, amongst others we have the Multicast Open Shortest Path First (Multicast OSPF, MOSPF), the Distance Vector Multicast Routing Protocol (DVMRP), the Protocol Independent Multicast (PIM) and Core-Based Trees (CBT). [10] Nevertheless, there are no means for direct multipoint-to-multipoint connectivity. If this type of connectivity is wanted, one has to introduce special mechanisms, like in the case of videoconferencing where a reflector is used absorbing multimedial data coming from multiple sources and distributing it to all members. [1]

Simultaneity

Simultaneity is the concept that at the same moment in time the network must be able to transport data belonging to several applications with possibly distinct quality of service requirements.

In theory there is no limitation for the number of services on the Internet. Originally the Internet was designed for best effort services, unspecified bit rate-like services, when using ATM terminology, where there is totally no guarantuee for performance (delay and bandwidth): data gets through when there is enough capacity on the net. In practice there is much too little bandwidth to allow for a large number of services, and since there are no traffic control mechanisms, like connection admission control or usage/network parameter control in the case of ATM, packet delays in the network and eventually congestion of the network resulting in packet losses are likely to happen. [11] [12]

Delay problems caused by access networks and provider sites can be relatively easily solved by upgrading the access networks and by upgrading the provider's equipment (hardware and software, such as a faster server) or by applying techniques such as IP clustering across multiple servers in order to achieve load balancing and fault resilience [7] [10]. Harder to solve are the delay problems caused by the Internet backbone. One solution offering more bandwidth and hopefully leading to less congestion and delays, is the use of faster network media, such as frame relay, FDDI and ATM networks, but this in itself still does not guarantee quality of service for Internet traffic. Arango suggests Guaranteed Internet Bandwidth (GIB) as a short term solution for providing predictable performance [7].

Moreover, since the start of the Internet a range of new services having other traffic characteristics, like constant bit rate, variable bit rate, and available bit rate services, putting forward a broad range of quality of service requirements, have emerged for which the Internet initially provided no solutions.

Gradually new protocols were and are being designed to be used with the common IP protocol, IPv4, to deal with quality of service parameters. We refer here to the Real Time Protocol (RTP) and the Real Time Control Protocol (RTCP), specified in RFC-1889 [3]. Other protocols are the Resource reSerVation Protocol (RSVP) and the STream Protocol Version 2+ (ST2+) providing resource reservation, and the Real Time Streaming Protocol (RTSP) enabling controlled, on-demand delivery of data with real time properties. [3] [10].

All of these protocols fit in the new Internet model, the 'Integrated Services Model', described in RFC-1633. This Internet Integrated Services Model supports as well best effort services, the traditional Internet services, as real time services by providing traffic control by means of controlled link sharing. [3]

Since the specification of this model a lot of research has been done in this area. Some of which led to the development of the above mentioned protocols. Other implementation possibilities for quality of service are also being studied. Preliminary research results can be found amongst the Internet Drafts. [10]

The evolution towards support for real time audio/video distribution and conferencing has even led to the development of a virtual overlay network, the MBone (Multicast Backbone) network. [13] [14] [15]

Also the new version of IP, IPv6, meets most of the shortcomings erected by the new generation of applications, such as support for quality of service by using priority and flow label fields in the IP header and better support for IP multicasting. [3] [9]

In order to support simultaneity of large numbers of all types of applications the Internet needs a large increase in bandwidth and a fast deployment of the newly developed protocols and techniques.

Applications

Some major key issues for applications on the information highway are the user-friendliness or ease-of-use, the transparency of underlying mechanisms and the availability of service directories, which provide the users with an easy navigation method towards the desired applications and information.

On the Internet we have transparent applications, hiding the lower layer protocols and mechanisms. For most applications there exists at least one version using a graphical user interface (GUI), which contributes greatly to the ease-of-use. With respect to service directories we can see on one hand individual attempts to categorise the available information into indexes, and on the other hand applications, like WWW browsers, capable of launching the necessary applications needed to perform certain tasks or represent information on the fly. Still, this is certainly an aspect of the Internet for which further study and development is required.

Interactivity

Interactivity translates into the support of a traffic/information flow in two directions, an up- and downstream, whereby both parties can interact on the information they receive. In other words the users must have some level of control over the information they exchange.

All services and applications on the Internet use one out of three client-server models. [16] In the classical model one client communicates with only one server. This model is used by the Telnet and FTP protocols. In the second model the client again communicates with only one server. Additionally this server can communicate with other simular (using the same application) servers, but only the server contacted by the client will communicate with the client. This is used for the SMTP and NNTP protocols. In the third model the server cannot communicate with other servers, but it can give directions to the client for starting a new communication with another simular server. This is used on the WWW by the HTTP protocol.

Hence there is interaction and two-way information traffic on the Internet.

Openness

Openness in this context means internetworking, the ability to interconnect the existing network media and the media yet to come, resulting in cooperation of applications on different networks and in overall exchangeability of information between those applications. In a wide scope including vendor and network provider independability. To reach openness open standards need to be developed, be easily available to all developers and be used. [5] [6]

The Internet, short for interconnected networks, is developed precisely to have this kind of openness. Theoretically the TCP/IP protocols can be used on any physical network medium, this is Internet offers a single platform for a wide variety of information infrastructure applications and services. [6] In practice however some improvements must be made to use (TCP/)IP efficiently over these new and mostly faster media, like frame relay, SMDS, FDDI, ATM, and wireless networks. Currently a lot of research is being undertaken in this area. [3] [10]

As for standardisation Internet's open standardisation process provides a fast delivery of proven standards based on consensus. [17] Proven, because for every standard to be accepted there needs to be at least one working application based on the implementation of the standard, freely available to all Internet users. Open, since on one hand literally everyone can make proposals for new standards, and on the other hand new standards must coexist with previously developed standards. Furthermore, standards are published as Requests For Comments (3) on the Internet, so that they are available to everyone. Before arriving at a standard, proposals have to be submitted to the Internet Architecture Board (IAB). A proposal document is called an Internet Draft and has a maximum life-time of 6 months. If no comments are given to a draft it may become a RFC. If comments are given by the Internet community, a new draft has to be prepared. [3]

Mobility

As discussed when handling openness, the highway must interconnect various networks, including the networks used by mobile users. Two kinds of mobile users can be further distinguished. Firstly, mobile users who from time to time move from one fixed site to another, but only use the network when they are physically connected to it, the migratory users, thus using wired networks. Then there are users who actually want to work on their computer while moving from place to place, the roaming users, thus using wireless networks. [4] A highway must therefore be able to cope with these mobile users.

On the Internet mobility is implemented by using home and foreign agents, and the techniques of host registration and tunneling, that is the host computer keeps the same IP address all the time but tells the network where it is located. [3] Implementations for mobile hosts on the Internet are the Internet Mobile Host Protocol (IMHP) and the Virtual Internet Protocol (VIP). [18] [19] In IPv6 we can also find support for mobility. IPv6 routing even has advanced capabilities for host mobility. [9] Although support for mobility on the Internet is not yet widely available, it is clear that it will be in the near future.

Universal access

Following the concept of openness, internetworking or the interconnection of various network media we need universal access. This translates into worldwide access requiring a universal addressing mechanism and efficient routing capabilities. [5] The Internet has a worldwide coverage of over 134 countries - over 186 countries when only counting email connectivity - has over 5 million hosts and has 20 to 50 million users according to estimations. The size of the Internet is said to double in between 6 to 12 months. [1] [2] [4] [20]

The Internet has a universal addressing mechanism by providing uniquely allocated IP addresses and fully qualified domain names (FQDN). Currently two IP address formats are in use: IPv4 and IPv6 addresses. IPv6 and IPv4 addresses can coexist, but IPv6 addresses offer several advantages: longer addresses (128 bits versus 32 bits) thereby providing an enormous amount of addresses and solving the scarcity of the IPv4 address space; better support for multicast; anycast capabilities and several different address types. For transferring the information from sender to receiver the network layer uses routing protocols. On the Internet there exists a wide range of routing protocols, one being more efficient than the other or meeting different network and service requirements. Again IPv6 routing protocols better meet new demands, such as provider selection, host mobility and auto-readressing. [9]

Hence, the Internet provides already universal access, but we still need a fast deployment of IPv6 to overcome some shortcomings.

Affordable access

Affordable access means in this context that everyone must have access to or must have the possibility to be provided with (a) basic service(s) at low costs; in other words universal service provisioning. [21] This also means the availability of different accesses at different cost and with different quality of service parameters for different users and application profiles.

Electronic mail, email, can be considered as the basic service for the Internet. Email connectivity puts low requirements on the access and transportation networks, making it very cost efficient. Also, every Internet user has at least the capability of performing email, which results in a large installed base, namely over 186 countries have email connectivity. On the Internet there exist various access methods - terminal access and host access via PSTN or ISDN, and network access - having different costs and offering different QoS parameters to a wide range of residential and business users, or to novices and very demanding users. Terminal and host access (via the telephone network, PSTN) are the cheapest ways to get access to the Internet.

Network management

Network management includes different functionalities: configuration, performance, fault, accounting and security management. [22]

On the Internet we have SNMPv1 and SNMPv2, two versions of the Simple Network Management Protocol. SNMP uses agents, managers and MIBs (management information bases). [3] SNMP offers some level of configuration, fault and performance management. For SNMPv2, a new and enhanced version of SNMPv1 which provides amongst others improved security features, there are no commercial applications yet.

At this time there exists no universal implementation for accounting on the Internet, but already several proposals have been formulated. [3] [10] It is clear that accounting will receive further attention in ongoing research. Accounting requires also good security functions. Security will be discussed in the next paragraph.

Security

Security demands a universal security policy [5], so that all users or applications can benefit from existing security implementations. Among the most important concepts of security are confidentiality (secrecy), integrity (accurateness), availability and/or reliability (no denial of service), authentication (identification of parties involved), non-repudiation (no denial of participating parties to having exchanged messages and to the content of those messages) [23] and privacy.

On the Internet the policy of universal security is met by the availability of security mechanisms at different layers of the TCP/IP reference model. Just beneath the IP layer, this is the network layer, thus in fact not part of the TCP/IP protocol suite, there are security implementations, such as Utimaco's Safe Guard IPN [24], on the network layer we now have the security extension headers which can be used in conjunction with IPv4 and IPv6: the authentication header (AH) and the encapsulating security payload (ESP) [3] [9], and on or just above the TCP layer and below the application layer we have the Private Communication Technology (PCT), the Secure Socket Layer (SSL), of which version 3 is included in the Transport Layer Security (TLS) protocol, and the SSH transport layer protocol. [10] [23] Non-universal mechanisms, in the sense that the provided security mechanisms can not be used by all services or applications, are integrated in applications, like in the case of Secure HTTP (SHTTP), S/MIME, MIME Object Security Services (MOSS), and Privacy Enhanced Mail (PEM); or are put on top of existing applications, like PGP to be used in conjunction with an email application, or like Secure Electronic Transactions (SET) which can be used on top of HTTP, SMTP or some other applications for securing the contents of transactions. [3] [10] [23] [25] All of the above mentioned mechanisms provide confidentiality, integrity and authentication. Although these methods now exist, they still have to be deployed on a wide scale. Non-repudiation by the sender can be achieved through integrity and authentication mechanisms. For non-repudiation by the recipient there are no real mechanisms at current, although some research is going on to have at least receipt acknowledgements, also called message disposition notifications, for email. [10]

Privacy issues can include secrecy of the content of information being exchanged, which can be reached by confidentiality mechanisms, or performing transactions in an anonymous way. For the latter there exist some ongoing developments and prototypes for electronic payments. [26]

Availability and reliability on the Internet is threatened by network hardware failures or by network congestion due to too little bandwidth being available or lack of traffic control. One solution is link or path redundancy for creating (fault) resilience and performing load-balancing. [16] Solutions for network congestion are putting less packets on the network or increasing the available bandwidth. The former is done by the TCP sliding window mechanism provided in standard TCP [10]. Since this mechanism offers a poor efficiency [4] new algorithms and extensions [3][10] were introduced in TCP and more efficient TCP-implementations have been developed, like TCP Vegas [27] and Indirect-TCP (I-TCP) [28]; but these implementations are not widely deployed yet. Increasing the bandwidth can be done by upgrading the networks, using faster network media and solutions for running IP more efficiently over these media, as already discussed previously.

Since the traditional Internet was not originally designed with security in mind, it was designed as a research environment, a lot of protocols, not to say practically all protocols contain serious security flaws, which may give rise to serious security breaches. [29] In addition to the security mechanisms mentioned above a lot of researchers have come up with security extensions for the existing protocols, like TCP, FTP, Telnet, DNS, SNMP, SMTP and the like. [3] [10] Another popular object of research forms the public key technology. [3] [10]

From this discussion of security technology, it should be clear that a lot of various security mechanisms have been developed. However, in general we can not see a wide-spread use or availability of the existing security mechanisms into commercial applications or services at this moment, but we can however expect that this security technology will become widely available and will be used in the near future.

4. Conclusion

Up to now the TCP/IP protocol suite and the Internet infrastructure were not capable of fulfilling the requirements put forward by the users and services of a highway. However it must be clear that the Internet certainly has future potential for becoming one. The development and deployment of IPv6, the adjustments made to run IP over new and faster media, thus upgrading the infrastructure to obtain higher bandwidths, the protocols added to support new services, and the introduction of various security mechanisms and security tools will make the Internet more into a highway, capable of supporting varying interactive multimedia services. New applications are continuously being developed and improved, turning the Internet in an integrated services digital network, approaching a highway.

Whether the Internet will or will not become or will be part of the information highway, it has provided and will be providing the community a large number of lessons about what technology works, what does not and why. Certainly many of the technologies ‘tested’ on the Internet will be part of every information highway.

References

Extended reference information with respect to RFCs [3] and Internet Drafts [10] can be found at URL: http://wcc.ruca.ua.ac.be/~didem/nocrefs.htm

[1] R. Kalakota and A.B Whinston., Frontiers of Electronic Commerce. ISBN: 0 201 84520 2. Addison-Wesley Publishing Company, Reading, 1996.

[2] Internet society, Frequently asked questions: what is the Internet? February 23, 1995, URL: ftp://ftp.isoc.org/isoc/faq/what-is-internet.txt

[3] An index of RFCs, URL: http://ds.internic.net/ds/rfc-index.html

[4] A.S. Tanenbaum, Computer Networks. ISBN:0 13 394248 1. Prentice-Hall Inc., Upper Saddle River, 1996.

[5] O. Farah (ed.), Framework for National Information Infrastructure Services. NIST Internal Report 5478, National Institute of Standards and Technology, Gaithersburg, MD, July 1994.

[6] B.G. Lee, M. Kang and J. Lee, Broadband Telecommunications Technology. ISBN: 0 89006 866 6. Artech House, Boston, 1996.

[7] M. Arango and M. Cortés, Guaranteed Internet Bandwidth, Conference Record of the 1996 IEEE Global Telecommunications Conference, London, 18-22 November 1996, 862-866.

[8] K. Maxwell, Asymmetric digital subscriber line: interim technology for the next forty years, *IEEE Communications Magazine* **10** (1996) 100-106.

[9] R.M. Hinden, IP next generation overview, *Communications of the ACM* **6** (1996) 61-71.

[10] An index of Internet Drafts, URL: http://www.ietf.cnri.reston.va.us/1id-abstracts.html

[11] J.A. Adam (ed.), Upgrading the Internet: extensive software improvements are coming to the worlds computer network, *IEEE Spectrum* **9** (1995) 24-29.

[12] ITU-T, Recommendation I.371, Traffic control and congestion control in B-ISDN. Perth, November 1995.

[13] S. Deering, IP Multicast and the MBone: enabling live, multiparty, multimedia communication on the Internet, December 95, URL: ftp://parcftp.xerox.com/pub/net-research/mbone/mbone-talk-dec95.ps

[14] H. Eriksson, MBone: the Multicast Backbone, *Communications of the ACM* **8** (1994) 54-60.

[15] M.R. Macedonia and D.P. Brutzman, MBone provides audio and video across the Internet, *IEEE Computer Magazine* **4** (1994) 30-36.

[16] P. Jurg and E. Zegwaart, Het Internet als Digitale Snelweg: de realiteit. ISBN: 90 71894 88 6. Otto Cramwinckel, Amsterdam, 1995.

[17] R. Rada, Sharing standards: consensus versus speed, *Communications of the ACM* **10** (1995) 21-23.

[18] C. Perkins, A. Myles and D.B. Johnson, IMHP: a mobile host protocol for the Internet, *Computer Networks and ISDN systems* **4** (1994) 479-491.

[19] F. Teraoka, K. Uehara, H. Sunahara and Murai Jun, VIP: a protocol providing host mobility, *Communications of the ACM* **8** (1994) 67-75.

[20] L. Landweber, International connectivity, version 15, 15/06/96, URL: ftp://ftp.cs.wisc.edu/connectivity-table/Connectivity_Table.text

[21] B.L. Egan, Information superhighways revisited, the economics of multimedia. ISBN: 0 89006 903 4. Artech House, Boston, 1996.

[22] ISO/IEC 7498-4:1989, Information processing systems - Open Systems Interconnection - Basic Reference Model - Part 4: Management Framework.

[23] A. Bhimani, Securing the commercial Internet, *Communications of the ACM* **6** (1996) 29-35.

[24] K. Schmid, Safe Guard IPN: Utimaco's solution for TCP/IP network security, Presentation held at Utimaco's Security Day, Diegem, Belgium, September 1996.

[25] P. Zimmerman, The official PGP user's guide. ISBN: 0 262 74017 6. MIT Press, Cambridge, 1995.

[26] P. Panurach, Money in electronic commerce: digital cash, electronic fund transfer, and ecash, *Communications of the ACM* **6** (1996) 45-50.

[27] L.S. Brakmo and L.L. Peterson, TCP Vegas: end to end congestion avoidance on a global Internet, *IEEE Journal on Selected Areas in Communications* **8** (1995) 1465-1480.

[28] A. Bakne and B.R. Badrinath, I-TCP: indirect TCP for mobile hosts, Proceedings of the Fifteenth International Conference on Distributed Computer Systems, IEEE, 1995, 136-143.

[29] S.M. Bellovin, Security problems in the TCP/IP protocol suite, *Computer Communication Review* **2** (1989) 32-48.

Core and ATM Networks
D.W. Faulkner and A.L. Harmer (Eds.)
IOS Press 1997 © AKM

Performance Comparison of Classical IP and LAN Emulation over the Mons-Valenciennes ATM Network

L. Meuleman*, J. Hanoteau**, N. Point***, L. Wuilmart*, M. Lamquin**

* *Multitel - Télécom,* ** *Service d'Electromagnétisme et de Télécommunications,* *** *CCTI*

Faculté Polytechnique de Mons, Boulevard Dolez, 31, B-7000 Mons (Belgium)

G. Bisiaux, A. Mayeur

Université de Valenciennes et du Hainaut Cambrésis (France)

Abstract. A performance comparison of the Classical IP over ATM and LAN Emulation approaches is presented in this paper. The network under test is part of the Mons-Valenciennes ATM network, an interconnection project at 155 Mbit/s between three universities in Mons (Belgium) and the University of Valenciennes (France).

1. Introduction

Two approaches are currently available to run TCP/IP applications over ATM networks : the *Classical IP over ATM* approach, proposed by the Internet Engineering Task Force (IETF) [1,2], and the *LAN Emulation* specification, proposed by the ATM Forum [3]. The Classical IP approach consists in the direct mapping of both IP datagrams and addresses into ATM equivalents. The LAN Emulation emulates a MAC service to higher-layer protocols (i.e. not only IP).

A quick comparison of these two solutions is presented in table 1. From a performance point of view, Classical IP over ATM is expected to provide better throughput figures due to a larger MTU (Maximum Transfer Unit) size : an emulated Ethernet indeed still uses 1500 byte-limited frames, whereas Classical IP over ATM defaults to a MTU size of 9180 bytes. Nevertheless, throughput figures are mostly relevant for file transfer-like applications. Other applications, like world wide web transactions, consist in many successive connection set-ups / tear downs. For these applications, connection set-up latency figures are more relevant.

Throughput and connection set-up latency measurements are presented and commented in this paper for both Classical IP over ATM and LAN Emulation approaches.

A rough comparison of Classical IP over ATM and LAN Emulation is described in the following table.

	LAN Emulation	Classical IP over ATM
Aim	Support of any network protocols and applications	Support of TCP/IP protocols and applications
Date	1Q95	1Q94
Document	LANE 1.0	RFC 1483&1577
Origin	ATM Forum	IETF
Broadcast domain	Emulated LAN (ELAN)	Logical Internet Subnet (LIS)
Interconnection with legacy networks	Layer-2 devices (bridges, switches)	Layer-3 devices (routers)
Address Resolution	MAC → ATM	IP ↔ ATM
ATM Adaptation Layer	AAL5	AAL5
Multicast	Supported	Not supported
Performances	Emulated Ethernet MTU size 1500 bytes	Default MTU size 9180 bytes
Quality of Service	None	None
Upgrade path	Smooth	Forklift

Table 1 : A rough comparison of Classical IP and LAN Emulation

2. The Mons-Valenciennes ATM Network

The ATM network under test is part of the Mons-Valenciennes ATM network, an interconnection project at 155 Mbit/s between three universities in Mons (Belgium) and the University of Valenciennes (France).

The main link is a 50-km optical fibre trunk at SDH STM-1 (155.52 Mbit/s) between Faculté Polytechnique de Mons and Université de Valenciennes. Distance-teaching, collaborative working, high-quality video-conferencing will be among the main applications. The network is also an ideal testbed for research and education.

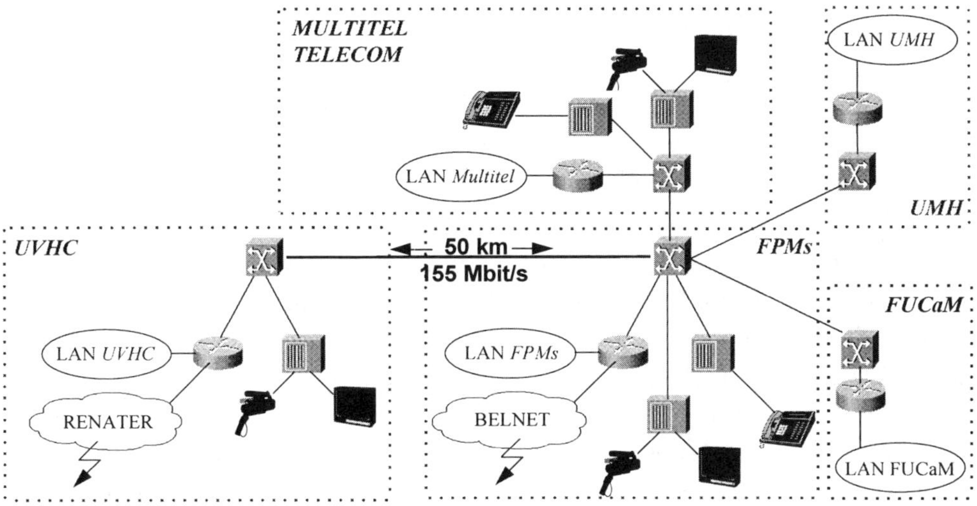

Figure 1 : The Mons-Valenciennes network

3. Logical Internet Subnetwork - Emulated LAN

The test configuration is shown on figure 2.

Two workstations were used in the test : a Hewlett-Packard C160 model running HP-UX 10.0 and a SUN SPARC running SUN OS 5.4. Both workstations were equipped with Fore Systems ATM interface cards and attached to a Fore Systems ASX200BX ATM switch.

The ATMARP (ATM Address Resolution Protocol) server was installed on the Mons ATM switch; whereras LANE servers (i.e. LAN Emulation Configuration Server - LECS, LAN Emulation Server - LES, and Broadcast and Unknown Server - BUS) were installed on a Cisco 7000 router located in Valenciennes. A *Logical Internet Subnetwork* (roughly, an ATM LAN according to the Classical IP over ATM specification) and an *Emulated LAN* (the LAN Emulation equivalent) were thus simultaneously present on the same physical network.

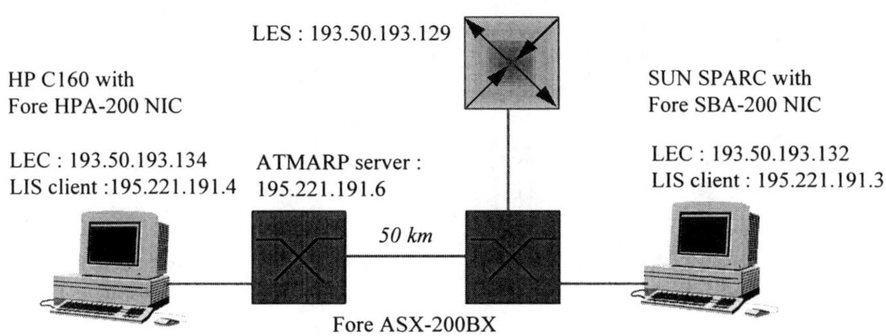

Figure 2 : Logical Internet Subnetwork and Emulated LAN under test

4. Throughput measurements

Throughput measurements were performed with *Netperf*, a well-known public domain network benchmarking software [4].

Netperf runs in a client/server fashion : the client *Netperf* requests a test to *Netserver* installed on a distant host. A control connection is first established between the two parties to exchange test parameters. A separate connection is then opened for the measurement itself.

The tests described here consist in bulk data transfer using TCP. The message size to be sent by the sending TCP is 8 kbytes, and send (S) and receive (R) socket buffers are modified to provide the following charts. Netperf allows also for the Nagle algorithm to be disabled (see, for instance, [5]).

Table 2 and 3 show that a 56 Mbit/s throughput can be obtained with Classical IP over ATM, whereas LANE reaches only 28 Mbit/s. This ratio of 2:1 is generally observed for any message size (cf. figure 3). Note that no dramatic decrease in TCP throughput, as mentioned in [6], was observed during the measurements. This shows that this deadlock situation is linked to the TCP implementation (i.e. to the UNIX flavour). Finally, disabling the Nagle algorithm does not have a significant influence on the results.

S \ R	4096	8192	16384	24576	32768	40960	49152	57344
4096	17.87 17.78	27.00 26.50	27.38 27.94	28.01 27.83	27.39 27.88	27.90 27.55	27.49 27.87	27.97 27.79
8192	17.98 17.86	26.93 26.88	27.35 27.44	27.62 27.98	27.69 27.95	27.60 27.73	27.99 27.85	27.83 27.85
16384	17.81 18.18	26.96 26.80	27.64 27.81	27.48 27.71	27.87 27.80	28.01 27.64	27.90 27.65	27.70 28.00
24576	18.12 18.24	27.11 26.80	27.30 27.84	27.75 27.81	27.86 27.77	27.82 27.69	27.93 27.68	27.74 27.50
32768	18.06 18.20	27.03 26.69	27.98 27.86	27.61 27.86	27.92 27.95	27.86 27.69	27.26 27.92	27.79 27.70
40960	18.16 18.27	26.92 26.72	27.84 27.49	27.71 27.80	26.90 27.85	27.90 27.97	27.70 27.82	27.75 27.67
49152	18.25 18.40	26.89 26.66	27.32 27.20	26.53 26.49	26.61 27.64	27.74 26.98	27.75 27.95	27.81 27.89
57344	18.32 18.16	26.46 27.09	27.37 27.85	28.00 27.92	27.93 27.82	27.57 26.77	27.91 27.18	27.85 27.85

Table 2 : Throughput measurements with LANE - upper row : Nagle algorithm enabled, lower row : Nagle algorithm disabled

R / S	4096	8192	16384	24576	32768	40960	49152	57344
4096	20.45 20.39	33.72 33.66	49.68 48.85	49.38 49.02	55.22 56.06	55.47 56.00	55.89 55.40	56.28 55.87
8192	20.51 20.32	33.64 33.47	48.81 49.12	49.41 49.38	55.63 56.09	55.88 56.00	56.27 55.92	56.46 56.39
16384	20.78 20.90	33.05 33.31	49.26 49.64	49.97 49.23	55.80 55.38	55.80 56.00	55.93 55.98	55.89 55.67
24576	20.87 20.87	33.40 33.27	48.66 48.39	48.74 49.40	54.73 55.86	55.90 55.97	56.18 55.88	56.41 55.78
32768	20.89 20.80	33.57 33.30	48.71 48.64	49.20 49.21	55.89 56.10	56.06 55.26	56.12 55.90	57.13 55.69
40960	20.84 20.72	33.42 33.57	48.65 48.40	49.37 49.04	55.91 55.65	55.12 56.04	55.94 55.82	56.38 55.15
49152	20.93 20.86	33.46 33.63	48.71 49.74	49.47 50.44	55.28 55.12	55.76 56.01	55.88 55.86	55.71 55.93
57344	20.88 20.90	33.51 33.34	48.47 48.65	49.49 48.98	55.65 53.75	56.04 55.13	55.86 54.42	53.08 54.24

Table 3 : Throughput measurements with Classical IP - upper row : Nagle algorithm enabled, lower row : Nagle algorithm disabled

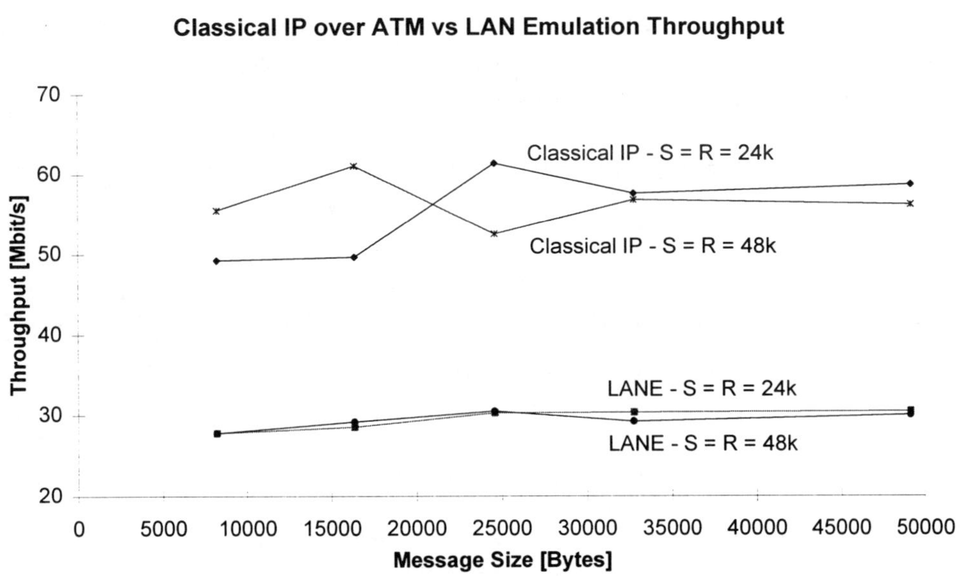

Figure 3 : Classical IP over ATM and LAN Emulation throughputs with respect to message size

5. Connection set-up latency measurements

Connection set-up latency figures can also be obtained thanks to Netperf. The kind of test run here is a so-called connect/request/response test which mimics the Hyper Text Transfer Protocol (HTTP) used by most web servers. In this test, a TCP connection is established prior to each request/response. This test was run for two minutes to reach a steady state behaviour. The request and response packet were 32 and 1024 byte-long, repectively.

The results are the following : 107.27 sessions per second occured with Classical IP over ATM, 116.22 sessions per second with LAN Emulation.

The connection set-up latency problem is indeed handled differently by Classical IP and LANE specifications. While waiting for the address resolution phase to be completed, any LAN Emulation Client willing to transmit can forward data through the BUS server. When the address resolution response is received, the flush protocol ensures successful switching between the default path and the direct ATM connection.

A direct IP to ATM resolution is performed in Classical IP. No default path is used for transmitting data while waiting for the address resolution completion.

In the case of the connect/request/response test, the first connection establishment makes the difference in the observed figures because it relies on the address resolution procedure. Subsequent connections are then established through a cache look-up.

The traditional *ping* tool was then used to estimate connection set-up latency. Figure 4 shows the kind of sequence of commands that can be used (in this case, for the LAN Emulation logical interface). The results are summarized in table 4. 10 connections were established for both Classical IP and LANE interfaces.

Connection set-up latency [ms]	Classical IP over ATM	LAN Emulation
minimum	67	2
mean	88	2.5
maximum	148	3

Table 4 : Classical IP over ATM and LAN Emulation connection set-up latency

```
# ifconfig el0 down
# ifconfig el0 up
# ./elconfig show -configured

ELAN(s) on Adapter 0
-----------------------
ELAN            NI        ATM Addresses
----            --        -------------
default         el0       LEC: 0x47.0005.80.ffe100.0000.f21a.2d07.002048067f5b.00
                          LES: 0x47.0005.80.ffe100.0000.f21a.21dc.00000c312904.01
                                direct vpi.vci=0.148
                                distribute vpi.vci=0.149
                          BUS: 0x47.0005.80.ffe100.0000.f21a.21dc.00000c312905.01
                                send vpi.vci=0.150
                                forward vpi.vci=0.151
                          LECS: 0x47.0079.00.000000.0000.0000.0000.00a03e000001.00
                          state: operational
```
At this state, the emulated LAN is operational (control connections to the LANE servers are established).
```
# ping -s 193.50.193.132
```
At this state, address resolution has to be performed prior to the first packet exchange.
```
# ./elarp show -all
00:20:D4:00:03:80 at 0x47.0005.80.ffe100.0000.f21a.21dc.0020d4000380.00
        ElanName=default flags=() vpi.vci=0.180
# ping -s 193.50.193.132
```
From this point, address resolution is performed by a cache look-up.

Figure 4 : Sequence to estimate connection set-up latency

6. Conclusions

A comparison of the Classical IP over ATM and LAN Emulation approaches was presented. Performances figures were obtained thanks to the use of public domain TCP/IP tools.

TCP throughput using Classical IP can reach twice the figures obtained on an emulated Ethernet. This is due to the larger MTU size used by Classical IP. On the contrary, LAN Emulation takes the lead when connection set-up latency is considered, since the LANE protocol allows for a default data path to be used before the completion of the address resolution phase.

From the performance point of view adopted in this paper, one might conclude that the Classical IP over ATM approach is the only choice. However, efficiency is not the only issue (see table 1): for instance, Classical IP runs TCP/IP applications only whereas LANE allows for other popular protocols to run over ATM.

Furthermore, LANE offers a smoother upgrade path. Firstly, because evolution of legacy LANs towards ATM can be progressive (legacy LAN workstations can be first attached to ELANs through ATM/LAN switches before replacing the Ethernet NIC by an ATM NIC). Secondly, because the future Multi Protocol Over ATM (MPOA) specification from the ATM Forum is built on LANE [7] and is the main candidate to route efficiently any layer-3 protocol over ATM.

Finally, it is important to note that all these results were obtained in a non-congested situation. TCP throughput in case of cell loss deserves a lot of attention. Large MTUs generally cause problems in this case (see for instance [8]).

Acknowledgements

Multitel - Télécom is funded under *Objectif 1* funds of *Région Wallonne* and *European Commission.*

References

[1] M. Laubach « Classical IP and ARP over ATM », IETF RFC 1577, January 1994, http://ds.internic.net/rfc/rfc1577.txt.

[2] J. Heinanen « Multiprotocol encapsulation over ATM Adaptation Layer 5 », IETF RFC 1483, July 1993, http://ds.internic.net/rfc/rfc1483.txt.

[3] ATM Forum « LAN Emulation over ATM version 2 - LUNI specification ».

[4] Netperf web page, http:// www.cup.hp.com/cgi-bin.

[5] W.R. Stevens « TCP/IP Illustrated - volume 1 : The Protocols », Addison-Wesley, 1995.

[6] D.E. Comer, J.C. Lin « TCP buffering and performance over an ATM network », Internetworking : Research and Experience, vol. 6, 1-13, 1995.

[7] ATM Forum « Multi-Protocol Over ATM Version 1.0, ATM Forum/BTD-MPOA-MPOA-01.11, Draft Version » February 1997.

[8] A. Romanow, S. Floyd « Dynamics of TCP traffic over ATM networks », ftp://ftp.ee.lbl.gov/papers/tcp_atm.ps.Z.

Core and ATM Networks
D.W. Faulkner and A.L. Harmer (Eds.)
IOS Press 1997 © AKM

An SDL model for simulation and evaluation of LAN emulation over ATM*

J. Ramón VIDAL, Jorge MATAIX
Universidad Politécnica de Valencia, 46071 Valencia, Spain
E-mail: jrvidal@dcom.upv.es, jmataix@dcom.upv.es

Abstract. In this paper we present a simulation model for evaluating data service over ATM with LAN Emulation. The aim of this model is to study the performance of data transfer services in ATM networks when offering best effort service. We are interested in the performance from the point of view of the upper layer protocols like TCP, and of the applications. To develop the simulation model, we propose a formal description technique approach based on SDL language. By using this formalism, we describe the components of the protocol stack in this system. A discrete event simulator can be generated from this model by an automatic tool. We also show some simulation results to illustrate some of the points under study.

1. Introduction

Future broadband ISDN networks will be able to carry traffic from different applications. To do this, the Asynchronous Transfer Mode (ATM), will provide several service classes: constant bit rate (CBR), variable bit rate (VBR), available bit rate (ABR) and unspecified bit rate (UBR). The two last classes are designed to carry traffic from data applications. This type of traffic is called 'best effort', and it is usually generated by host end systems running TCP/IP protocols.

TCP is the most widely used transport protocol in data networks. It was designed for low and medium speed networks, but it will not disappear with the introduction of broadband networks. The study of TCP performance in ATM networks is therefore interesting. This study is mainly concerned with how the TCP congestion control mechanisms interwork with ATM best effort service classes.

To allow existing TCP/IP networked applications to run over ATM, some interconnection solutions have been proposed. One of the most straightforward is LAN Emulation (LANE), which provides a service identical to the MAC service of standard IEEE 802.3 and IEEE 802.5 networks. This allows standard bridges to connect through an ATM network as if it was a LAN segment, and so allows standard protocols like TCP to run over ATM without changes.

In this paper we describe the problems related with this situation and present a formal model of the protocol layers intervening in interconnection of TCP hosts through ATM, with ABR service class and LANE.

* This work is financed by CICYT project number TIC96-0680

2. TCP over ATM performance

Service classes UBR and ABR do not provide a quality of service guarantee. So users are exposed to bandwidth fluctuations depending on network resource availability. This can lead to congestion in the ATM layer, and if this occurs some upper layer segments can be lost. If the ATM service user is TCP, it can only detect congestion in the network after the loss of some packets. This feedback is very slow, and can provoke an arbitrarily low throughput, as has been shown in some studies [1].

Over UBR service class, an added problem is the segmentation of TCP packets into ATM cells. In case of network congestion, some ATM cells can be lost, and the packets to which this cells belong will arrive incomplete and be discarded by the receiver. But some cells of these corrupted packets will be unnecessarily transmitted, making the congestion worse. To alleviate this problem, some mechanisms in the ATM switches are proposed. One is the Partial Packet Discard (PPD), in which, once a cell has been dropped by a switch, all the subsequent cells of the packet are then dropped. Another is Early Packet Discard (EPD), in which entire packets are dropped when the switch buffer reaches certain thresholds. By doing this, packets partially accepted into the buffer are protected from being dropped.

PPD and EPD improve TCP throughput, but require large buffers and are not efficient under severe congestion. UBR service has these limitations because PPD and EPD cannot solve the feedback problem, so congestion probability is not sufficiently reduced.

A better performance is achieved with ABR traffic class [3]. ABR provides a congestion control mechanism, so the traffic source can adjust its traffic rate to the network availability – and this avoids unnecessary frame looses. In LAN environments, with low transmission delays, ABR achieve this goal without large switch buffers. But in WAN environments, TCP throughput depends of the switch buffer size [6]. Other aspects of TCP performance, like fairness between connections, are also improved with ABR .

Some studies have been made on TCP over ABR, showing that TCP performance can be improved by tuning ABR source parameters [4] [5]. These parameters determine the source behavior in response to the network [2]. Nevertheless, ABR can be implemented in switches with. The potential benefits and disadvantages of each different algorithm from the point of view of TCP performance is a point to be investigated.

3. Congestion control in TCP and in ATM

When TCP is running over ATM with ABR service class, two mechanisms for flow and congestion control are simultaneously active: the TCP end-to-end sliding window [8] and the ATM layer feedback control [7], based on a resource management cell loop.

TCP implements flow control by sliding windows, whose size is the maximum amount of correlative bytes sent without acknowledgment (ACK). Two windows are used: a transmission window, specified by the receiver in the ACKs; and the congestion window, that is adjusted to recover from segment loss. Data segments are sent until either the transmission or the congestion window end is reached. When a data segment is lost, TCP reduces the congestion window size to one segment (this is called slow start); and increases this window with every received ACK. This increase is made in two phases. First, the size doubles every round trip time; and secondly when it reaches a threshold, it increases linearly. The threshold is set to half the congestion window size at the moment when a segment loss is detected.

Segment loss is detected in two ways: by time-out or by out-of-order ACKs. To set time-outs, the sender estimates the round trip time from the ACK reception times, and sets a timer for every segment transmitted. If a timer expires, the segment associated with this timer is considered lost. The other possibility is that several non-correlative ACKs are received, meaning that at least one preceding segment is lost. This mechanism is called fast retransmission, and is predominant when running TCP over ATM.

At the same time, the ATM layer implements a feedback control between end systems (ABR users) and the network. The ABR source end system periodically sends control cells, called resource management (RM) cells. These cells go along the ABR connection and are returned by the destination end system. On the way back, the switches write some information into them allowing the source to dynamically adjust its rate.

The ABR end system behavior is standardized by the ATM Forum [2], specifying how the end systems must send RM cells, return RM cells to the peer system, and adjust cell emission rate depending on the information received in the RM cells.

The ATM switches must calculate the feedback information and write it in RM cells. They can operate in two modes: binary mode or explicit rate (ER) mode. In binary mode, the switches only write some flags in the RM cells to indicate some congestion conditions. In ER mode, the switches have to estimate periodically what cell rate can be allowed for the connections, and write this in the RM cells. This estimation can be implemented with different algorithms, each of which has certain advantages and drawbacks [7].

There are several factors that limit the performance of TCP over ABR. The most important is the reaction time of the ABR congestion control algorithm. When congestion occurs in a switch, because bandwidth availability changes, the ABR sources have to adjust their cell rates. But informing the sources about the new network conditions, by means of the next RM cell, requires a certain time. During this time, the sources can emit cells at the formerly allowed rate, and so the switch buffers are filled with cells. If the switch buffer limit is reached, cells are discarded, and so TCP segments are lost.

The TCP recovery from segment loss is slow. This means that after segment loss, TCP needs a relatively long time to adjust its congestion window to the optimal size. During this time the available bandwidth is not used. This is because TCP connections react to segment loss, caused through congestion, by reducing its congestion window to a minimum.

There are other factors concerning ABR performance, like fairness between connections, and link usage during transitory periods. But, from the point of view of TCP, these are not as important, because the main problem with TCP is its slow response to congestion. This means that, when a single cell is lost, TCP needs a long time to recover and use the available bandwidth again.

4. Network interconnection through ATM: LAN Emulation

One way to run TCP applications over ATM networks is with LAN Emulation (LANE) over ATM. LANE is specified by the ATM Forum [9], and emulates the service offered by IEEE 802.3 and IEEE 802.5 networks. It provides a connectionless service with multicast capability over the ATM connection oriented service.

LANE service is offered by clients (LECs). To participate in the emulated LAN, the LECs must register in a LANE server (LES). A LES has a table with the ATM addresses of participating LECs, and uses it for the resolution of LAN addresses into ATM addresses. When a LEC is requested by the user to send data, it has to send it by the appropriate ATM connection. If the connection is not set, the LEC has to set it; and if the LEC destination address is not known, it must resolve it together with the LES. During the address resolution time, the LEC may send data through a Broadcast and Unknown Server (BUS), that is also used to send data to non-registered LAN addresses and to multicast addresses.

At the start time, the LECs set ATM connections with the LES and the BUS, and then they register in the LES all the LAN addresses to be served. LECs may also operate in proxy mode, needed when the LANE user is a bridge, and then they serve any LAN address.

The ATM service is accessed by LANE layer through the ATM Adaptation 5 (AAL5) layer. This provides non-real time service for data applications. The AAL5 is made up of: the Segmentation and Reassemble (SAR) sublayer, for segmenting frames into ATM cells and reassembling them at the destination; and the Convergence Sublayer (AALCS).

The connections between LECs use a best effort class service. This could be of UBR or of ABR class. A low probability of congestion at the switches is desirable, and so best performance will be achieved with ABR connections operating in explicit rate mode.

5. Simulation methodology with SDL descriptions

To study by simulation the interoperation of all the mechanisms mentioned, it has been necessary to develop a complex model. This was intended to fulfill several conditions:

- It had to be structured in the same way as the layered protocol stack. So it was necessary to separately describe each layer, in a way that the layer entities could be configured to model end systems.
- The descriptions had to be easily readable, so that the system behavior could be easily understood. This saves work in the test and validation phases. To accomplish this, a formal language with real-time features was necessary.
- Because several parts of the description will be changed very often, an easy to maintain description was is desirable, in spite of its complexity.
- Several parts of the system are standard protocols. Therefore it was desirable that this parts (mainly protocol layers) could be easily reused in further models.
- The model description must be simulated in an efficient way, by automatically generating executable code from the description.

All these requirements are accomplished by the Specification and Description Language (SDL) standardized by the UIT-T [10], together with an automatic tool able to generate high level language code from the SDL descriptions.

SDL is a process-based language, in which processes are modeled as extended state machines, communicating with each other by asynchronous signals. Processes have states and variables, and receive and send signals through defined routes. The processes can be structured into blocks, and the blocks can be nested. The latest SDL version (SDL-92) is object-oriented, so it is possible to define a class for any layer protocol entity. With these classes, end systems can be easily described.

An automatic tool can generate C code from the descriptions, which can be compiled and linked with the needed libraries. These kind of tools have other features, like graphical editors, analyzers, simulators, validators, etc.

In SDL, abstract data types and their operators can be specified. The operator behavior is not easily described in SDL, but they can be implemented in C and linked with the generated code. To run a discrete-event simulation of the system, there are some abstract data types that have to be used in the description; like random numbers, types for file manipulation and queue-processing. These types are implemented in external libraries.

The random number library is used to generate random times for modeling, for instance, the cell jitter at the physical layer, or the message interarrival time from the applications. The file manipulation library is needed to read the simulation parameters from configuration files at the start time – and to write into files traces of the events under study. The queue-processing types are needed to model buffers in the communicating systems.

6. The SDL simulation model

The model consists of a set of classes. Each of them describes a protocol layer entity, an end system, or a network configuration. They are the following :

- LAN user : includes an application traffic generation model and the TCP layer.
- ATM layer : describes the behavior of the user side of the ATM layer.
- SAR sublayer : describes the behavior of the SAR entity.
- AALCS sublayer: describes the behavior of the AALCS entity.
- SVC sublayer: describes the behavior of the Switched Virtual Circuit (SVC) entity.
- LEC, LES and BUS: which form the LANE layer. Each of them contains a specific LANE entity, and instances of ATM, SAR, AALCS and SVC.
- ATM host : includes an instance of the LEC and an instance of the LAN user. A host models a LEC and one or several TCP traffic sources.
- ATM network : modeling the network side of the ATM User Network Interface (UNI). It contains processes modeling the switch behavior and link delays. This class can be configured with several nodes linked in chain, for studying the most common configurations used for performance evaluation.
- Simulation configuration block: used to configure the simulation scenario.

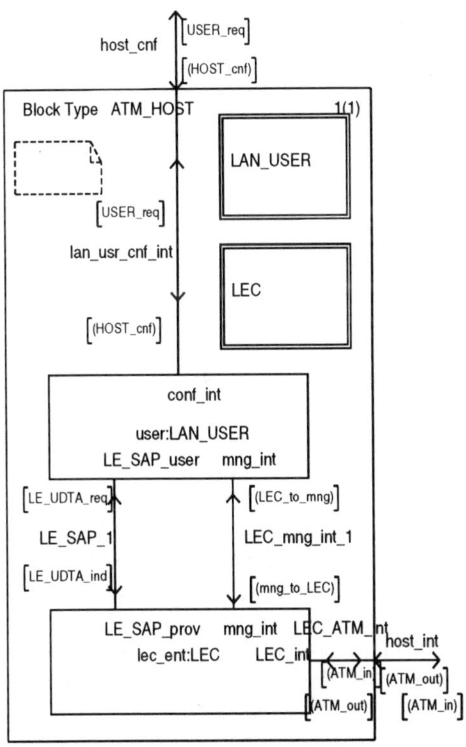

Figure 1. ATM Host block type

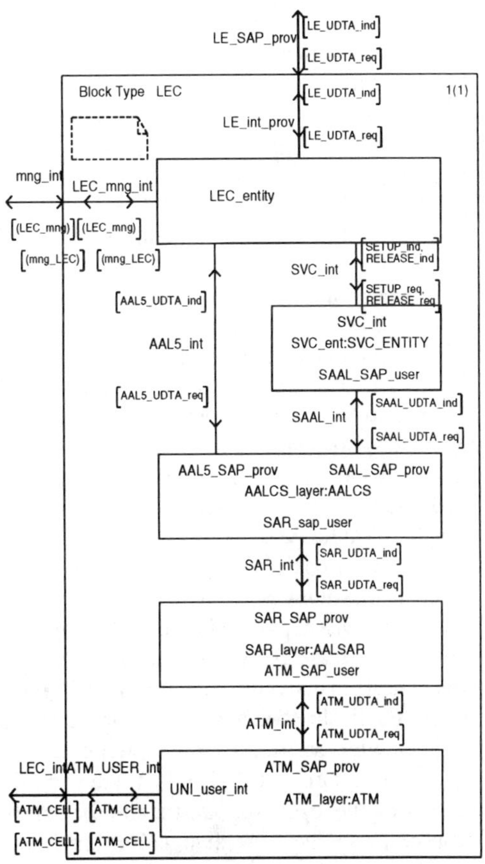

Figure 2. LEC block type

The system to be simulated is composed of an instance of the ATM network, the BUS, the LES, the simulation configuration block, and several instances of ATM hosts. At start time, the simulation configuration process reads the simulation parameters from a configuration file and configures the hosts (addresses, operation mode, traffic sources, etc.).

Some parts of the model are shown in figures 1, 2 and 3. In figure 1, the host structure can be seen, containing a LEC instance and a LAN user block instance. Figure 2 shows the SDL description of the LEC class (in SDL called block type) with the protocol stack used in the LEC end system, and where each layer is an instance of another block type. Figure 3 shows a simulation scenario with two hosts.

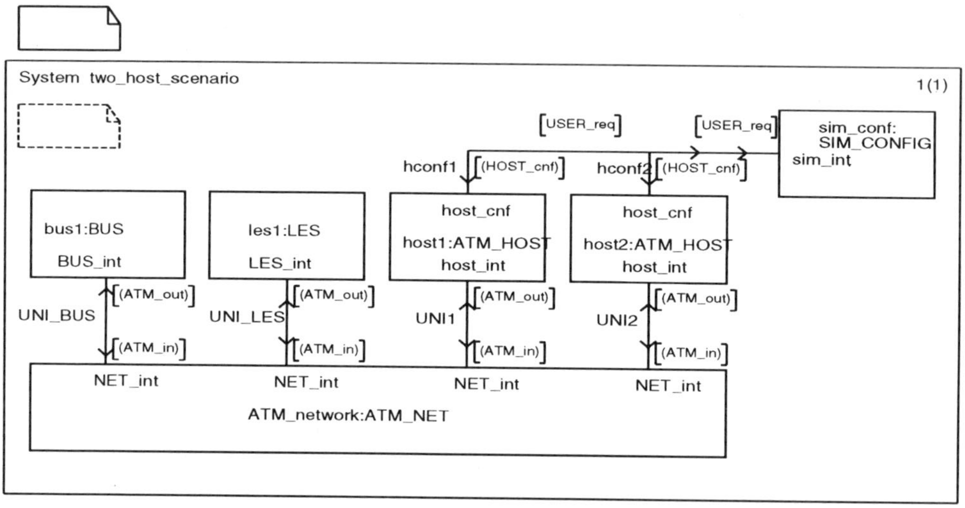

Figure 3. Simulation scenario with two hosts

7. Some results

As an example of the use of the model, we present the simulation results of a simple configuration, in which some of the described dynamics can be seen. This configuration consists of two ATM switches, connected by a link of 100 Km at 150 Mb/s. Each switch has five hosts connected by 100 Km at 150 Mb/s links. These link lengths mean a delay of 0.5 ms each, and allows as to study the problem of feedback delay in ABR service, and its influence on TCP. The switches implement the ERICA algorithm to provide ABR service in Explicit Rate mode.

Each host contains a LEC, operating in host mode (not proxy), and a TCP traffic source. The hosts connected to one switch send traffic to the host at the other switch. So the trunk link will be crossed by cells belonging to five unidirectional TCP connections.

To illustrate the problem of switch congestion, we simulate a change in the link bandwidth, occurring when all the TCP connections have reached the steady state. This is achieved by instantaneously reducing the bandwidth by one half (75 Mb/s). This could happen when some connections with other traffic classes (CBR or VBR) are set.

In figure 4, the five TCP connections throughput is plotted. This is measured in b/s of correlative bytes received at the TCP receiver entity (goodput). It can be seen that after a transitory period, the five connections fairly share the available bandwidth. When the bandwidth is reduced, there is a new transitory stage, and the goodput is adjusted to the new conditions.

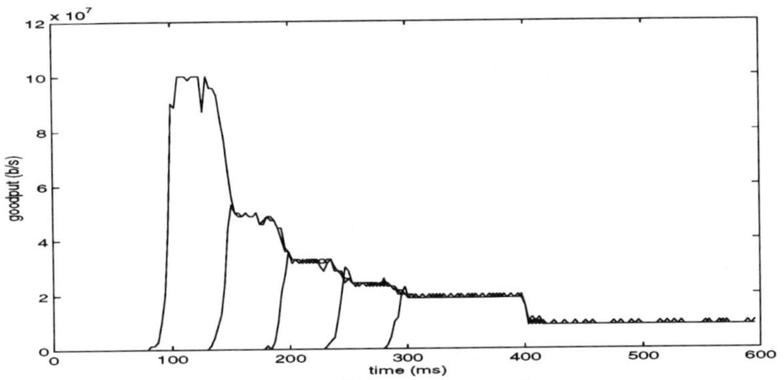

Figure 4. TCP connections goodput

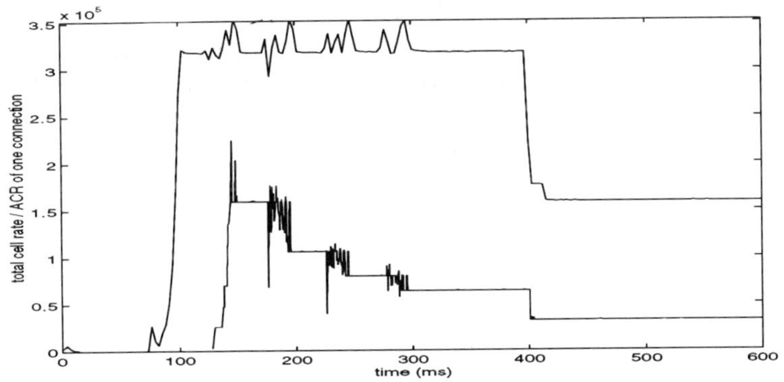

Figure 5. VCC cell rate and link cell rate

Figure 5 shows how the source cell rate is adjusted by the ABR feedback mechanism. Here the available cell rate (ACR) of one of the sources is plotted in b/s. The ACR is communicated to the sources by the RM cells returned in the feedback loop.

Figure 6 shows the number of cells queued at the first switch output buffer of the trunk link. There is a bottleneck is in this link between the switches. There are peaks in the buffer level when a new connection enters, and when the bandwidth is reduced. These peaks occurs because, under changing network conditions, the sources send cells at the former rates until the sources ACRs are adjusted. If this peak reaches the buffer limit, some cells are dropped, causing some TCP connections to enter the slow start phase.

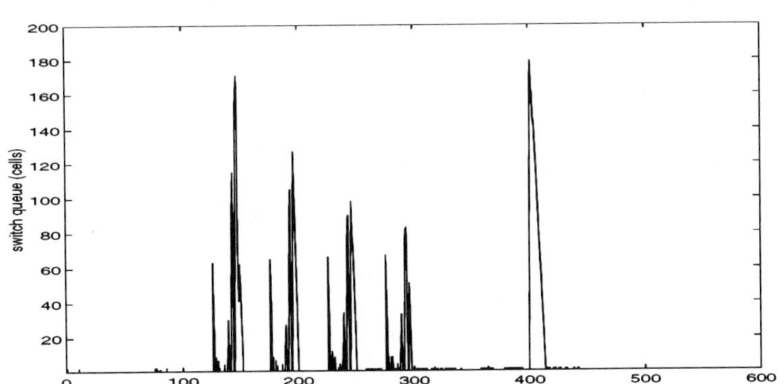

Figure 6. Number of cells in the switch buffer

In this example we can see that is worthwhile studying switch buffer dimensioning with the objective of achieving cell-loss free operation. Another possibility is to define a dropping strategy to minimize the effect on TCP performance.

8. Conclusions and future work

We have confirmed that the simulation model description in SDL formal language is a very useful technique for performance evaluation. It allows as to describe complex models that are very easy to test, validate, maintain and reuse. These models consist of sets of abstract classes, that can be instanciated to easily specify multiple simulation scenarios. Efficient discrete-event simulators can be generated automatically from these specifications.

The objective of the described model is to study the performance of the ATM best effort service (UBR and ABR) – from the point of view of the transport layer and the applications. In this paper, we have described some of the aspects under study, and we have shown some illustrative only simulation results. This study will go on to compare the different possibilities to implement the ATM services described. This possibilities include different algorithms for implementing ABR service; the switch buffer dimensioning; the ABR source parameters election; as well as others.

Acknowledgment

The authors would like to thank the members of the InterATM** group of the UPV for their contributions to this work.

References

[1] A. Romanow, S. Floyd, Dynamics of TCP over ATM Networks. *IEEE Journal on Selected Areas in Communications*, vol. 13, num. 4, 1995, pp. 633-641.

[2] R. Jain *et al.*, Source Behavior for ATM ABR Traffic Management: An Explanation. Department of Computer and Information Science. Ohio State University, Columbus. 1996.

[3] H. LI *et al.*, A Simulation Study of TCP Performance over ABR and UBR Services in ATM LANs. *IEICE Transactions on Communications*, vol. E79-b, no. 5 1996, PP 658-667.

[4] G.Hasegawa *et al.*, Performance of TCP over ABR Service Class. *Proceedings of the GLOBECOM'96*, 1996.

[5] G. Hasegawa *et al.*, Performance Evaluation and Parameter Tuning of TCP over ABR Service in ATM Networks. *IEICE Transactions on Communications*, vol. E79-b, no. 5 1996, pp. 668-683.

[6] S. Kalyanaraman *et al.*, Performance of TCP/IP over ABR Service on ATM Networks. *Proceedings of the GLOBECOM'96*, 1996.

[7] A. Arulambalan, X. Chen, Allocating Fair Rates for Available Bit Rate Service in ATM Networks. *IEEE Communications Magazine*, November 1996, pp. 92-100.

[8] D. E. Commer, Internetworking with TCP/IP, vols. I and II. Prentice Hall, 1991.

[9] ATM Forum Technical Committee, LAN Emulation over ATM. 1995.

[10] UIT-T Recommendation Z.100, Specification and Description Language (SDL). Helsinki, 1993.

** Information about the group in http://www.giatm.upv.es

Core and ATM Networks
D.W. Faulkner and A.L. Harmer (Eds.)
IOS Press 1997 © AKM

Strategic Network Planning for ATM Deployment - A Bell Canada Perspective

Stephen G. Bartlett and Ahmad Sultan
Bell Canada - Network Technology Planning
160 Elgin Street, Suite 540, Ottawa, ON, K1G 3J4 Canada
Email: sbartlet@post.bell.ca

Abstract. Network providers are facing pressures associated with large-scale ATM switch deployment and its impact on network architectures and investment priorities. Methods and findings of target architecture and transition planning research for ATM deployment in Bell Canada are described. Insights gained through this work span a broad range of issues and provide a compelling rationale for a phased switch deployment to form a distributed concentrator architecture, and for coordinated planning of ATM and transmission system architectures.

1. Introduction

ATM offers the potential to standardize on one network architecture defining the multiplexing and switching method, with SONET/STM providing the basis for the physical transmission standard for very high-speed rates. ATM also supports multiple Quality of Service (QoS) classes for different application requirements on delay and loss performance. Thus, the vision of ATM is that an entire network can be constructed using ATM and ATM Application Layers (AAL) switching and multiplexing principles to support a wide range of applications such as voice, Internet Protocol (IP), Frame Relay, video, imaging, and Circuit Emulation.

This paper addresses issues associated with strategic planning for ATM deployment in large metro inter-office networks where market drivers are first expected to occur. The metro core network referred to here may best be defined as the service non-specific inter-office infrastructure and technologies performing the ATM switching and transport functions for traffic leaving the serving area of a Central Office (CO). Adaptation to ATM takes place at the edge of the core network for a range of business and residential service traffic. This traffic is switched through the core network to its destination over high-speed inter-switch links provisioned on, for example, SONET transport systems.

Our vision is to develop the ATM core network to accommodate the existing and emerging services (e.g., analog Internet and Work at Home/Teleworking, high-speed Internet and Work at Home/Teleworking, high-speed LAN, DS3-UNI, etc.) with ATM adaptation generally taking place closer to the residential and business customers (see Figure 1). The network vision comprises all of the services and technologies, planned, deployed and evolved to optimize the interworking of each layer with the others to increase the efficiency of the network as a whole.

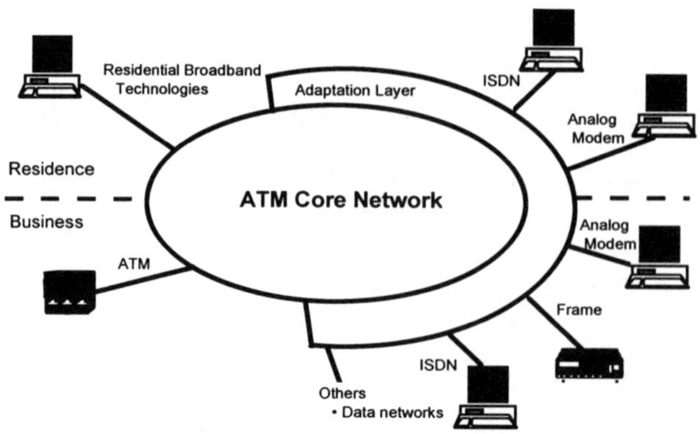

Figure 1. ATM adaptation evolves towards the access

2. Objective

Large regional business and high density (urban) residential markets are recognised as being the most at risk to competition. Research of network evolution scenarios indicates that ATM technology can be effectively employed as a means both of acquiring market share for emerging services and of reducing costs to ensure existing services remain competitive.

The evolution from circuit to cell-based metro core networks will present challenges and opportunities for network planners. If costs can be contained while ATM presence in the network is established, telcos can capitalize on the promise of cheaper service development and provisioning on a fully integrated network.

The objective of the research undertaken over the past two years by the Network Technology Planning group in Bell Canada is to develop the ATM deployment direction and strategies required to take best advantage of the benefits offered by this technology.

3. Method

An extensive modeling process is employed to gain perspective on ATM architectures, the associated switching and transport requirements, and planning costs.

The process begins with a series of forecasts for services expected to be ATM-adapted. These typically take the form of a total number of terminations per year, but an end-to-end view of demand (from CO to CO) is required. To distribute terminations for residential Internet access via ADSL, for example, demographic information about CO serving areas, forecasted service penetration and bandwidth requirements, Extended Area Service (EAS) boundaries, and Internet Service Provider market share and locations are all applied to build an end-to-end view of all virtual circuits requiring switching and carriage through the core to service anticipated demand. For other services the distribution may be based on communities of interest of existing data services. If details regarding traffic characteristics, such as bandwidth requirements, delay tolerances or communities of interest (COI), are unavailable, assumptions must be made and sensitivity analysis performed.

A similar end-to-end view is built for each forecasted enterprise or residential service for each year of the forecast for the study area, which in this case comprises eighty-five CO's serving a million households. These are combined for input to the *ATM Sizer*- a switch and trunk dimensioning tool developed by the Network Performance group of Stentor, a Bell Canada research affiliate. This step permits the evaluation of a given architecture under loads we might reasonably expect to carry and ensures that the network is scaled to accommodate cell delay and cell loss tolerances for all forecasted virtual circuits.

The next step is to take the inter-switch links output from the sizing exercise and model the provisioning of these facilities on the core transport network. Transport capacity requirements and the need for new transport technologies and architectures can then be assessed.

Finally, planning costs can be developed as a measure of the efficiency of designs across both the ATM switching and transport systems architectures of the core network.

4. Analysis

4.1 Terminology

Edge Switch: a CO or access ATM network element usually with a relatively small matrix size and low/native speed interfaces performing the ATM adaptation function for multi-service demand.

Concentrator Switch: with a larger matrix size and predominantly high speed interfaces, this ATM network element provides concentration of traffic from edge switches to improve network efficiency.

Concentrator Site: a CO designated for concentrator switch deployment.

Gateway Switch: concentrates traffic and provides gateway functionality to the national or regional Inter-Exchange (IX) network.

4.2 Architecture Alternatives

A range of connectivity options have been evaluated using the modeling method outlined above. The main characteristics differentiating the architectures studied are the distribution of the concentration and tandeming functions among the switches in the network and the nature of the impacts on the core transport systems of provisioning the inter-switch links. These architectures are:

1- Hub Architecture

In the fully hubbed architecture (Figure 2) the concentration and tandeming functions are performed at a single site connected to all switches in the network. There is no interconnectivity between switches at the edge. Facilities are transported longer distances and tend to have lower utilization on average.

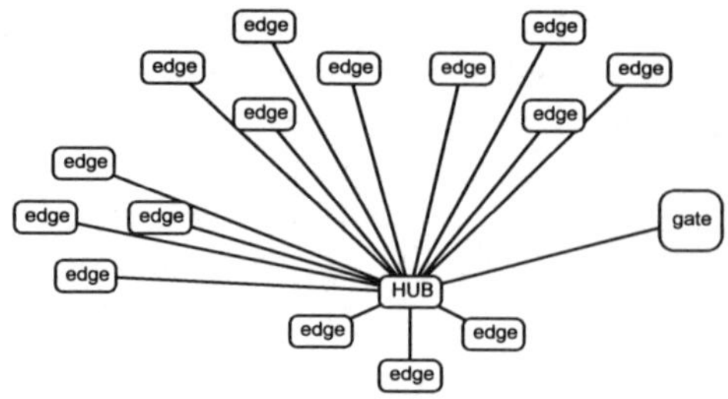

Figure 2. Hub Architecture

2- Distributed Concentrator Architecture

Concentrator sites are employed so that low-fill facilities coming into the core from the edge are carried shorter distances. This architecture (Figure 3) consists of edge switches (relatively small matrix size and low/native speed interfaces) performing the ATM adaptation function, connected to concentrator switches (larger matrix size and predominantly high speed interfaces) performing the tandeming function among other highly inter-connected concentrators.

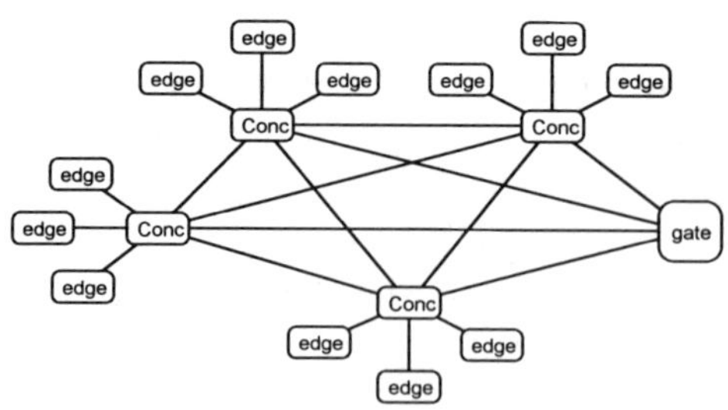

Figure 3. Distributed Concentrator Architecture

3- Mesh Architecture

The mesh architecture (Figure 4) has the highest utilization on the fewest and shortest (least cost) facilities for a given traffic matrix. There is no defined switching hierarchy and facilities ideally reflect the communities of interest in the traffic. This connectivity would result from following the provisioning rule that switches, with sufficient traffic between them to justify a dedicated facility, should be directly connected.

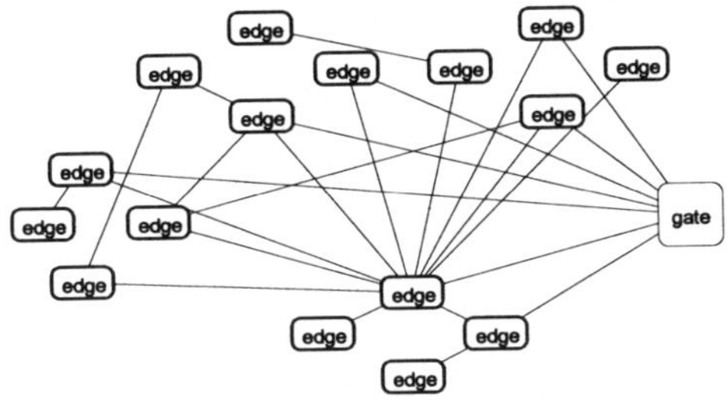

Figure 4. Mesh Architecture

4.3 Architecture Direction

Ideally suited for ordered growth and simplicity in planning, a distributed concentrator architecture is superior in all aspects analyzed. This holds true given any practical set or distribution of ATM connections sufficient to justify more than a single hub switch and is expected to scale well to regional networks. This architecture facilitates ATM-layer survivability with a highly inter-connected core of concentrator switches and a framework for diversity from the edge that can be implemented as required. For example, increasing the number of concentrator sites (CO's in which a concentrator switch is deployed) over time, and adjusting to new homing arrangements while maintaining existing links, will result in the connection of each edge switch to at least two concentrator switches. Furthermore, congestion avoidance is facilitated by having the tandeming function distributed among several switches rather than a single hub.

Given its structured framework, with destinations for backhauled accesses or inter-switch links from the edge to the core clearly defined and rationalized in advance based on broad and forward-looking criteria, this architecture is the most tolerant of churn. This is a significant advantage given that, with the potentially large bandwidth of a single service connection, the justification for switch deployment and inter-switch links may be associated with a single customer. The distributed concentrator architecture is robust, simpler to administrate and scale and also allows switching products to be selected and configured to suit particular networking functions (e.g., adaptation or path switching).

4.4 Transition of the Transport Network Layer

Carriage requirements associated with the inter-switch trunking of the distributed concentrator ATM core architecture have the best correspondence with the evolution envisioned for the transport network layer. ATM switch placement can be expected to skew former communities of interest and concentrate what were once widely dispersed and low speed facilities into hubbed backhauling patterns and high speed inter switch links. Increased trunk sizes will trigger high-capacity transport technology on cross sections between concentrator sites, leading to relatively lower carriage costs and earlier opportunities for optical networking. A fully meshed or hubbed ATM architecture, on the other hand, would result in higher volumes of lower-speed links spanning longer distances. Designing an appropriate transport layer architecture would be difficult. Multi-system transits would be practically unavoidable and increased reliance on cross connects assured, with triggers to next generation low-cost transport technologies certainly delayed.

It is highly recommended that planning for the deployment of ATM and transmission system design be coordinated. While the architectures of choice will be network-dependent and based on many criteria, the problems associated with a mismatch are clearly best avoided. The provisioning of ATM inter-switch links on transport systems not designed with that in mind can be expected to undermine their effectiveness and advance further investment.

The ultimate transport layer architecture, accommodating as it must the carriage requirements of ATM, cannot be designed independent of the technology selected or the timing of its availability. Where SONET remains the transport technology of choice, the way it is deployed and planned (ring design, linear versus ring thresholds, etc.) will require adjustments. To further complicate matters, it is unlikely that the ATM cell-based approach will entirely supplant the existing circuit-based approach to transporting traffic, so that interfaces to both ATM and STM technologies will have to be maintained well into the future.

Given the prominence of ATM in the network vision, transport architectures which ignore switch placement and trunking requirements can be expected to prove inadequate and unnecessarily costly to augment. Inefficient use of technology (e.g., SONET ring exhaustion due to extensive OC-12c provisioning requirements) and expenditures inconsistent with the objective (e.g., increased rather than decreased reliance on circuit-based technologies) are a likely result.

4.5 Deployment Strategy

Preliminary findings of research modeling backhaul scenarios in a large metro network indicate that, given current life-cycle expectations of ATM switches versus SONET transport network elements, it may be economical to backhaul multiple accesses at any rate from the serving area of a CO to a concentrator site rather than deploy a switch. When adaptation is centralized in this fashion, fewer edge switches are required and utilization improves dramatically. The associated burden on the transport network will force and/or advance increases in transport capacity, but this is considered a better investment given that the technology is more mature and has a significantly longer life expectancy. Installed First Cost (IFC) analysis is not adequate to prove in switch deployment over backhauling access(es) directly to a concentrator site, given that the cost of the switch may recur in as little as three years.

There are also benefits in phasing concentrator switch deployment. Backhauling or trunking initially to fewer than the anticipated number of concentrator sites will result in eventual diverse links from edge switches and in significantly greater trunk speeds earlier between the initial concentrator sites. This is encouraged given that the use of higher speed interfaces is considered a means of extending the benefits of SONET.

The difference in total cost between backhaul scenarios can be small; however, the technology invested in varies significantly. Delaying switch deployment tends to perpetuate dependency on muxes and cross-connects and extend the need for low speed interfaces on transmission network elements. Such expenditures are inconsistent with the network vision of an ATM cell-based core network. Rules for switch deployment should therefore consider not just access-side interface requirements, but also opportunities to avoid continued large investment in circuit-based technology.

5. Conclusion

Ongoing research investigating strategies for the deployment of ATM adaptation and switching elements, and for transporting the associated traffic, is providing insight and direction for metro core network planning. Much of the information required to model ATM networks, however, is not readily available. Anticipated services are either at a too premature stage of development to consider or else have forecasted market and traffic characteristics which change frequently and dramatically.

Rather than planning to accommodate a specific long range forecast, we should select the technology and architectures that will efficiently adapt, switch and transport immediate/near term demand for which confidence is relatively high. They will also best position the network to handle the unexpected (i.e., as yet undeveloped services or unanticipated levels or distribution of service take). This does not necessarily translate into maintaining unreasonably large spare capacities. The objective is to enable ordered, modular growth with a minimum of risk and investment. Coordinated planning of the ATM switching and transmission system architectures in metro core networks is a necessary starting point.

References

[1] Uyless Black, ATM: Foundation for Broadband Networks, Prentice-Hall, 1995.

[2] D. McDysan and D. Spohn, *ATM: Theory and Application*, McGraw-Hill, 1995.

[3] Network Planning and Engineering group, Stentor Resource Centre Inc. (SRCI), *Network Evolution Plan*, December 1996.

[4] Christian Emond, Status and Evolution of SONET Benefits (A Review of the SONET Situation), SRCI document number: 12703-95-0003, April 1995.

List of Acronyms

AAL	ATM Adaptation Layer
ADSL	Asymmetric Digital Subscriber Line
ATM	Asynchronous Transfer Mode
CO	Central Office
COI	Community Of Interest
EAS	Extended Area Service
IP	Internet Protocol
LAN	Local Area Network
QoS	Quality of Service
SONET	Synchronous Optical NETwork
STM	Synchronous Transport Mode

Core and ATM Networks
D.W. Faulkner and A.L. Harmer (Eds.)
IOS Press 1997 © AKM

Efficiency and usability of equivalent bandwidth algorithms for ATM network dimensioning

M. Naldi, P. Pelusi
Telecom Italia, Teletraffic Engineering Dpt., Via di Val Cannuta 250, 00166 Roma, Italy

Abstract. A particularly critical issue in the dimensioning of ATM networks is the treatment of variable bit rate connections. The prevailing approach is to use a single descriptive parameter, named equivalent bandwidth. In this paper four algorithms devised for the determination of the equivalent bandwidth are compared under both aspects of efficiency and usability. Source models as well as real data are used for this purpose. While the performances are acceptable for ON-OFF sources, MPEG-coded sources represent a very critical load, as the attainable efficiency is very low in all the cases examined and all the algorithms heavily underestimate the bandwidth needed. In addition, the traffic data that can be reasonably expected to be available to the network operator are shown to limit quite strongly the range of usable algorithms.

1. Introduction

As the activity on ATM shifts from the purely experimental and research phase to the realization of an operating network, a growing importance is attached to the establishment of an effective dimensioning procedure. A particularly critical issue in the allocation of network resources is the treatment of variable bit rate (VBR) connections, typically accounted for by capturing their overall need of bandwidth through a single descriptive parameter, named equivalent bandwidth (EB). By using the equivalent bandwidth paradigm, the VBR connections can be considered, for dimensioning purposes, as constant bit rate ones; the dimensioning process can therefore be carried on by resorting to the well established tools for multi-rate circuit-switched networks.

A number of algorithms have been proposed for the derivation of the equivalent bandwidth. The aim of this paper is to compare the most relevant ones under two aspects: *efficiency* and *usability*. Both real data and models are used to evaluate the efficiency of the algorithms under consideration; usability is instead evaluated by establishing a qualitative metric based on the availability of the input data needed by each EB algorithm

2. Algorithms for the equivalent bandwidth computation

Four algorithms (hereafter briefly described) have been considered in this paper:

1. Lindberger's algorithm;
2. Tidblom's algorithm;
3. Guerin's algorithm;
4. Fan and Mars' algorithm.

Lindberger's algorithm

In this algorithm [1] the actual cell arrival process is replaced by an equivalent one in which bursts of cells (emitted at a uniform rate within the burst) arrive according to a

Poisson process. The equivalence between the two processes is obtained through moment matching. Under this hypothesis, and assuming buffers of negligible length with respect to the average size of a burst, Lindberger arrives at a very simple formula which relates the equivalent bandwidth B_{eq} to the mean (m) and the variance (σ^2) of the actual bit rates, and to the capacity C of the virtual path (VP) which has to accomodate the source:

$$B_{eq} = a \cdot m + b\frac{\sigma^2}{C} \qquad (1).$$

The two parameters a and b depend on the desired cell loss probability B according to the following formulas, derived heuristically:

$$a = 1 - \frac{\log B}{50} \qquad (2),$$

$$b = -6a \cdot \log B \qquad (3).$$

If the source is known to be of the ON-OFF kind, the expression (1) must be more appropriately reformulated as

$$B_{eq} = a \cdot m + b \cdot m\frac{(h-m)}{C} \qquad (4),$$

where h is the peak bit rate (i.e. the bit rate during the ON period).

A sample parametric representation of the relation between the equivalent bandwidth and the source burstiness for a VP capacity of 150 Mbit/s is given in Figure 1.

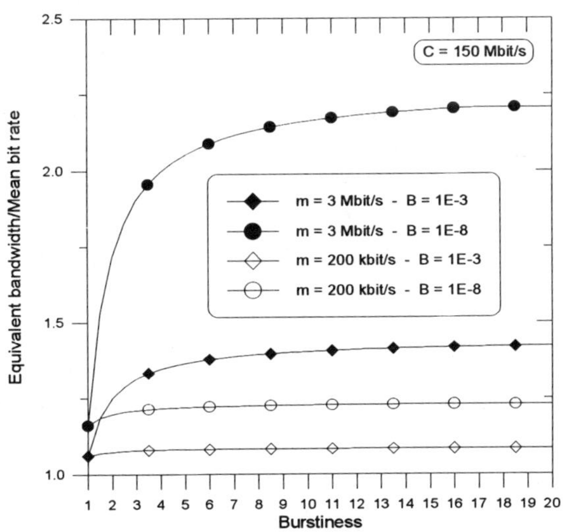

Figure 1 - Equivalent bandwidth according to Lindberger's formula

Tidblom's algorithm

Though possessing the appealing quality of being very simple, Lindberger's formula leads to an under- or over-estimation of bandwidth needs when the desired cell loss rate is larger than 10^{-9}. The most dangerous case of underestimation happens when the peak cell rate of each source is comparable to the available capacity (i.e. when the ratio C/h is low). Tidblom [2] noticed this shortcoming and provided the following correction formula

$$B_{eq} = a \cdot m - \frac{2h \cdot b \cdot m(h-m) \log B}{C^2} \qquad (5),$$

to be used instead of the expression (4) when

$$C/h < -2 \log B \qquad (6).$$

An additional shortcoming of Lindberger's formula is that for smooth sources it may lead to an equivalent bandwidth exceeding the actual peak cell rate. In order to limit such overestimation, for those sources Tidblom suggested to use the following maximum value

$$B_{eq} = a \cdot h \qquad (7),$$

so that for the most stringent case ($B = 10^{-9}$) the overestimation factor is 18%.

Guerin's algorithm

In Guerin's algorithm [3] the equivalent bandwidth is computed from the combination of two different approaches, one based on a fluid-flow model, and the other on an approximation of the stationary bit rate distribution. The first approximation is expected to provide an accurate estimate when the impact of the individual connection characteristics is critical; the second approximation is representative of situations in which the effect of statistical multiplexing is relevant. The equivalent bandwidth is obtained as the minimum of the values B_{eqf} and B_{eqs} given by the two approaches:

$$Beq_f = \frac{\alpha d(1-r)h - R + \sqrt{[\alpha d(1-r)h - R]^2 + 4R\alpha dr(1-r)h}}{2\alpha d(1-r)} \qquad (8),$$

$$B_{eqs} \cong m + \alpha' \sigma \qquad (9),$$

where R is the buffer size, d is the mean burst duration, r is the activity ratio, and the parameters α and α' are

$$\alpha = \ln\left(\frac{1}{B}\right) \qquad (10),$$

$$\alpha' = \sqrt{-2\ln(B) - \ln(2\pi)} \qquad (11).$$

Fan and Mars' algorithm

In Fan and Mars' algorithm [4] the superposition of a number of sources is approximated as a two state MMPP process (Markov modulated Poisson process). The mean arrival rates and the mean duration of each state can be derived on the basis of the knowledge of the mean (m), the peak (h), and the variance (σ^2) of the actual bit rates, and of the covariance coefficient.

3. The set of sources

In order to compare the earlier described algorithms a set of sources representing all the three kinds (voice, video, and data) has been used.

For the voice and the data cases ON-OFF sources were considered; in this paper the results are reported for one voice and one data source, whose characteristics are shown in Table 1.

Table 1 - Characteristics of voice and data sources

	Voice source	Data source
Peak bit rate	33.125 kbit/s	11.041 Mbit/s
Mean bit rate	11.639 kbit/s	1.104 Mbit/s
Variance	250.075 (kbit/s) ^{2}	10.97 (Mbit/s)^2
Burstiness factor	2.846	10
Mean activity duration	352 ms	13 ms
Mean silence duration	650 ms	117 ms

For the video sources we used instead a number of real sequences emitted by an MPEG coder. In this paper just the results for three of them (whose names are ATP, Bond, and Dino, and whose characteristics are shown in Table 2) are reported.

Table 2 - Characteristics of video sequences

	ATP	Bond	Dino
Peak bit rate [Mbit/s]	5.278	6.752	3.307
Mean bit rate [kbit/s]	609	676	366
No. of frames		40000	

4. Efficiency of the EB algorithms

Each of the algorithms so far considered allocates an amount of bandwidth to each connection of a given kind; the maximum number of connections that the link is capable of accomodating is determined by summing their equivalent bandwidth as far as this sum doesn't exceed the overall link capacity. As the bandwidth assigned to a given sourc depends on the EB algorithm, the various algorithms differ as to the number of connection they envisage to admit on the link. In the following we consider homogeneous aggregates of sources; this is not a limiting assumption, as the segregation of services is an ever more recommended policy.

At this point the efficiency of each EB algorithm can be evaluated by computing the ratio between the mean bit rate of this aggregate of sources and the link capacity. This measure alone is not however extremely reliable. In fact a generous EB algorithm, underestimating the bandwidth needed by each source, could allow for too many connections on the same

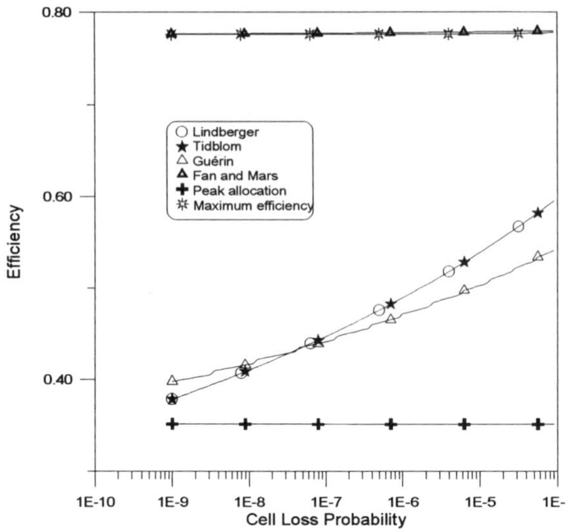

Figure 2 - Efficiency of the EB algorithms for the voice source (C=2 Mbit/s)

link and result in a lower than expected quality of service. Hence the efficiencies of the EB algorithms have to be compared against a lower and an upper bound. The lower bound is given by the efficiency obtained by assigning each source its peak bandwidth. The upper bound is instead obtained by determining in an accurate fashion the number of sources that can be accomodated on a link. This is done either by simulation (in the case of the ON-OFF sources) or by a specific model (for the MPEG sources [5]), whose performance had been previously tested through simulation. The results are shown in Figure 2 through 6.

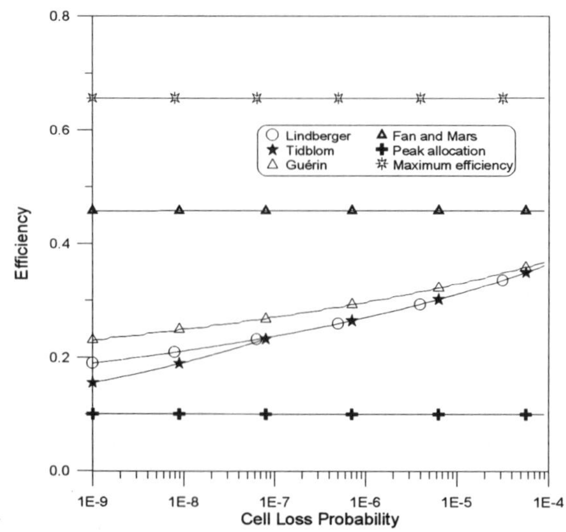

Figure 3 - Efficiency of the EB algorithms for the data source (C=155 Mbit/s)

As can be seen, for the voice and data sources the maximum efficiency is very high (ranging from 65% to 80%), but the algorithms are not equally capable of exploiting the link capacity. In both cases Fan and Mars' results are by far the best, achieving the optimal efficiency in the voice case. There are not appreciable differences between the other three algorithms, as all of them result overly conservative (especially for the data source).

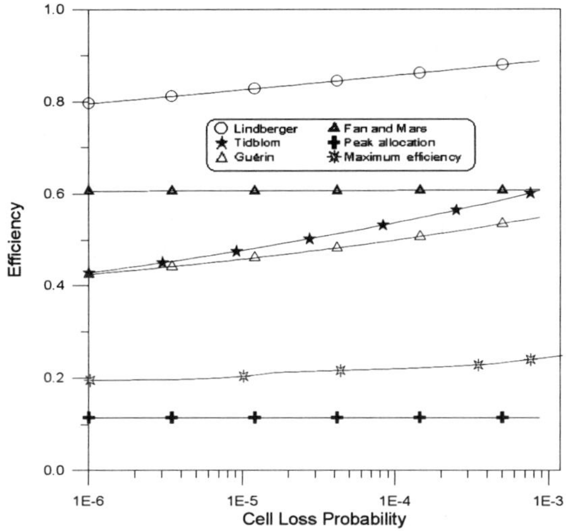

Figure 4 - Efficiency of the EB algorithms for the MPEG "Atp" sequence (C=155 Mbit/s)

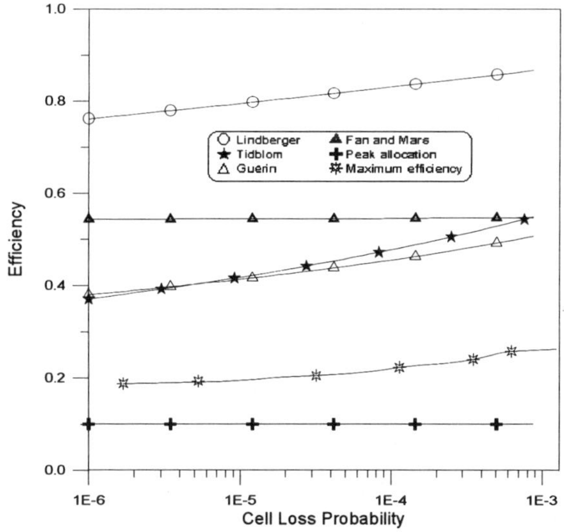

Figure 5 - Efficiency of the EB algorithms for the MPEG "Bond" sequence (C=155 Mbit/s)

Things are radically different in the MPEG case. First, the allowable range for the efficiency is very limited (from 10% to 20% in the best of cases); the room for improvement is not so much as in the voice or data case (though a 20% efficiency represents a 100% increase over a 10% figure). What is more dangerous is that all the algorithms (Fan and Mars' is the worst to this regard) heavily underestimate the bandwidth needed; that's the reason for the apparently high efficiency values appearing in Figures 4, 5 and 6.

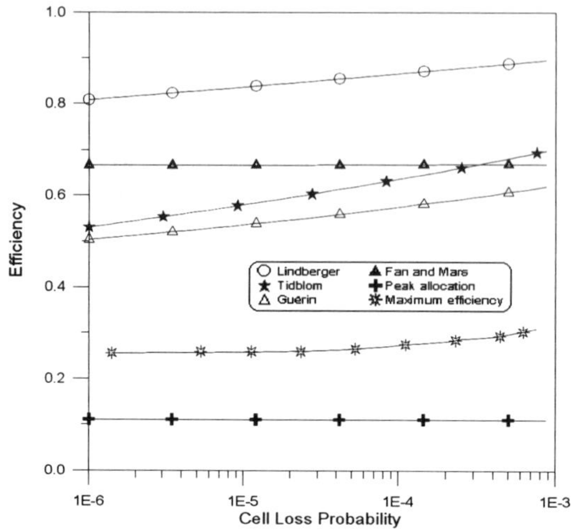

Figure 6 - Efficiency of the EB algorithms for the MPEG "Dino" sequence (C=155 Mbit/s)

5. The usability aspect

Though efficiency is an important evaluation parameter, it is not the only one. Any EB algorithm must be usable, in the sense that it must be capable to operate in the real environment under the available knowledge. In this context an algorithm can be considered to be usable if its input parameters are actually available. For the EB algorithms considered in this paper the required parameters are reported in Table 3.

Table 3 - Parameters required by the EB algorithms

	Mean bit rate	Peak bit rate	Variance	Mean burst duration	Autocovariance coefficient	Burstiness
Lindberger ON-OFF	X	X				
Lindberger	X		X			
Tidblom	X	X				
Guérin		X		X		X
Fan e Mars	X		X		X	X

The first critical point is to define who has to provide the values for these parameters. Two possibilities may be foreseen: the parameters can be either extracted from the users' declarations at the connection set-up or determined through network measurements.

In the first case the list of desired parameters must be standardized so that every ATM terminal is able to provide them. At the present the main ITU Recommendation on traffic control, I.371 [6], has standardized just two source traffic descriptors: the Peak Cell Rate and the Sustainable Cell Rate (which can be assimilated to the average cell rate). Under these conditions the only algorithms that meet the usability definition are those due to

Lindberger and Tidblom. Though the standards can be enriched to this regard, a question naturally arises: aren't we asking too much of the user ? While an accurate determination of the source bandwidth needs is feasible in some applications (e.g. Video-on-Demand or CBR sources), the source emission cannot be accurately predicted in most cases.

If it is up to the network operator to build and maintain a database of bandwidth information, then a rich set of measurements must be defined and routinely performed. As the impact on the switch processing load can be disturbing, a trade-off must be established. This means that too demanding measurements will have to be avoided. By taking a bref look at Table 3, we can note that the algorithms are strongly different to this respect. While the average and the peak cell rate can be thought to be easily available, the variance, the mean burst duration and the autocovariance coefficient are (in order of increasing difficulty) certainly not.

These latter considerations would lead us to exclude Fan and Mars' algorithm from the list of feasible algorithms, though this is tantamount to accept a far lower efficiency than would otherwise be possible.

6. Conclusions

All the algorithms considered for the determination of the equivalent bandwidth perform acceptably in the case of the ON-OFF sources (which is often their main assumption) with the remarkable dominance of Fan and Mars' algorithm, which attains the maximum efficiency.

In contrast, all algorithms, with no exception, heavily underestimate the bandwidth needed to cope with MPEG sources, where the maximum efficiency is anyway low (of the order of 20%).

As to the usability aspect, all the algorithms, with the notable exception of those due to Lindberger and Tidblom, appear to be too demanding on the user or to require an excessive load on the switch measuring system.

Three lessons can therefore be learnt:

1. no algorithm is a jack-of-all-trades;
2. MPEG sources are particulary critical, and a heuristic allocation based on some fraction of the peak bit rate could be envisaged;
3. the network operator requirements on traffic characterization and the actual user capabilities to this regard must be somewhat reconciled.

References

[1] K. Lindberger : "Analytical methods for the traffical problems with statistical multiplexing in ATM networks", 13th International Teletraffic Congress, Copenhagen, 19-26 June 1991, pp. 807-813

[2] S.E. Tidblom : "Complete equivalent bandwidth formulae for various cell loss", COST 242 TD(92)036, August 1992

[3] R. Guerin, H. Ahmadi, M. Naghshineh : "Equivalent capacity and its application to bandwidth allocation in high speed networks", IEEE Journal on selected areas in communications, vol. 9, no. 7, September 1991, pp. 968-981

[4] ITU Recommendation I.371 : "Traffic control and congestion control in B-ISDN"

[5] P. Skelly, M. Schwartz, S. Dixit : "A histogram-based model for video traffic behaviour in an ATM multiplexer", IEEE/ACM Transactions on networking, vol. 1, no. 4, August 1993, pp. 446-459

[6] Z. Fan, P. Mars : "Effective bandwidth approach to connection admission control for multimedia traffic in ATM networks", IEE Electronics Letters, vol. 32, no. 16, August 1996, pp. 1438-1439

Core and ATM Networks
D.W. Faulkner and A.L. Harmer (Eds.)
IOS Press 1997 © AKM

Designing ATM Networks Using LINDA*

Jochen Frings
Lehrstuhl für Kommunikationsnetze, Technische Universität München, 80290 München, Germany
Email: frings@lkn.e-technik.tu-muenchen.de

Abstract

When ATM equipment is ready for commercial use, cost efficient design of ATM networks becomes important. Because many requirements are unknown today, the network planning tools must have a very flexible software architecture to make their redesign easier.
We present the integrated network design tool LINDA and its completly modular architecture. To show some of its features, a complete network planning example including node placement, traffic matrix generation, topology optimization and the VP network design is done.

1 Introduction

Today, as ATM equipment is almost ready for commercial use in public networks, the cost efficient design of ATM networks is becomming increasingly important.
To solve these design problems, network planners need the help of integrated network planning tools (NPT), which cover the whole planning process. This, because network optimization problems have an incredible high problem complexity.
But despite the fact, that the NPT development has to be started now, any assumptions about the lateron introduced telecommunication services, as well as their corresponding traffic modells, must be inaccurate. Thus many future requirements are unknown today and each NPT will have to undergo several major redesigns as long as new services and networking techniques are introduced into the networks.
The only way to save the tool investments and to allow rapid prototyping (an important feature, if there are several competing carriers) is to design the NPT with an extremly flexible software architecture, that makes integration and combination of planning algorithms easier.
In contrary to that, many tools are designed for very special planning cases at this moment and often they cover only a single step of the whole planning process. Hence several different tools have to be used to design a single network.
To overcome this problem and because a flexible NPT is also an ideal research environment for new network planning methods, we decided to create LINDA, the *I*ntegrated *N*etwork *D*esigner for *A*TM, at our institute, the *L*KN. Its architecture and some features will be discussed in section two into detail.
Although LINDA is not yet (and will not be) a finished tool, it is possible to solve

*This work was supported by the central research and development department of Siemens AG

some typical network planning problems. As an example we modell the case of a power supplier wanting to become also a carrier, in section three. In section four we discuss the planning methods actually integrated into LINDA and show how to design an ATM network for the example case. Section five concludes and directs to further research.

2 LINDA

To start with, here is a short list of conditions, a modern NPT has to meet. Altogether they require the flexibility stated above.

1. It should be very easy to integrate a new network planning algorithm regardless of its data structure, programming language or programming paradigma. Thus it is possible to use the most qualified technique to implement the algorithm and to integrate even third party algorithms.

2. A NPT will be used for a long time. Thus, to be able to use the the best available computers, it should be designed very portable.

3. Most NPTs are (for many reasons) no mass products. Hence, they should be designed as prototypes, with full custom design only for the planning algorithms and of the shelve modules wherever possible, to reduce the implementation expense.

4. Because network planning sometimes is a very challenging problem, the architecture should allow to run algorithms without to much modification on remote special hardware (e.g. vector or parallel computer).

5. The network planner has to deal with lots of different parameters and algorithms. To work effectively and faultless, thats why he needs an interactive graphical user interface, which guides him through the network planning process and does some plausibility checking.

The architecture of LINDA, shown in figure 1, was designed in consideration of these conditions. The tool is modularized into several loosely coupled programms, to allow easy integration of algorithms (C1) and rapid prototyping (C3). The interfaces are ASCII flat files (e.g. control file, data file), unix commands with and parameters to start algorithms or TCP/IP sockets for closer couplings (e.g. topology editor - database) and they are system independent (C2) and easy to implement or convert (C1).
The planning algorithms can be run as stand alone programms, after they got the required data. Thus they can be run off line or on remote computers (C4), if their startup procedure in the process control was modified accordingly.
The process control is written in Tcl/Tk [6], an easy to learn script language for graphical user interfaces (GUI), which is available on many computer systems. The GUI development can be supported by the interactive GUI design tool called *xf* [3]. Hence the process control can be easily extended to new algorithms (C1) an is as comfortable as required (C5). To make plausibility checks possible, a finite automate of the planning process was implemented. Its actual status is animated with the colour and layout of the algorithm start up buttons (C5) (see figure 3).
Another important aspect is the data management concept, especially, because different data formats have to be supported (C1). In LINDA, the process control knows only about the configuration parameters and startup sequences of the algorithms. The consistence of the network data is guaranteed by the network database, which also has the

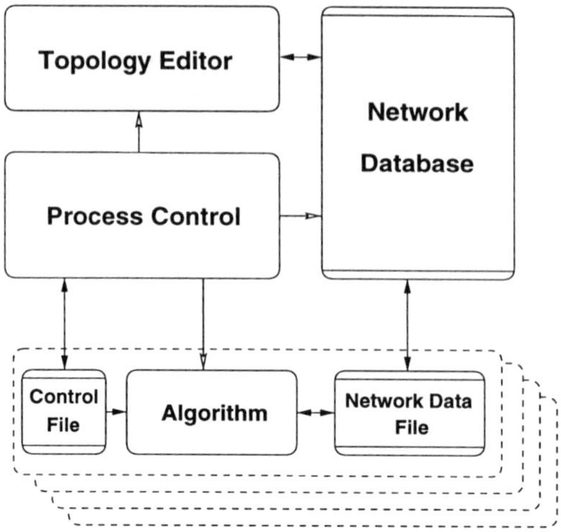

Figure 1: Software Architecture of LINDA

integration feature to have a meta data format. All other data formats are translated to and from it. While the process control is the network planners interface to the process data, the topology editor is the interface to the network data (R5).
The database system and the topology editor should be standard tools with standard programming interfaces, like SQL in the case of the database (R2) (R3).
The algorithms integrated into LINDA at present were developed to design wide area ATM backbone networks. Therefore the design process shown in figure 2 is used. This design process is divided into five succeeding steps, because the whole problem is too complex to be solved by a monolithic algorithm. For some steps, there are alternative algorithms. This is needed in steps, in which most algorithms are heuristic or solve only approximations of the original problem, because each algorithm works best for its parameter range.
It is possible to go back to any step at any time, if another variant (parameters, design algorithm) of the network has to be generated. To step forward is also possible, if the data produced in the step leaped over are not needed as an input for the choosen algorithm. Only these allowed actions can be done by the designer, because of the finite design process automate included into the process control and GUI.
The GUI (figure 3) animates this design process.
In section four more details about the currently implemented algorithms are given.

3 An ATM Network Planning Problem

To give an example of LINDAs current network planning features, we modell and solve a typical design problem.
A consortium of German power suppliers is willing to become also a national telecommunication services provider.
The conditions are good, because most of the consortiums power lines can be equiped with fiber without much effort and city carriers in all large towns are willing to connect

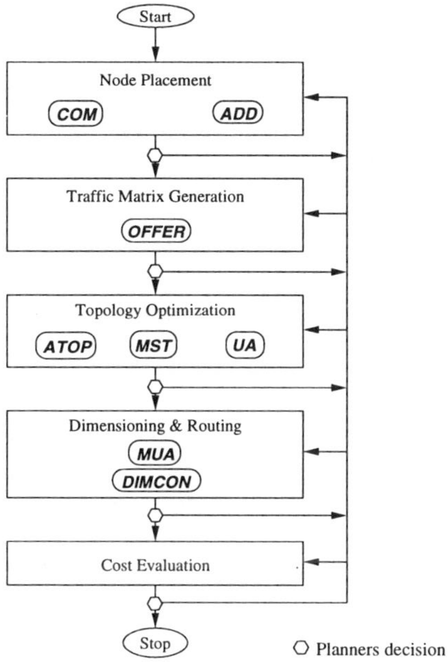

Figure 2: Currently implemented design process

to the new wide area network (NET).
It is possible to place a NET node in any town, but the optimal number and locations are unknown. They have to be a result of network planning. City carriers, which as a result do not host a NET node, are connected to NET with a direct link to a NET node.
In the next step the NET nodes have to be connected. It is required, that each pair of nodes must have at least two completly disjunct connections, because of network availability.
A telecommunication marketing survey has been made, which allows some assumptions about the offered traffic. As a result the typical users telecommunication habits are put together in figure 4. In addition to that, it can be assumed, that about half of all

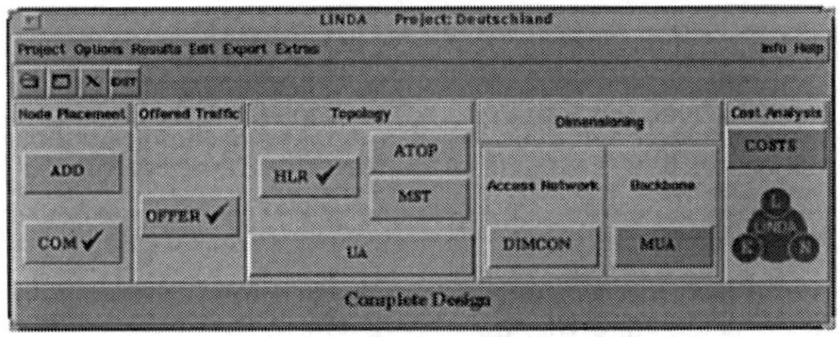

Figure 3: Graphical user interface of LINDA

Class	Service	Bitrate in Mbit/s		Traffic
		mean	max	in Erlang
CBR1	Videophone	1	1	0.1
CBR2	Television/VOD	3	3	0.03
CBR3	Video Conference	10	10	0.001
VBR1	Data1	1	5	0.03
VBR2	Data2	5	10	0.001

Figure 4: A typical users offered traffic

inhabitants of each town will use NETs services.
The planners are not restricted to use only ATM crossconnect nodes. Rather the used switching modes (VC/VP (permanent) switching) and the routing strategy shall be a result of the network planning phase.

4 A network planning example

In figure 5 the 42 participating city carrieres, which have to be connected with NET, are shown. To start with, the NET node locations have to be choosen. For this purpose two

Figure 5: 42 participating city carriers

algorithms are available. One is the centre of mass algorithm (COM), which clusters the two node clusters, whose weighted centers are nearest neighbours, as long, as some boundaries are adhered to. The other algorithm, ADD, chooses an additional node location, as long, as costs are decreasing. Both are simple heuristics taken from [5], but give good results. In figure 6 some COM results for different node numbers are given. Now the traffic matrix can be generated. We assume to have three CBR and two VBR services as given in figure 4 and that the communication interest between two persons decreases exponentially with their distance [4]. In addition to that, it is assumed, that the communication interest is depending on the number of inhabitants of the sink town. Now, after normalization of the interests and weighting them with the offered traffic of the typical user and the number of users, the traffic matrix can be filled.

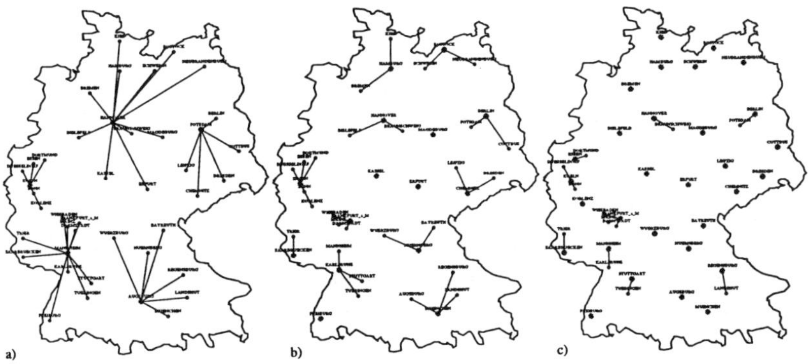

Figure 6: Node locations and clusters for a) 5 nodes b) 15 nodes and c) 30 nodes

In the next step the link topology has to be generated. As assumed in section three at least two totally disjunct paths are needed for any NET node pair. Again more than one algorithm is available.
First a minimum spanning tree algorithm (MST) can be used to connect all nodes. To get alternative paths, the MST can be applied again on the remaining unused links. Of course this algorithm can not pay attention to the later traffic streams. Therefore a new algorithm, called ATOP, was introduced. It approximates the NP complete topology optimization problem with a series of linear programms. ATOP is able to do the optimization under several restrictions, like percentage of remaining capacity after a network element failure, maximum link distance for nodepairs (equal to maximum transmission time), maximum capacity of network elements, maximum number of intermediate nodes (equal to maximum delay jitter) and so on. To transform the originally nonlinear optimization problem into a linear one, we use Siebenhaars [7] superpath concept. In figure 7 the influence of the maximum number of intermediate nodes on the link topology design is shown. A third algorithm for topology design is the Unified Algorithm (UA)

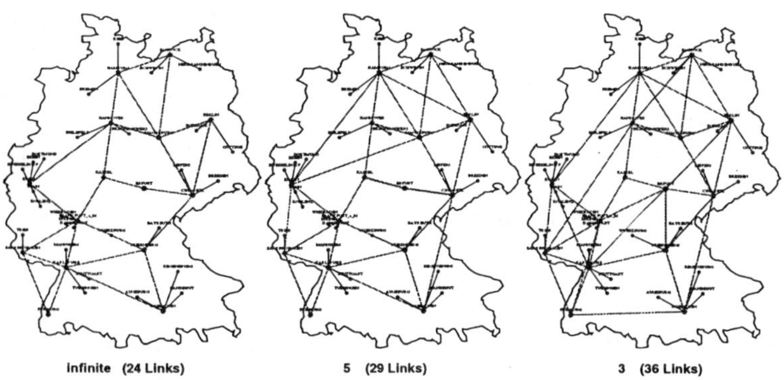

Figure 7: Topologies for different numbers of allowed intermediate nodes

of Ash [1]. Because it was originally designed for telephone networks, the traffic matrix dimension has to be reduced. The implemented version does not guarantee a certain connectivity.
The last planning step in the design process is dimensioning and routing.

In general, it is not possible to perform dimensioning and routing optimization separately, because the results of each of these steps depends heavily on the other step. Especially in the case of ATM network planning routing optimization is a demanding task, because of the logical VP network. The VC routing problem has to be solved together with the logical VP topology problem.

In [2] Bauschert describes an extension of the UA (see above), the *M*ultiservice UA (MUA). He uses the link topology optimization feature of the UA to optimize the logical VP network. Because the two problems, dimensioning and routing, are to complex to be solved in a monolithic optimization problem, in the MUA they are both solved, until convergence, in an iterative process. Dynamic nonhierarchical alternate routing is used for VC routing optimization. MUA is also a part of LINDA.

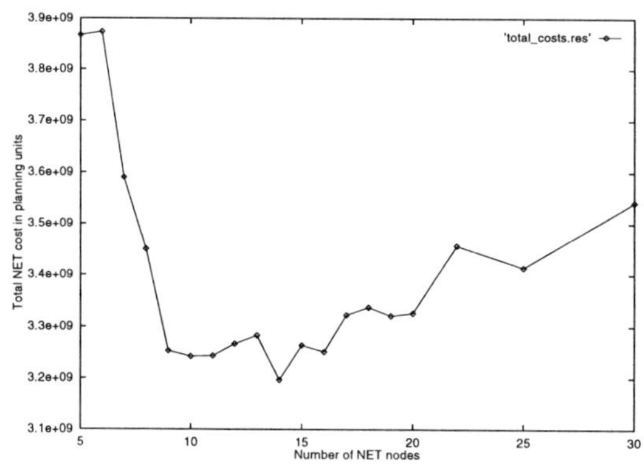

Figure 8: Dependency of total network costs from number of nodes

Initially it was unknown, how many NET nodes will be needed to get a cost effective network. Therefore the network planner has to evaluate several alternatives for different node numbers. This was done for the planning example under the assumption of a maximum call blocking probability of one percent for all services and node pairs.

The result can be seen in figure 8. Two tendencies are noticeable. First, starting from very few nodes, additional nodes lead to rapid decreasing of the network costs, because the star network (one node) becomes a meshed network, which is more effective. But then the additional node costs are higher than the savings by the star to mesh conversion. In addition to that, it becomes clear, that there is a large spectrum from about eight to 20 nodes, in which the difference between the solutions is very small, compared to the total costs. This is an often documented result.

For our network the minimum is reached at 14 nodes. The jitters can be declared with the discrete structure of the link topology and the nonsystematic node locations. The final 14 node network is presented in figure 9.

5 Conclusion

It was shown, that modern network planning tools need a very flexible software architecture, that is adopted to the special conditions of the network planning environment. Only in this case the investments into the tool can be saved for a longer period.

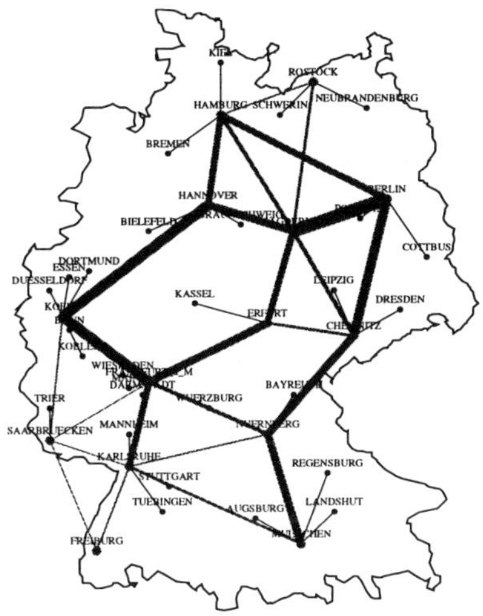

Figure 9: The final 14 node network

The software architecture of LINDA was presented and the concepts behind it were introduced.
A network planning problem was developed and solved with the help of LINDA. Finally it was shown, that a network planning tool is only a decision support system. Thus e.g. to find the optimal number of nodes is an iterative process.
Several planning aspects, like maximum capacity and extensioning of already running networks, were not discussed, but can be calculated with LINDA.
The next versions of LINDA will contain an interactive topology editor and the integrated network database. Currently we work on the discrete network element planning problem, which transforms planning results into real network elements.

References

[1] G.R. Ash, R.H. Cardwell, and R.P. Murray. Design and optimization of networks with dynamic routing. *The Bell System Technical Journal*, 60(8):1787–1820, October 1981.

[2] Thomas Bauschert. Design of VP based ATM networks with nonhierarchical routing. In *Networks'96, 7th International Network Planning Symposium*, pages 523–528, 1996.

[3] Sven Delmas. Information on xf. *http://panther.cimetrix.com/sven/xf.html*, 1996.

[4] R. Jäger. Zusammenhang zwischen Gebühren und Einnahmen im Telekommunikationsbereich. *Der Fernmelde-Ingenieur*, 44(1/2):1–41, 1990.

[5] A. Kershenbaum. *Telecommunication Network Design Algorithms.* McGraw-Hill, New York, 1993.

[6] John K. Ousterhout. *Tcl and the Tk Toolkit.* Addison-Wesley Publishing, 1994.

[7] Rainer Siebenhaar. *Verkehrslenkung und Kapazitätsanpassung in ATM-Netzen mit virtuellen Pfaden.* Thesis, Lehrstuhl für Kommunikationsnetze, Technische Universität München, 1996.

Core and ATM Networks
D.W. Faulkner and A.L. Harmer (Eds.)
IOS Press 1997 ©AKM

Hybrid STM/ATM Network Architectures - A First Assessment

Ahmad Sultan
Bell Canada - Network Technology Planning
700 De La Gauchetiere W, Office 18W2, Montreal, QC, H3B 4L1, Canada
Email: asultan@qc.bell.ca

Abstract. Hybrid STM/ATM is an emerging technology promising to be a cost effective evolution path to ATM-based networks. Efficient transport of ATM traffic in the existing synchronous network may be achieved by deploying hybrid network elements capable of processing both STM and ATM traffic.

In this paper we address the emerging hybrid STM/ATM technology and its impact on network architecture from an end-to-end perspective. We show how a SONET network can evolve to handle the increasing ATM traffic. We investigate the opportunities of this new technology in both access and transport networks and provide some results to show the benefits that can stem from it. Our results show that hybrid network elements and architectures provide good opportunities to support increasing ATM demands within the existing SONET network. For instance, significant savings in bandwidth and interface requirements can be achieved by deploying hybrid STM/ATM rings. Opportunities exist in both access and transport networks and depend on the available multiplexing technology of ATM traffic. These results were obtained from an economic analysis based on a real network.

1. Hybrid STM/ATM Principles

A hybrid SONET/ATM network element (NE) is a SONET transport NE enhanced with ATM cell processing capabilities. As the growing ATM services are utilizing existing SONET networks, it may be beneficial to add ATM functionality to SONET NEs, permitting a more efficient transport of ATM traffic over existing SONET networks.

Hybrid NEs provide centralized points for handling STM and ATM traffic, and may provide a cost-effective evolution path from STM to ATM networks for broadband applications that require flexible bandwidth management and high capacity, such as Local Area Network (LAN) interconnection, multimedia, and video services. By incorporating ATM functionality into SONET networks, an overlay network to handle ATM will be unnecessary. This will result in fewer network elements and facilities to manage [1,2,3].

Hybrid STM/ATM traffic refers to the transport of both STM and ATM services on the same network, where ATM traffic is carried in ATM cells and STM traffic is carried in STM format (e.g., DS1, or DS3 mapped to a SONET payload).

There are three different techniques for combining STM and ATM traffic over a single transport network [1]:

1. The first technique uses a conventional SONET transport and ATM switching network to combine the STM and ATM traffic onto the same SONET stream but over different

STS-Ncs. SONET NEs are used for transport and ATM edge and hub switches are used for ATM cell switching. ATM cells are mapped to SONET payloads (e.g., into an STS-Nc Synchronous Payload Envelope) but the SONET transport NEs only process the bit stream at the SONET layer; they do not process it at the ATM layer. The SONET transport network provides transport of ATM traffic from the ATM Customer Premise Equipment (CPE) to the ATM switch and back to the ATM CPE. With this technique, each ATM stream requires its own STS-Nc.

2. Hybrid NEs (e.g., STM/ATM ring, hybrid DCS) may be used to aggregate ATM traffic to achieve better fill of the SONET payload. Different ATM traffic streams can be aggregated and carried in a single STS-Nc, allowing for more efficient transport. The hybrid STM/ATM ring is a SONET-based ring in which the Add-Drop multiplexers (ADMs) have ATM cell processing capabilities. This allows sharing of a SONET payload among several ring nodes. The hybrid DCS provides ATM cell cross-connection, grooming and management functions. An ATM Service Access Multiplexer (SAM) can be used at the edge of the public ATM network to provide ATM interfaces and adaptation for customer services.
3. The third technique for combining STM and ATM traffic on the same SONET stream is by converting the STM traffic (via ATM Circuit Emulation Service) into ATM cells and carrying all STM and ATM traffic as ATM cells in a single STS-Nc. While this last technique may provide the more efficient bandwidth utilization, it requires circuit emulation for the STM traffic, which is costly today.

2. Hybrid SONET/ATM NEs and Architectures

2.1. Hybrid SONET/ATM Rings

A Hybrid SONET/ATM ring is a SONET-based ring (either Unidirectional Path-Switched Ring [UPSR] or Bidirectional Line-Switched Ring [BLSR]) enhanced with ATM cell processing capabilities (i.e., adding/dropping/cross-connecting of ATM traffic, policing, etc.) as shown in Figure 1. An STS-N(c) on the ring is allocated and can be shared by all ATM traffic that is generated by ATM nodes connected to the ring. To provide this capability, an upgrade of existing SONET ring ADMs will be necessary (e.g., ADMs that need to drop ATM traffic should be upgraded to provide ATM cell processing, whereas ADMs that need to only drop STM traffic do not need to be upgraded) [1].

Hybrid SONET/ATM rings use the ATM Virtual Path (VP) layer for ATM traffic management on the STS-N(c) reserved for ATM traffic. A path is provisioned to only carry ATM traffic, utilizing an STS-1, STS-3c, or STS-12c synchronous payload envelope (SPE). This path can be shared among several ATM ADMs, with multiple virtual path connections terminating at different nodes, and bandwidth is provisioned on the ring according to the bandwidth needed for a particular VP. This allows better utilization of the provisioned bandwidth for ATM traffic.

Consideration should be given to include statistical multiplexing and traffic policing functions in ATM ADMs, if they are deployed in architectures in which they are the first NEs in the network that perform ATM processing. More detailed information and requirements for the ATM ADM functions can be found in reference [1].

Finally, the BLSR appears to be more appropriate for an ATM ring rather than the UPSR. The BLSR is more efficient in terms of capacity for ATM network applications, and hence, more economical. In UPSR, each STS-N(c) reserved for ATM traffic must be duplicated with a protection path in the opposite direction around the ring. Ring bandwidth would thus be quickly exhausted if UPSR protection is extended to the shared ATM STS-N(c) and thus an alternative ATM layer protection mechanism should be developed. With BLSR, no major changes to the STM protection layer are needed when applied to hybrid rings [1,4]. Hybrid SONET/ATM ring protection requirements are currently under consideration for standards work in T1X1.5 and/or T1S1.5.

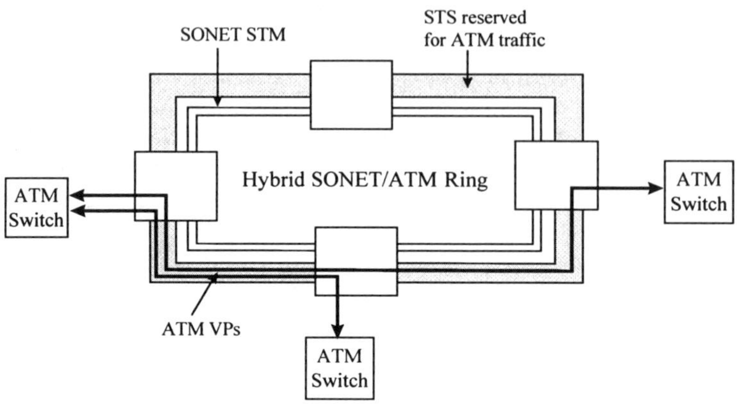

Figure 1. Hybrid SONET/ATM Ring

2.2. Hybrid DCS

A hybrid DCS is a conventional SONET DCS with ATM functionality, allowing the gradual addition of growth in ATM services over existing SONET networks. It provides ATM cell cross-connection, grooming, and management based on virtual paths. ATM functionality is present in a hybrid DCS via an ATM cross-connect matrix. The ATM cross-connect matrix along with an STM cross-connect matrix are the two major components of a hybrid DCS [2].

Hybrid DCS interfaces may contain a mixture of STM and ATM tributaries; they are internally demultiplexed/multiplexed, and forwarded to the appropriate cross-connect matrix. ATM cell processing takes place in the ATM VP matrix and is based on Virtual Path Identifier (VPI) processing.

A hybrid DCS may be deployed to interconnect hybrid STM/ATM rings. It can also be used to connect multiple rings to a Broadband (ATM) Switching System (BSS). In addition, a hybrid DCS can serve as a hub, connecting to various other STM, ATM, or hybrid NEs in other locations. Other applications of hybrid DCS can be found in [2].

3. A Network Evolution Scenario

A potential evolution strategy of a SONET transport network into an ATM transport network, capable of efficiently carrying ATM traffic, consists of the following three phases:

1- Deployment of SONET systems (rings):

This first phase of the evolution consists of deploying a SONET-based platform in the case of a new application and relieving the existing Asynchronous platform with a SONET-based platform when exhaust occurs, in the case of an existing application. This phase is aimed at introducing a broadband stage so that new advanced capabilities are introduced due to SONET inherent characteristics such as add/drop capability and self-healing ring architecture. In addition, this SONET stage will also serve as the physical layer to transport ATM-type traffic. This scenario is based on conventional SONET ADMs and hub ATM switches in the transport network to perform ATM processing, with possible edge switches at the access level for localized ATM processing. All traffic entering the network from the customer premise is mapped into STM SPEs for transport.

2- Deployment of hybrid SONET/ATM equipment and architectures:

Deploying only conventional SONET NEs leads to inefficient use of bandwidth when transporting ATM traffic. As the amount of ATM traffic increases, ATM cell processing capabilities can be added to SONET NEs, resulting in hybrid STM/ATM network elements and architectures.

3- Deployment of pure ATM NEs and architectures:

This final phase consists of deploying pure ATM NEs when the ATM demand far exceeds the STM demand. All STM traffic entering the network is converted to ATM cells, and converted back to STM traffic when exiting the network (via circuit emulation).

Note that phase 1 and/or 2 could be skipped, depending on the ATM technology evolution, the hybrid equipment availability, and the growing demand of ATM traffic. For example, if ATM demand is low and cell-based traffic is a small portion of the total traffic, we may chose to incorporate limited ATM functionality into existing SONET transport equipment and cross-connects as a cost-effective way to provide ATM presence. If, on the other hand, ATM demand is high and cell-based traffic is expected to quickly overtake traditional circuit-based services, then we will likely move directly to pure ATM network elements and carry legacy services in circuit emulation mode.

4. Network Analysis

In the previous section, we showed the potential evolution scenario from the existing SONET network to the pure ATM network. The hybrid STM/ATM NEs and architectures represent an intermediate stage in this evolution. In this section, we provide some results to show the benefits of the hybrid STM/ATM technology.

In order to address and quantify the perceived benefits of the hybrid technology, we performed an economic analysis to assess the benefits of hybrid STM/ATM rings. In this analysis, we considered the network shown in Figure 2. This network consists of two transport rings (T1 and T2) with 30 COs and 5 access rings (A1 - A5) with 20 business and residential sites. The dashed lines represent the traffic demand.

We have considered business and residential traffic. Business traffic includes Advanced-LAN (ALAN) and DS-3 UNI; residential traffic includes Internet and teleworking. Table 1 summarizes the characteristics of these services.

4.1. Assumptions

- All traffic is provisioned on rings; traffic provisioned on linear systems has been removed.
- Maximum of four sites per access ring.
- Traffic includes inter/intra access rings, inter/intra transport rings.
- DS-3 interfaces in the access and the transport.

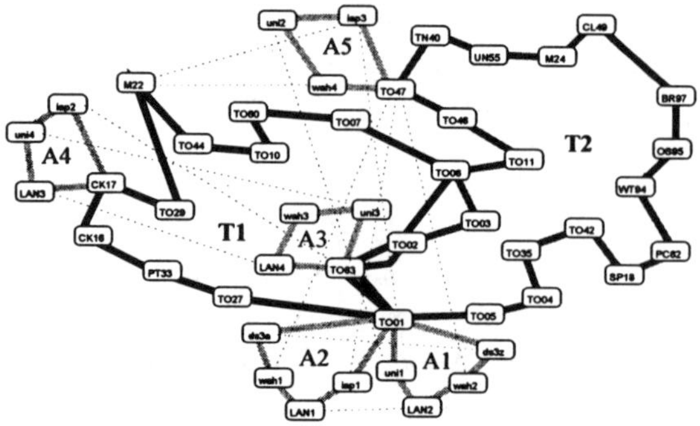

Figure 2. Test network

Table 1 Service characteristics

Service	Peak bandwidth (Mbps)	Average bandwidth (Mbps)				
		96	97	98	99	00
ALAN	10	5	5	5	5	5
DS-3 UNI	45	45	45	45	45	45
High-speed Internet	2	0	.02	.04	.08	.16
High-speed Teleworking	2	0	.04	.07	.14	.28

4.2. Analysis Scenarios

In this analysis, we focused on investigating and quantifying the opportunities of Hybrid SONET/ATM rings. Three scenarios were considered:

1. Conventional SONET and ATM network consisting of conventional SONET rings (BLSR); this is scenario 1 (Reference).
2. Hybrid SONET/ATM network consisting of hybrid SONET/ATM BLSR rings. There are two possibilities:

   2-a. Hybrid ADM with service multiplexing; this is scenario 2 (Alternative 1).

   2-b. Hybrid ADM with statistical multiplexing; this is scenario 3 (Alternative 2).

We considered scenarios 2 and 3 to assess the two multiplexing options [1]:

1. *Service multiplexing* of the inputs into a fixed size VP. It is also referred to as channel merging or service concentration. This is a simple concentration of all the incoming cells into a fixed-size VP that is sized based on the sum of the peak bandwidth (i.e., peak rate of VC services within VPs) used on each low-speed interface for ATM traffic. For example, a node with 3 interfaces, 2 DS1s (<=1.5 Mb/s) and 1 DS3 (carrying <=10 Mb/s) of traffic with all traffic destined to a single node on the ring, would require a VP of 13 Mb/s bandwidth.
2. *Statistical multiplexing* of the inputs into a fixed-size VP on a per ring node basis. In this case, the VP could be sized based on the sum of the mean bandwidth (i.e., sustained rate of VC services within VPs) used on each low-speed interface for ATM traffic. For example, a node with three interfaces, each a DS3 with an average rate of 5 Mb/s and a peak rate of <= 10Mbs, may require only a VP of bandwidth 15 Mb/s (3 x

5 Mb/s). It may be sized somewhere between the sum of the mean bandwidths and the sum of the peak bandwidths, depending on the desired multiplexing gain. The VP bandwidth requirement would be 30 Mb/s if the first option of service multiplexing was used.

4.3. Results Summary

4.3.1. Bandwidth Analysis

Figure 3 illustrates the bandwidth savings of hybrid rings. It shows the bandwidth requirements in terms of the number of DS3 needed in the overall network (access and transport) for the three scenarios. Five different demands are considered, based on the service forecast for a five-year study. The traffic volume increases every year. This figure shows that scenarios 2 and 3 (Alt1 and Alt2) provide bandwidth savings of 16% and 32% respectively, relative to the reference model. Note that the benefit from statistical multiplexing increases for higher traffic volumes.

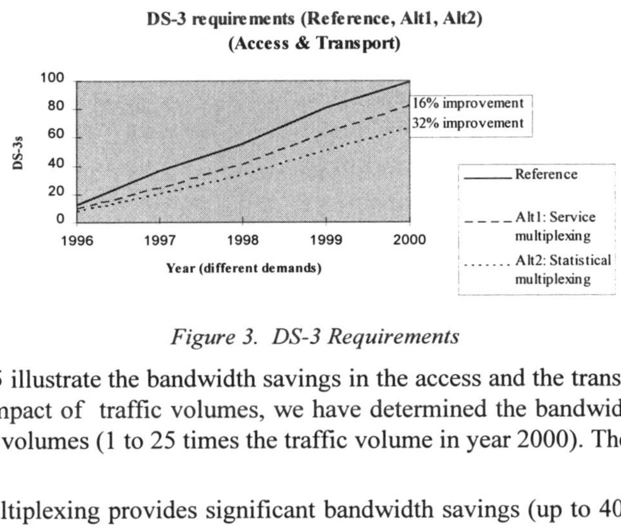

Figure 3. DS-3 Requirements

Figures 4 and 5 illustrate the bandwidth savings in the access and the transport, separately. To show the impact of traffic volumes, we have determined the bandwidth requirements for high traffic volumes (1 to 25 times the traffic volume in year 2000). These results show that:

1. Service multiplexing provides significant bandwidth savings (up to 40% in the access and 20% in the transport) for low traffic volumes.
2. Service multiplexing provides less or no benefits for higher traffic volumes. The reason is that higher traffic volumes produce more efficiently filled facilities.
3. Benefits from statistical multiplexing are ensured with both high and low traffic volumes, in the access as well as in the transport network.
4. Opportunities for statistical multiplexing improve when combining multiple service types.

4.3.2. Network Interface Analysis

Figure 6 shows the network interface requirements in the three scenarios. Improvements are in the order of 5% and 11% for service multiplexing and statistical multiplexing respectively.

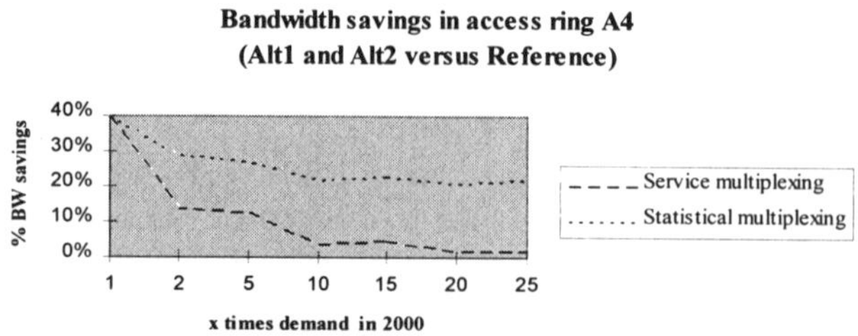

Figure 4 Bandwidth savings in access ring A4

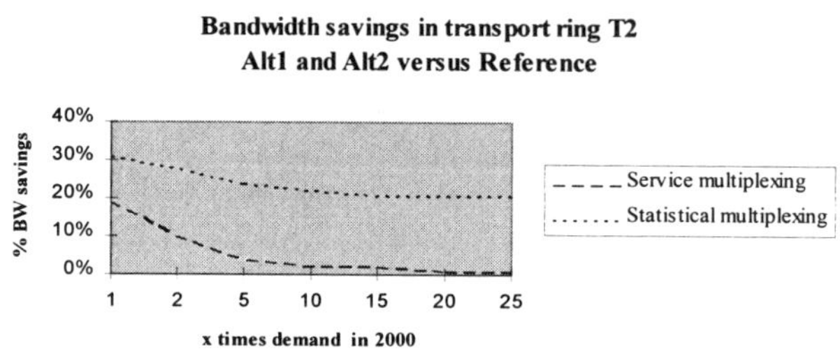

Figure 5 Bandwidth savings in transport ring T2

	Ref	Alt 1	Alt 2
DS-3	92	86	76
STS-1	134	128	114
Total	226	214	190

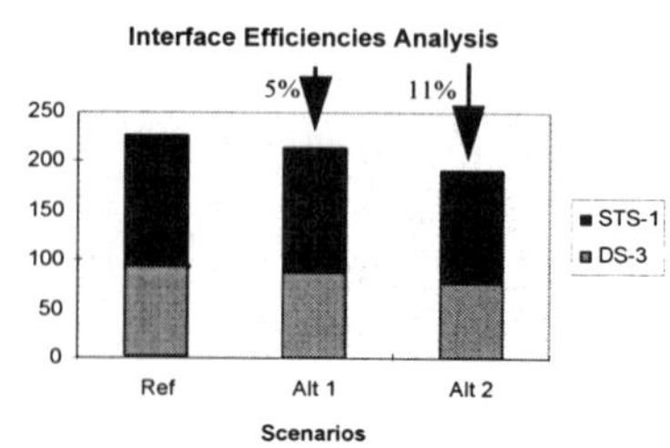

Figure 6 Network interface results

4.3.3. Impact of Interface Rates

As we mentioned earlier, DS3 interfaces were assumed in the previous results. We show now the impact of statistical multiplexing when higher interface rates are assumed. The interface rate at which ATM and terminal technologies are inter-connected influences the hybrid ring opportunities. The following table summarizes some preliminary results:

Interface	Potential statistical multiplexing benefits
OC-3	20% improvement
OC-12	30% improvement

In addition, other benefits associated with larger interfaces include: provisioning efficiencies (a 70-85% filled OC-12 is more efficient than multiple OC-1 pipe), flexibility in planning, and operational simplicity.

5. Conclusion

Hybrid SONET/ATM technology offers additional network efficiencies. Integration of SONET and ATM services within the same network provides consolidation of both network and operations management and may potentially reduce the overall cost of the network by allowing the same equipment to be efficiently used for both STM and ATM services. By incorporating ATM functionality into SONET equipment, an overlay network to handle ATM may be unnecessary, potentially resulting in fewer network elements and facilities to manage.

Hybrid STM/ATM rings lead to significant bandwidth savings when compared to existing conventional SONET rings. Improvements are mainly due to service multiplexing and statistical multiplexing of ATM traffic. With service multiplexing, there are more benefits in the access than in transport. With statistical multiplexing, the hybrid rings provide more opportunities in both access and transport.

A detailed analysis is needed to determine the hybrid STM/ATM operations economics and whether the additional operations costs can be offset by equipment economic saving. Another analysis is also needed to quantify the potential benefits of the hybrid DCS. These are the subject of an ongoing research.

References

[1] Bellcore, GR-2837, *ATM Virtual Path Functionality in SONET Rings*, Issues 2 and 3, Dec 95/Oct 96.

[2] Bellcore, GR-2891, *SONET Digital Cross-Connect Systems with ATM Functionality*, Issue 2, June 1996.

[3] Bellcore, SR-3945, *Evolving your SONET network to ATM*, May 1996.

[4] J. Sosnosky, "Potential Network Applications for Virtual Path SONET ATM Rings", National Fiber Optic Engineers Conference, NFOEC'94.

List of Acronyms

ADM	Add-Drop Multiplexer
ALAN	Advanced LAN
ATM	Asynchronous Transfer Mode
BLSR	Bidirectional Line-Switched Ring
BSS	Broadband (ATM) Switching System
CES	Circuit Emulation Service
CPE	Customer Premise Equipment
DCS	Digital Cross-connect System
NE	Network Element
SAM	Service Access Multiplexer
SPE	Synchronous Payload Envelope
STM	Synchronous Transport Mode
UPSR	Unidirectional Path-Switched Ring

Core and ATM Networks
D.W. Faulkner and A.L. Harmer (Eds.)
IOS Press 1997 © AKM

Comparison of IP and PNNI Routing for the Use in Intelligent Routing with Respect to Survavibility, Congestion and Mobility

Ilse Lievens, Piet Demeester
University of Gent - IMEC, Sint-Pietersnieuwstraat 41, B-9000 Gent, Belgium
ilse.lievens@intec.rug.ac.be

Abstract. In this paper, the concept of intelligent routing is introduced, with the integration of survivability, congestion control and support of mobility. All these aspects focus on rerouting of connections. Since the success of the Internet and the effort put in ATM routing, the routing in IP and PNNI is looked at to see for interesting reusable features. A newly developed algorithm, the Superposition DRA, is then presented.

1. Introduction to intelligent routing

Efficient usage of the resources in a communications network is of great importance as well for the user as the service provider and network operator. In an ATM network, also the Quality Of Service (QOS) will play an important role. Therefor the idea of intelligent routing is introduced, which can be interpreted as to use the available network resources as efficient as possible. In a network environment, this corresponds to integrating the aspects of network survivability, congestion control and support of mobility.

Survivability looks at the ability of the network to survive network element failures. We primarily focus on algorithms that find alternative routes to replace the interrupted ones, by making use of distributed restoration. For congestion control in intelligent routing, we look at the connection level, rerouting new and existing traffic around the congested area is the key issue. With mobility it is important that users can keep their connections without interruptions and that routes to mobile users are routed as efficiently as possible. Summed up, in all these fields a rerouting of an affected connection has to be done, which comes to searching for alternative routes in the network.

In this paper, a new algorithm is presented, aimed at providing survivability in the network, in particular in areas of the network which are heavily loaded or nearly congested. For that development, we have studied the routing mechanisms in IP and PNNI with the purpose of reusing some interesting features. The most interesting items in IP and PNNI are discussed in Section 2. The state of the art in survivability or more specific distributed restoration is given in Section 3. Section 4 presents the principles of the new algorithm. We conclude with future work and some conclusions in Section 5 and 6.

2. IP and PNNI routing techniques

2.1 Internet Protocol

Application domain. The success of the Internet is enormous, its size is said to double every year. In this datagram environment, the routing protocols provide a basic function for packets to be delivered to destinations. This packet delivery service is a best-effort one.

Hierarchy. The Internet architecture [1] consists of a set of Autonomous Systems (ASs) interconnected through gateways. An AS is a set of routers and networks under the same administration, with internal connectivity. The routing within an AS is provided by Interior Gateway Protocols (IGP), routing between ASs is done by Exterior Gateway Protocols (EGP).

Algorithms. The recommended IGP is the Open Shortest Path First (OSPF) protocol. It is a link state protocol, which implies that a large topology map containing an overview of links with related information of the AS is built in the nodes, which can then compute the shortest path to each destination. This map must be identical in all nodes, and is built by exchanging states of links between each other. For the route computation in the nodes, a shortest path algorithm, e.g. Dijkstra's, is used. Several metrics can be used: largest throughput, lowest delay, lowest cost, highest reliability. However no real-time load information can be taken into account.

The recommended EGP is the Border Gateway Protocol (BGP). The main function is to exchange reachability information between ASs. This provides full connectivity between the ASs. This implies that no full topology of other ASs is known within an AS, the border routers only know to which destinations they can relay packets.

2.2 Private Network-Network Interface

Application domain. For ATM networks, PNNI (Private Network-Network or Network-Node Interface) [2] has been recently specified for routing between private networks and switches, it includes a routing and a signaling protocol.

The goals of PNNI are scalability since large ATM networks are expected, simplicity to ensure that even small networks are able to apply PNNI, policy supporting for enabling routing cross network or provider boundaries, and QOS support.

Routing hierarchy. The routing architecture (Figure 1) is made hierarchical for scalability reasons. At the lowest level of the hierarchy, the ATM network is organized as a collection of Peer Groups. A Peer Group (PG) is a set of nodes, which all exchange information to maintain an identical view of their group, implying identical databases in each node of a PG. One node is PG Leader, which collects and aggregates the information that makes it possible to represent the PG as a single logical node one level up in the hierarchy. At that level, there is a collection of logical nodes, each representing an entire PG of the lowest level. At their turn these logical nodes are organized into PGs, in which the same information exchange takes place as at the lowest level. This hierarchical layering can then be repeated recurrently to complete the full PNNI routing hierarchy. It is important to notice that each node has only full knowledge of the topology within its own PG.

Apart from the internal topology of the PG, the nodes also need to know how nodes in other PGs can be reached. Therefore the concept of Uplinks is introduced (see Figure 2).

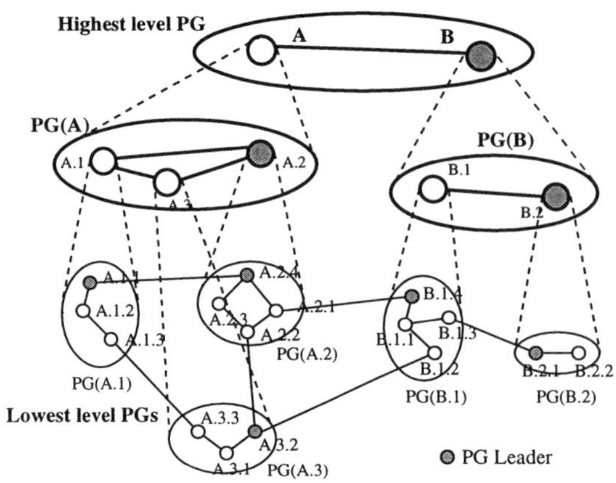

Figure 1 PNNI routing hierarchy

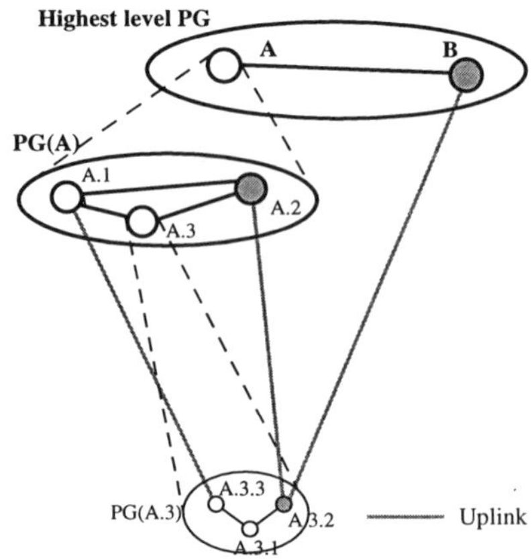

Figure 2 Uplinks for PG (A.3)

An Uplink is required when two nodes belonging to different Peer Groups (e.g.A.3.2 and B.1.2) have a link between them. Then an Uplink is established in the hierarchy from one node (A.3.2) towards the representation of the other node (B) in the higher level, which is the one in which both nodes are represented for the first time in the same PG.

Algorithms. For collecting the routing information and building the topology databases in the nodes, a link state based protocol is used. The database forms a topology map, containing full topology information on the own PG and information on the parent PGs, through the Uplinks. The map also stores resource information for enabling selection of 'best' paths. The amount of descriptive information is reduced for scaling reasons.

Since ATM is connection-oriented, path selection has to be done carefully. Therefor PNNI applies source routing, in which the source is responsible for selecting the entire path to the destination. Since PNNI has a hierarchical architecture in which the nodes do not know all the details of the entire topology, they cannot layout the path till the tiniest detail. Therefore PNNI specifies a hierarchically complete source route. The path is indicated as a list of sub-paths, each representing one hierarchy. When entering another PG, the route through that PG is learned at that moment from examining the border node's database. By this, the full path can be established, although the source node has no full overview.

2.3 Interesting features for use in new algorithm

After having discussed some important characteristics of IP and PNNI routing, the aspects that seem useful in the process of searching an alternative route are listed.

First of all, the hierarchy in the network - certainly in PNNI and less formal in IP - is an attractive feature, considering the importance of scalability. Together with this scalability, also the Uplinks are important, for a restricted, though complete view on the network's topology. Hierarchically complete source routes can be used in gathering alternative routes for intelligent routing, while also the notion of gateway nodes can play an important role.

3. State of the art in distributed restoration

Several techniques exist for providing survivability in the network. Basically they all search for an alternative route to replace an affected route. Here distributed restoration is considered, in which the network nodes themselves will search by exchanging messages between each other. The advantage of having no central control is the accuracy as well as the robustness since the actual state of the network is taken into account.

For the exchange of the messages a protocol is required for start, process and stop criteria. Several protocols can be found in literature, mostly distinguishing in the way the search is started. In the Sender-Chooser approach [3], the search is started by one end-node of the failed link or connection, towards the other side of the failure. In the Two-Prong protocol [4], a route is searched from both sides of the failure. This is more complex, but also faster. Another complex issue, due to the distributed nature, is resource reservation on candidate routes. When applying reservation during search, a solid release procedure is required.

Research has already been performed in the area of distributed restoration. This has resulted in algorithms for SDH networks [3,4,5,6] which are currently operationally used. Algorithms for distributed restoration in ATM networks [7,8,9,10] are still in a phase of research and development. Yet, ATM [11] offers several features, which can be successfully applied for a better restoration. E.g. the bandwidth allocation policy of ATM is of great help for the efficiency of the global use of the network: restoration messages will only take bandwidth when they are required. In addition, there is also the independence between a connection's route and its bandwidth: the route can already be downloaded in the translation tables of the node, without any bandwidth being assigned to it. When needed, the route can take in its bandwidth. This has already been used for restoration purposes [7].

4. The Superposition Distributed Restoration Algorithm

4.1 General design choices

The Superposition DRA is a distributed algorithm, searching for alternative routes. It applies path rerouting, which means that the entire connection will be replaced. It is intended to search long alternative paths in a reasonable time, to avoid crowded network areas. Because of the distributed real-time search, accurate real-time information like network load can be taken into account when searching for spare capacity.

It is based on the Two-Prong protocol, implying that both endnodes of the affected connection will start with sending of messages to look for spare capacity in the network.

4.2 Restoration hierarchy

The basic idea is to make use of a two-level hierarchy in the network (Figure 3), in which the lowest level, being the local level, is partitioned into groups of nodes, called local groups. These local groups are represented as global nodes in the higher level, being the global level, thus having a logical network in the global level which is much smaller in size. The local groups in the local level are isolated from each other, the inter-local group connections occur in the global level. Hierarchy can already be present in the network - refer to PNNI where a hierarchical routing structure is used. In that case the partitioning in local groups can be based on address prefixes. If the network has a flat structure, the network has to be partitioned, using some arbitration based on local connectivity rules available in graph theory.

A route in the layered network can be represented as a concatenation of route parts in the local groups, connected by global level links (Figure 4). Therefor, the alternative route

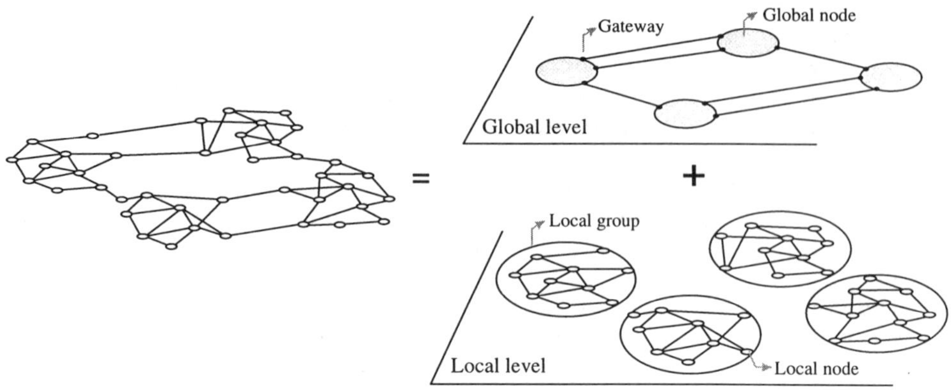

Figure 3 Network representation

is searched at the two levels. At the global level a route is searched between the global nodes, in which the end nodes of the affected connection are situated. At the local level, routes in the local groups must be found mainly between gateways, being nodes at the border of a local group.

4.3 Restoration process

The gateway nodes play an important role in the restoration process. They are visible as well in the global level (as ports of the global node) as in the local level (being border nodes of local groups). To speed up the restoration process these gateway nodes are activated as soon as possible after the occurrence of a failure. Therefor HyperLinks (Figure 5) are pre-established between every local node and all the gateways of this local group, and between all the gateways of the same local group. These HyperLinks are in fact pre-established ATM Virtual Paths (Figure 6), set-up in a pre-restoration phase, to be low-bandwidth but permanently available channels. When a failure occurs, the end nodes of the affected connection(s) inform the gateway nodes of their local group through these HyperLinks. Then the gateways forward the activation towards the other global nodes. This is a fast process since the connections are already established, only switching must be performed in the intermediate nodes. It must be stressed that these HyperLinks are only used for the algorithm's "signaling" requirements, they are absolutely independent of the

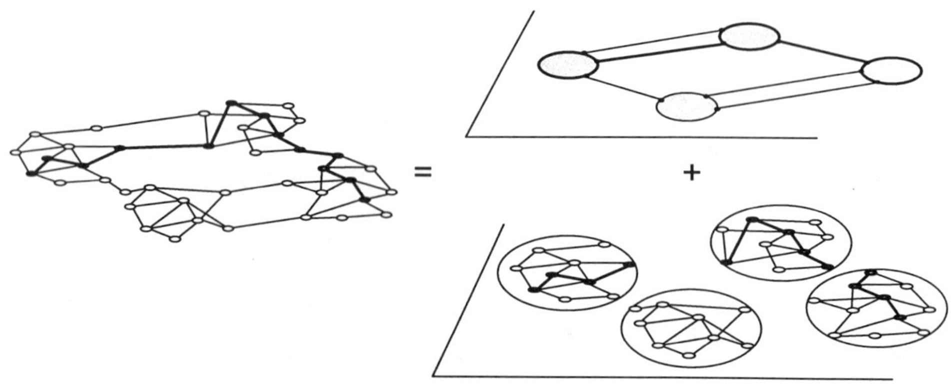

Figure 4 Route representation

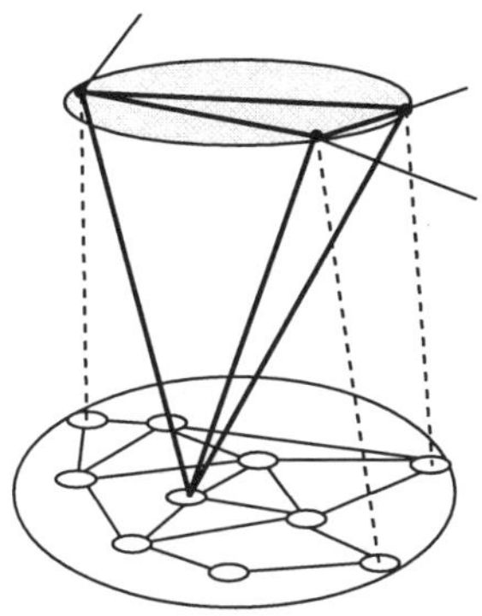

Figure 5 HyperLink view

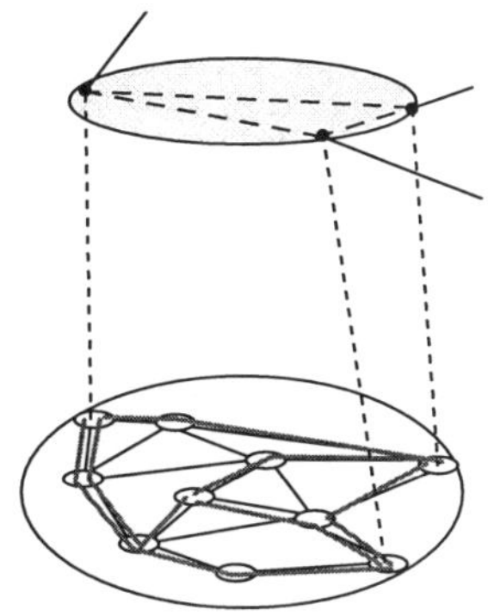

Figure 6 VP view

actual route a restoration path will follow once it has been found.

Once gateways are activated they also start a local level flooding process in the local group they belong to, to find an actual route through the local group. In this way, different simultaneous search engines are started at different places in the network, aiming at accelerating the search process (Figure 7). The restoration route will thus be found in pieces and it is the task of one node to collect all these pieces and create one valid route out of it.

In the algorithm there are three levels of search messages: search messages inside an local group (FL from Flooding), search messages through the HyperLinks to gateways in the same local group (HL from HyperLink) and search messages between two global nodes (FW from ForWard). On all these levels hits can occur, indicating that a candidate route part has been found (Figure 8).

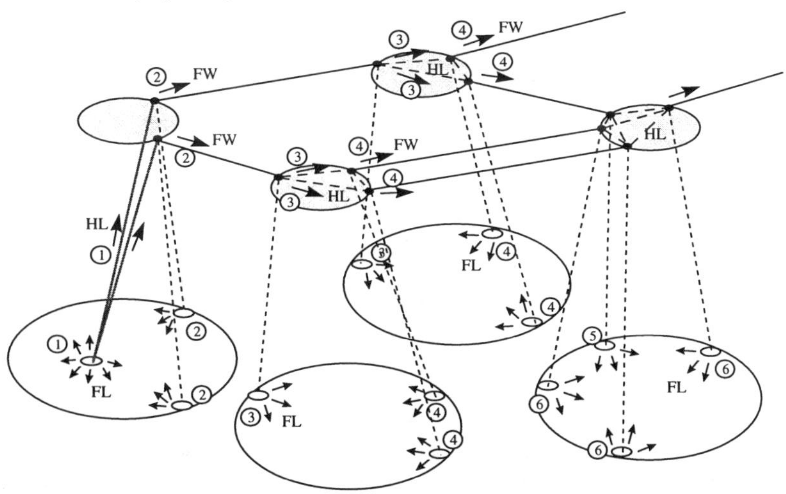

Figure 7 Two-level flooding

Information on these hits is stored in particular databases in the network. Part of the information is stored in the gateways, and the rest in the Chooser (CR) node, i.e. one of the connection's endnodes. The Chooser node builds a hierarchically complete overview of the restoration area, so that he can choose the actual restoration route(s) only with information visible to him. E.g. in Figure 9, the route specification by the CR will be [CR-g-d-e]-[GN4-GN2-GN1], being a hierarchically complete route. This implies that to the Chooser node a

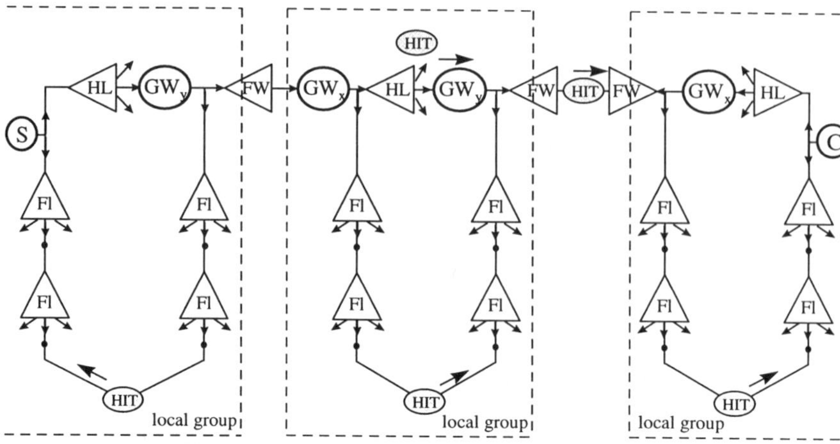

Figure 8 Generic representation of the algorithm

global node is merely a switching function between two ports (i.e. gateways) with a certain throughput. The actual information on the route between these gateways inside the local group is not relevant to the Chooser, and is only stored in the gateways of that local group.

During the searching phase, no reservation of the spare capacity is done. This non-reservation policy has the advantage of avoiding complex release protocols in case of reservations which are no longer needed since other routes were found. Since the Chooser has an overview of the local group in terms of available capacity, it is the only node which will actually allocate bandwidth to routes and thus pick out the restoration route among several possible candidates. By keeping a single overview, over-allocation problems are also avoided.

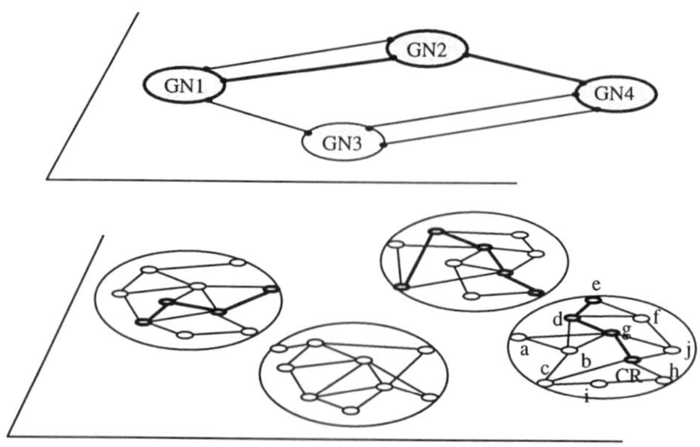

Figure 9 Hierarchical complete route from Chooser (CR)

To conclude this section, we can say that a two-layered algorithm has been developed that searches for alternative routes on both levels, with one node of the affected connection assembling the different parts into one route.

5. Future work

So far no implementation of the algorithm has been performed. This implies that no exact data on the algorithms' performance are available. Yet a good performance is expected because of the simultaneous searches in different parts of the network, as opposed to the classical way of a distributed restoration algorithm, where the search starts from one or two points in the network.

Typically this algorithm will perform well for finding long routes within reasonable search time. This implies that it can be used for routing around areas in the network which are in a congested or nearly-overloaded state, making an integration between survivability and congestion control in intelligent routing.

Moreover, the nature of the algorithm is generic, which means that reuse of functionalities and procedures at the different levels is possible.

The algorithm will be implemented in the simulation tool OPNET, with use of a realistic reference network already used for assessing the performance of other restoration algorithms.

6. Conclusions

In this paper, an overview of the main routing functions of IP has been given. The ATM PNNI routing has been discussed as well. From this study the most interesting features from both fields have been transferred into the area of rerouting. Out of this, a new algorithm that merges issues in IP, PNNI and distributed restoration has been developed, with application in intelligent routing for survivability and congestion control.

Acknowledgment

This work has been supported by the Flemish Government through the IWT project ITA/950214/INTEC.

References

[1] Christian Huitema, 'Routing in the Internet', Prentice Hall PTR, New Jersey 07632, 1993.

[2] The ATM Forum Technical Committee, "Private Network-Network Interface Specification Version 1.0 (PNNI 1.0)", af-pnni-0055.000, March 1996.

[3] W.D. Grover, "The Self-Healing Network: a fast distributed restoration technique for networks using digital cross-connect machines", Globecom 1987, pp. 1090-1095.

[4] C.E. Chow, J. Bicknell, S. McCaughey, S. Syed, "A Fast Distributed Network Restoration Algorithm", International Phoenix Conference on Computers and Communications, March 23-26, 1993.

[5] C. Han Yang, S. Hasegawa, "FITNESS: Failure Immunization Technology for Network Survivability", Globecom '88, pp. 1549-1554.

[6] H. Komine, T. Chujo, T. Ogura, K. Miyazaki, T. Soejima, "A Distributed Restoration Algorithm for Multiple-link and Node Failures of Transport networks", IEEE 1990, pp. 459-463.

[7] R. Kawamura, K. Sato, I. Tokizawa, "Self-healing ATM Networks Based on Virtual Path Concept", Networks '92, pp. 129-134.

[8] K. Struyve, P. Demeester, L. Nederlof, L. Van Hauwermeiren, "Design and Evaluation of distributed link and path restoration algorithms for ATM meshed networks", Lecture Notes in Computer Science, Vol. 1044, Procs. International Zurich Seminar on Digital Communications, ETH Zurich, Switzerland, February 1996.

[9] L. Nederlof, H. Vanderstraeten, P. Vankwikelberge, "A new distributed restoration algorithm to protect ATM meshed networks against link and node failures", Proceedings of International Switching Symposium '95, April 1995.

[10] I. Lievens, P. Demeester, "Design of a new high-performance restoration algorithm for survivability in ATM networks", to appear in NOC'97, June17-20, 1997, Antwerp, Belgium.

[11] M. De Prycker, "Asynchronous Transfer Mode, solution for broadband ISDN", 2nd edition, Ellis Horwood Limited, 1993.

Core and ATM Networks
D.W. Faulkner and A.L. Harmer (Eds.)
IOS Press 1997 © AKM

Routing in Integrated SDH-ATM networks

Graham Knight, Rong Shi

Computer Science Department, University College London, UK

Abstract : Many efforts have been made to manage homogenous B-ISDN networks across multiple domains. However, less effort has been made for hybrid B-ISDN network management. The ACTS **MISA**(**M**anagement of **I**ntegrated **S**DH and **A**TM networks) project addresses the issue of hybrid ATM and SDH management, aiming to automate the end-to-end **GBC** (Global Broadband Connectivity) service in the integrated ATM and SDH network environment by realisation of the Global Broadband Connection Management (**GBCM**) service according to the TMN(**T**elecommunication **M**anagement **N**etwork) management architecture. The routing issue is fundamental to automating GBC so as to choose an optimal or nearly optimal route for the purpose of meeting various constraints imposed by customers and, at the same time, achieving good network performance. In this paper we discuss the routing issues, within a domain and across domains, in the MISA environment. By analysing the characteristics of the MISA network and MISA management system, the MISA routing requirements are identified. Through reference to ideas from the existing datagram network routing approaches and an understanding of the distinction between MISA routing and datagram network routing, a MISA inter-domain routing scheme is proposed for use in the pan-European demonstration. The scheme is based on a link-state algorithm tailored to allow for QoS constraints and simple policy routing.

1. Introduction

A great deal of effort has been made towards achieving global homogenous ATM or SDH network management and considerable progress has been made. However, in today's increasingly deregulated telecommunications world, it may be that not all domains support the same B-ISDN technology, furthermore in certain domains, instead of having a homogenous ATM or SDH network, a hybrid with ATM over SDH fulfils B-ISDN tasks. The management of such a global hybrid network is understood to be very complex regarding integrated management of ATM and SDH networks as well as co-operation between administrative domains. In particular, end-to-end **G**lobal **B**roadband **C**onnections (**GBC**) are set up and configured today by manual actions using faxes and telephone calls across different telecom operators. A network management system which enables automation of **GBC** formation is highly desired. One benefit of such a management system is that the heavy burden of manual actions for setting up and configuring the end-to-end (**GBC**) is released. Another benefit is that it pioneers in the direction of managing different inter-connected networking technologies in a unified way.

The ACTS **MISA**(**M**anagement of **I**ntegrated **S**DH and **A**TM networks) project aims to address the above issues by building a TMN(ITU-T M.3000) compliant **G**lobal **B**roadband **C**onnection **M**anagement (**GBCM**) service in a multi-provider and multiple domain environment[Ref 7]. A routing mechanism is essential in the GBC formation procedure and this mechanism must make efficient use of network resources as well as satisfying customer QoS and policy requirements. The GBC is perceived as being able to support 'a wide range of audio, video and data applications' with different traffic and QoS requirements. Besides these, there may be policy related requirements regarding the GBC customer's preference, for example, in choosing a network. Therefore the MISA routing can be seen as *a process of selecting the best end-to-end paths for the purpose of satisfying various GBC requirements, minimising the cost, whilst at the same time meeting network provider's various constraints and achieving good network performance in the multi-provider and multi-domain environment.*

In this paper, we mainly discuss MISA inter-domain routing issues. In addition, the QoS-based intra-domain routing is described. The remainder of the paper is organised as follows. In Section 2 the MISA system architecture is briefly introduced and the nature of MISA routing is analysed. In Sections 3, the MISA routing requirements in terms of inter-domain routing are identified. In Section 4, the proposed inter-domain routing scheme in the MISA is introduced at functional architecture level. Finally, some unsolved problems and open issues are discussed in Section 5.

2. The MISA System Architecture

The MISA system[Ref. 4] can be subdivided into two separate but tightly related systems - the *MISA network* and the *MISA management system*(see Figure 1). Obviously, the MISA network is a global broadband ATM and SDH hybrid network on which the GBC is established. The MISA management system performs management actions on the MISA network through configuration, performance, fault, accounting and security management functions. The MISA management system is designed according to the TMN architecture and contains as many co-operating MISA management systems as the number of administrative domains. Here the administrative domain (**AD**) is defined in terms of the PNOs' administrative authority.

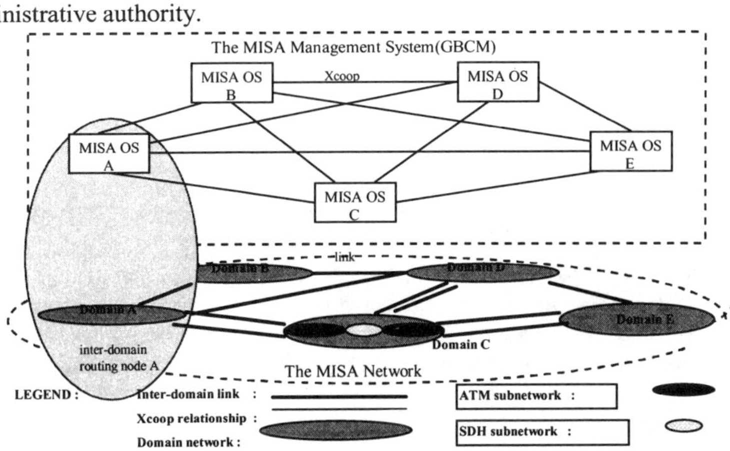

Figure 1. The MISA system

2.1 Inter-Domain Routing and Intra-Domain Routing

In the MISA system, GBC set-up is realised by the GBCM in which the routing mechanism is embedded. Suppose, a customer requests a GBC from London to Athens for a real time Variable Bit Rate service, with source traffic descriptor, QoS and domain policy constraints. The GBC request firstly is submitted to a British telecommunication operator which checks the destination address' integrity and finds an optimised route from London to Athens which satisfies the GBC QoS and policy constraints according to the network resources available.

As described above, each domain is only visible to the other domains as a black-box and does not expose its internal topology or ATM (SDH) sub-network capabilities. When sources and destinations of a GBC request fall in different domains, routing operated at the level of end-to-end GBC involves **inter-domain routing** which makes routing decisions based on the domain network capacity and domain inter-connection information for the purpose of meeting the various constraints given by both customer and network provider. The route achieved at this level is the best or near best set of ADs with ingress(on-ramp) and egress(off-ramp) access points. Then each GBC request can be broken into a series of GBC sub-network connection requests each of which is passed to the MISA OS residing in the selected AD across *Xcoop* interface[Ref. 4]. During the whole process of the inter-domain routing, a MISA OS plays the role of an inter-domain routing node Figure 1), which is responsible for collecting and propagating network state information via *Xcoop*, maintaining the routing information in the Management Information Base(MIB) and calculating routes for a GBC request.

In the case of a domain with two or more ATM islands connected by a SDH sub-network (See Figure 1 sub-network C) or accessing ATM via a SDH circuit, the GBC sub-network connection cannot be supported by a single ATM or SDH sub-network. Therefore, an **intra-domain routing** mechanism is essential at the level of GBC sub-network connection to choose a route. This route is defined by a set of access points which includes the termination access points of the GBC sub-network connection as well as the access points inter-connecting the ATM and SDH sub-networks. A GBC sub-network connection request is again broken into ATM sub-network connection requests and SDH sub-network connection requests. Intra-domain routing is invoked by inter-domain routing to form the domain network capacity information based on the ATM sub-network and SDH sub-network capacity information as well as the ATM and SDH inter-network topology information. In the case of a domain with only a single ATM or SDH network, **intra-domain routing** performs a direct mapping function between the GBC sub-network connection and the ATM or SDH sub-network connection.

The **inter-domain routing** and **intra-domain routing** are tightly coupled since **inter-domain routing** relies on the domain network capacity information that **intra-domain routing** provides. Further, inter-domain routing resolves each hop on the route to a pair of end-points which can then be presented to the intra-domain algorithm.

3. MISA Inter-Domain Routing

3.1 QoS and Policy Requirements

As stated in Section 1, MISA inter-domain routing must be QoS and policy routing based. In order to support a wide range of B-ISDN services, the GBC must accommodate both

ATM end-to-end semi-permanent paths (ATM Path Provisioning - APP) as well as SDH end-to-end leased lines(SDH Path Provisioning - SPP). The set of QoS parameters defined at the ATM layer are Peak Cell Rate, Sustainable Cell Rate, Cell Delay, Cell Delay Variation and Cell Loss Ratio. SDH network connections are only characterised by throughput, Bit Error Ratio and end-to-end delay. In order to make the GBC accommodate both APP and SPP, the QoS and traffic metric of the GBC must be the union of both traffic and QoS parameter sets of ATM and SDH [Ref 2]. In fact a sub-set of these parameters including Peak Bit Rate, Sustainable Bit Rate, delay and jitter has been chosen as the first stage GBC QoS parameter set. The other ATM QoS parameters will be considered in the next stage.

The policy factors have two aspects. On the one hand, ADs may have different administrative policies concerning which traffic may be transported on their network or which customer's traffic is allowed to terminate etc.. On the other hand, GBC customers may express their preferred PNOs in a GBC request. So the MISA inter-domain routing is expected to support simple policy routing which only concerns the policy factors that are relatively stable. Both *stub control* (which specifies who can send traffic to my internal resources) and *transit control* (which specifies who can send transit traffic through my domain) must be supported by the MISA inter-domain routing mechanism.

3.2 Explicit or Implicit Paths

In the classic datagram-based network, although the network holds sufficient information to be able to route datagrams from A to B it does not retain knowledge of "connections" or "paths" as such. In a virtual-circuit based network connections, paths are explicitly part of the state information maintained by the network and there is a clear separation between the connection set-up phase (during which route selection takes place) and the data transfer phase. The networks used in MISA are of the latter type and we expect that, normally, once an inter-domain GBC has been reserved between A and B, the reserved network resources will be dedicated for the GBC use during its lifetime. Thus we do not expect to see the highly dynamic re-routing associated with classic datagram networks. MISA routing will adapt, but only to relatively long-term load trends or major topology changes.

3.3 Frequency Requirement of Routing Information Update

Routing information and route computation are the essence of routing in general. In order to avoid GBC QoS degradation and network congestion, MISA inter-domain routing must be able automatically to adapt to inter-domain network topology changes as well as domain network capacity and policy changes. Ideally, optimal routing decisions rely on up-to-date routing information which is obtained at the moment a GBC request is made. However, given that our objective is to adapt routing to long-term trends and major topology changes, it is not worth paying too heavy a price to have timely information. In fact a quasi-static model can be used in the MISA inter-domain scheme, in which the routing information remain unchanged in a certain period or until significant changes occur, such as inter-network topology change, domain policy change and tremendous domain network capacity change. It must be remembered that there will always be a route set-up phase following the route calculation. Occasionally we will find that a domain cannot satisfy a request to provide a GBC segment at the required QoS because its circumstances have changed since it last provided routing information. When this happens an alternative route must be found. Our aim in setting our update strategy must be to keep the frequency with which these recalculations are required within acceptable bounds.

3.4 Centralised or Distributed Route Calculation

In a *fully centralised* scheme, a single entity is responsible for pre-calculating all routes for every pair of source and destination and distributes results to the constituent nodes. There is an assumption that all information relevant to the calculation of a routes is simultaneously available in one place. In such a situation it is possible to guarantee loop-free routing and that the routing information distributed to nodes is consistent. It is difficult to see how such a scheme could work in the MISA context. Firstly, there is an administrative problem of multiple independent domain administrations agreeing on the setting up of such a scheme and then jointly managing it. Secondly, QoS and policy based routing means that there are many "best" routes between a particular source/destination pair depending on the particular values of QoS and policy parameters in the customer request. It is not feasible to pre-compute and distribute multiple routes just in case they may be needed later.

In a *fully distributed* scheme each node or domain performs routing calculations for its local environment and exchanges routing information with other nodes(domains). No single entity has knowledge of any entire route. This means that inconsistencies and routing loops may arise unless precautions are taken. Although it is mainly associated with datagram networks there is no reason why this "hop by hop" approach should not be considered for MISA.

Mid-way between the fully centralised and fully distributed schemes is classic "source routing" in which the initiating node(domain) is responsible for negotiating a route through the network. In datagram networks the source route has to be explicitly stated in all datagrams. In a virtual-circuit network the initiating node (which need not necessarily be the source node) would select the route and then negotiate with all the nodes along the route in order to reserve resources and set up the connection. Knowledge of the connection is then held in the per-node state information. We call this scheme "*initiator controlled*" routing. Since the whole of the route selection is performed by a single entity, it is possible to guarantee loop-free routing within an initiator controlled scheme.

In the MISA environment, when a domain agrees to set up an inter-domain GBC for a customer, a contract is established under which the domain undertakes to set up and maintain the GBC for a particular price. This is the only contract the customer knows about ("one-stop shopping"). All negotiations with other domains are hidden and are the responsibility of the initiating domain(InD). When the InD sets up the GBC it negotiates with other domains and agrees sub-contracts with them. The negotiations may conform to the distributed or initiator controlled schemes or perhaps to some combination of the two.

Initiator controlled routing is attractive in a multi-domain environment as it keeps the route selection exclusively in the control of the domain which has the contract with the customer. This allows the domain great flexibility in the sort of contract and guarantees it offers to its customers. Further, it is straightforward to avoid loops in an initiator controlled scheme.

3.5 Routing Information Collection and Route Calculation

Dynamic routing composes *routing information collection and dissemination protocols* and route calculation algorithm. Regarding routing protocol, A number of inter domain routing protocols such as EGP (Exterior Gateway Protocol), BGP (Border Gateway Protocol) and IDRP (Inter Domain Routing Protocol) etc. [Ref 6] have been extensively researched in the context of *datagram networks*. Each of these approaches addresses the problem of distributing routing information and it is worth considering what relevance this work has for MISA. It is evident that MISA routing and datagram network inter-network routing protocols have many common requirements such as both must have the mechanism for collecting and maintaining network state information across domain. However datagram networks maintain no connection state and try to adapt to load changes on a comparatively short time-scale. A more subtle distinction is that in typical datagram networks the links used for exchange of routing information are precisely those which carry the data. This means it is relatively easy to exchange routing information with neighbours, relatively hard to do so with distant nodes. In the MISA context routing information is exchanged between OSs which are on a conceptually independent control network; it may be as simple for an OS to consult the OS of a distant domain as it is for it to consult the OS of a neighbour.

Dynamic routing can be classified as *distance-vector* or *link-state. A distance-vector* requires that each node maintains the "distance" from itself to each possible destination and sends this information to its neighbours at intervals. New distance estimates are computed using the information in neighbour's distance vectors. In a *link-state*, each node has complete information on the network topology so that each node can carry out route computation independently. Nodes send out updates to all other nodes by broadcasting either regularly or when a node detects any changes in local link distances.

The chief problems of *distance-vector* can be categorised as :

- Slow convergence of network state information: if the maximum diameter of a network is N hops, N routing updates are needed to propagate changes across the network.
- Routing information inconsistency: due to the slow convergence it is possible that not all domains will receive network state updating information timely. This leads to inconsistent views of the networks state and may result in routing loops.
- *Distance-vector* typically only allows a node to find the next hop for a specific destination. Therefore, they are not suitable for an initiator-controlled scheme.
- QoS and policy routing : the distance-vector algorithm is not suitable for QoS and policy routing as it is impossible for each routing node to maintain distance information to each potential destination in terms of different QoS parameters.

Link-state algorithms avoid the defects above. They also have the advantage that each node accumulates complete information about the network. This is required if we adopt the pure initiator controlled routing scheme. All these suggests strongly that a link-state-like algorithm would be better choice for MISA. However, conventional link-state algorithms employ a flooding strategy to propagate routing information. Whereas, setting up semi-permanent CMIP associations between all the OSs (via Xcoop) for forwarding routing update messages is sufficient to the MISA system.

4. The Functional Architecture Of MISA Inter-Domain Routing

Having analysed the MISA inter-domain routing requirements, initially for use in the pilot MISA system, we propose an inter-domain routing scheme with *link-state based routing information collection by using periodic fetching and notifying significant network state change mechanisms, initiator-controlled routing mechanism and QoS and policy based route calculation algorithm.*

This implies that each inter-domain routing node(MISA OS), should obtain full knowledge of the inter-domain network topology in its MIB by querying other routing nodes using statically configured identification information of the remote MISA OSs after the local MISA OS is installed; should notify all other domains of major state changes; should periodically collect network state information from other domains; should maintain local domain network state information in its MIB, which includes the local network capacity, inter-domain links and domain policy; should calculate the route based on full knowledge of the inter-domain network topology by using the QoS based routing calculation algorithm. Accordingly, five fundamental functional elements(shown in **Figure 2. MISA inter-domain routing functional model**) are essential to keep routing information consistent and to compute optimal inter-domain route for a GBC.

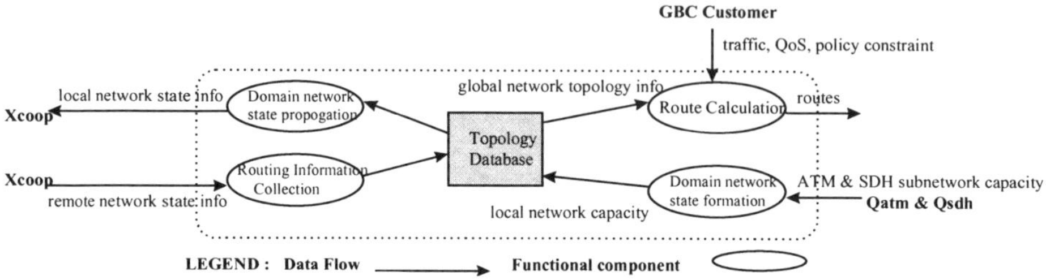

Figure 2. MISA inter-domain routing functional model

Domain Network State Formation : This element contains the intra-domain routing mechanism and is responsible for measuring local network capacity in the form of domain network traffic load allowance and QoS tolerance. A *domain network state formation* action occurs only when there is a serious ATM or SDH sub-network state change. In the QoS-based routing case, local network capacity is measured as potential domain network connections with the perceived traffic allowance and QoS tolerance synthesised from ATM and SDH sub-network capacities. The local domain network state is stored as local routing information in the *Topology Database.*

Domain Network State Propagation : As the routing is link-state based, every MISA OS will have full knowledge of the MISA network topology and capacity. Each OS is obliged to propagate any serious changes regarding its local network capacity, inter-domain link status and domain policy to all other routing nodes by using CMIS notification primitive via *Xcoop*. Different notifications represent different local network capacity change, such as a serious domain network capacity change, domain policy change, inter-domain link status change(topology change) and inter-domain link capacity change.

Routing Information Collection : This function is mainly responsible for collecting the remote network state change information by periodically fetching remote network state information or being notified of the remote network state change.

Route Computation : The best routes for a GBC is determined by using shortest-path algorithm based on the full network topology information stored in the *Topology Database* under the various constraints given in a GBC request. Since QoS based route calculation is a NP-complete problem, a heuristic solution based on human-knowledge has to be applied. Also due to the fact that the GBC is end-to-end based, the traffic and QoS required by a GBC must be jointly fulfilled among the domain networks crossed. Suppose the final route computed contains n domain networks each of which is identified as i, the criterion considering the traffic and QoS should be :

PBR[end-to-end] = Min(PBR[i]); SBR[end-to-end] = Min(SBR[i]); i = 1 to n

jitter[end-to-end] = Σ(jitter[i]); delay[end-to-end] = Σ(delay[i]); i = 1 to n

The best routes are based on a selected shortest path criteria which can be cost against traffic and QoS parameters or number of hops that a GBC traverses.

Topology Database : In addition to the full network topology information for inter-domain routing, the *Topology Database* maintains the local domain network state as well. The *Topology Database* is part of MISA OS MIB.

5. Conclusion

Inter-domain routing in the integrated ATM and SDH networks has been extensively studied in the MISA project under the TMN context. An inter-domain routing scheme has been proposed for use in the MISA pilot system by absorbing experience from various routing approaches in datagram networks and adapting these to the rather different MISA environment. In this paper, we have only briefly introduced the MISA inter-domain routing at architectural level. Various detailed routing issues under study in the MISA are not included, such as details of the algorithm for calculating the inter domain routes based on a set of QoS metrics and scaling issues in the routing information update mechanism.

In order to have an inter-domain routing system for the MISA robust enough for the real world with optimal functionality, further research work within various MISA trials and a simulated environment has been planned to address the following issues :

- QoS based route computation : issues regarding how to combine the multiple QoS graphs to form a single tree as basis of the final route selection for the purpose of QoS routing; what is the proper mechanism in solving the NP-complete problem; how to divide or recalculate the end-to-end QoS parameters such as end-to-end delay and the error bit ratio etc.
- scaling issue in routing information update : issues include the setting-up of the time-period for routing information collection and the threshold for determining the notification of serious network state change, etc..

All these work requires a great deal of effort to be spent in the next stage of the project.

6. References

[1]. Brijesh Kumar, "Policy Routing in Internetworks", Ph.D thesis, Dept. of Computer Science, UCL; 1993.

[2]. Celine Verdier and Magda Chatraki, "Quality of Service Requirements", MISA Consortium internal document; July 1996.

[3]. Graham Knight and Rong Shi, "Inter-domain routing in the MISA", MISA Consortium internal document; August 1996.

[4] MISA-Deliverable 3, "Initial MISA High Level Design", MISA Consortium, September 1996.

[5] ATM Forum User Network Interface ver.3.1, September 1994.

[6]. Martha Steenstiup, "Routing in Communication Networks", Prentice Hall, 1995.

[7] A.Galis et. al, "Towards Multi-domain Integrated Network Management for ATM and SDH Networks", Conference BSTW'96, Berlin, Germany, Oct. 96.

Core and ATM Networks
D.W. Faulkner and A.L. Harmer (Eds.)
IOS Press 1997 © AKM

The AMUSE Residential Multimedia Trials: Network Architecture and First Results[†]

Joachim CHARZINSKI, Johannes FÄRBER, Jürgen ENSSLE[‡]

University of Stuttgart, Institute of Communication Networks and Computer Engineering
Pfaffenwaldring 47, D-70569 Stuttgart, Germany
e-mail: {charzinski|faerber|enssle}@ind.uni-stuttgart.de

Matthias FRANK

University of Bonn, Institute of Computer Science IV
Römerstraße 164, D-53117 Bonn, Germany
e-mail: matthew@cs.uni-bonn.de

Abstract. The European ACTS project AMUSE (Advanced Multimedia Services to Residential Users) carries out ATM-based multimedia field trials in a heterogeneous environment. The AMUSE infrastructure is tested in six European countries under different technical and regulatory circumstances.

The network architecture, starting at a generic system configuration, adopts different options in the four domains of Server, Core Network, Access Network and Customer Premises Equipment. A media server provides Video on Demand as well as other interactive multimedia services. Additional gateways are provided for high-speed internet access and digital access to broadcast TV. The range of access networks used includes Hybrid Fibre-Coax (HFC) and Asymmetric Digital Subscriber Line (ADSL).

Service usage and user acceptance have been evaluated by automatic monitoring tools and by user questioning after the trial. In addition, traffic measurement results including packet size distribution and packet interarrival distributions are available from the first trials.

Keywords: Field Trial, ATM, Multimedia, Network Architecture, Access Network, Measurement

1. Introduction

1.1. Multimedia Trials in the European Environment

Several Interactive Multimedia field trials involving real residential users have been and will be conducted within the European ACTS project AMUSE (*Advanced Multimedia Services to Residential Users*). The trials, located in different European countries, all depend on ATM as the basic transfer mode. Parts of the trial equipment and some applications are shared between trials whereas others are site dependent.

Section 2 explains the general reference configuration and the specific HFC-based trial configuration. Section 3 lists the services offered in the trials and section 4 presents service usage and traffic statistics obtained from automatic monitoring from two trials, carried out in Munich, Germany and in Basle, Switzerland.

1.2. The AMUSE Consortium

The AMUSE consortium comprises several companies, telecom operators, research institutes and universities, combining their knowledge and experience from hardware

---

[†] This work was funded by the European Community ACTS project AC011.

[‡] The author is now with the Alcatel SEL Research Division, D-70499 Stuttgart

development to content provision and from theoretical studies to the implementation of user interfaces.

Equipment and Content Providers		**Telcos and Cable Operators**	
ITALTEL Spa (prime contractor)	I	IDEA S.C.	B
McCann Erickson Italiana Spa	I	Swiss Telecom PTT	CH
Siemens Nixdorf Information Systems	D	Deutsche Telekom AG	D
Siemens AG	D	Portugal Telecom / CET	P
Siemens Switzerland Ltd	CH	Telecom Italia Spa	I
ATEA	B	Post and Telecom Iceland	IS
Acorn Computers Ltd	GB		
GPT Ltd	GB	**Universities**	
Videotime	I	NTUA Institute of Communication	
Nyherji Ltd	IS	and Computer Systems	GR
Orckit	IL	University of Bonn	D
Research Institutes		University of Stuttgart / IND	D
Sirti Spa	I	University of Iceland	IS
Instituto de Engenharia de Sistemas e Computadores	P		

Table 1: The AMUSE Consortium

2. Network Architecture

The AMUSE trials implement different options for residential access to ATM-based broadband services. Figure 1 shows the Reference Configuration, which indicates the options used in the different network parts and the DAVIC reference points [1] in between. The server side consists of a *Video on Demand* (VoD) and interactive multimedia server, which will have a billing system attached in future trials. Optionally, there is a gateway to cable television offering *Switched Video Broadcast* (SVB). Internet access is provided either through a WWW proxy server in the Multimedia Server or by a separate Internet Gateway.

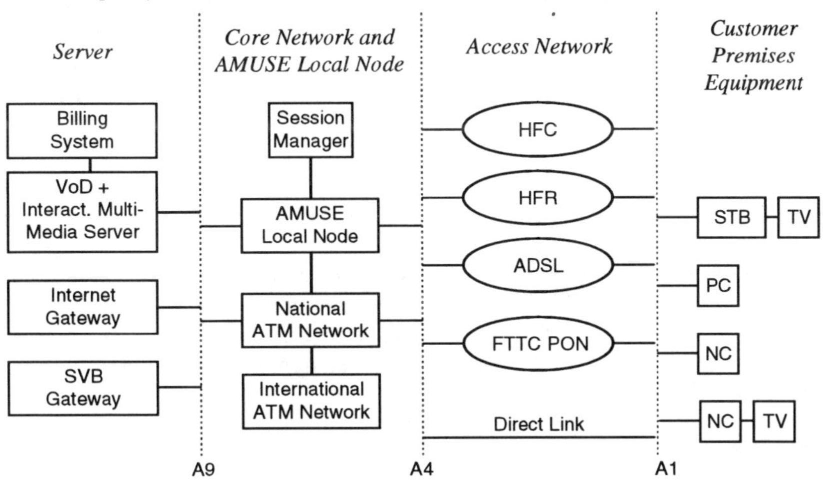

Figure 1: AMUSE Reference Configuration

The core network consists of a core switch (the *AMUSE Local Node*) and can optionally use parts of the national and international ATM network infrastructure. For the implementation of DSM-CC UN control functions, a session manager will be attached to the Local Node.

The access networks used in the different trials are based on existing infrastructure like HFC (*Hybrid Fibre Coax*, using TV cable networks) or ADSL (*Asymmetric Digital*

Subscriber Line, using the customer's telephone line) or use new infrastructure like HFR (*Hybrid Fibre Radio*, replacing the HFC coax network by a radio link) or PON (*Passive Optical Network*). Another option is not to use a special access network but to place a small ATM switch into the basement of each home to be connected ("direct link").

At the customer side, the usual equipment is a Set Top Box with a remote control, connected to a TV set, but the standard ATM interface allows also multimedia PCs or NCs (*Network Computers*) to be connected.

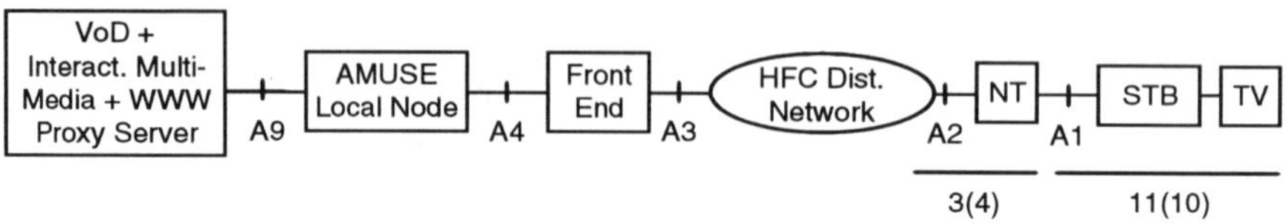

Figure 2: Munich and Basle Trial Configuration

The detailed results from the Munich and Basle trials reported in section 4 are based on the configuration shown in Figure 2, consisting of a video server with Multimedia server and WWW proxy server connected to an ATM switch at A9. The switch interfaces with an HFC access network serving three (four) houses with a total of 11 or 10 residential homes in Munich (Basle), which use the STB/TV combination. International interworking experiments have been conducted which show that this ATM-based platform can be flexibly used also for long distance connections, except for the control protocols, which may have to be modified to allow for long round trip delays without impairment of the perceived reaction speed.

3. Services

A whole variety of services are or will be offered in the framework of the AMUSE trials. Some of the corresponding applications are still being developed; others have been in use since the first trial. The most heavily used services in the trials so far were

- Internet Access: the set top box WWW browser gives access to the world wide web using a remote control for navigation
- Video on Demand (VoD): a choice of several movies and other films
- News on Demand (NoD): recorded TV news broadcasts of the past seven days. This service is going to be improved by news classification and indexing, so that personalised news services can be offered.

Other services offered are home shopping, home banking, tele games, info commercials on demand, travel information, travel agency and tele-voting services.

The amount and range of contents offered in the trials is a vital point also in the discussion of service usage results due to the limited number of different videos which could be offered in such a trial.

4. Trial Results

4.1. Service Usage Observation from Automatic Monitoring

One of the methodologies used within the AMUSE trials to evaluate the service demonstrations is the automatic monitoring of user actions. The objective of the automatic monitoring is to establish a usage profile to get an exact picture of when the participant uses the system and what services he uses most frequently. All relevant actions of the users are permanently monitored and evaluated automatically without additional effort for the user. Each action (e.g. selection, start, stop of a video, retrieval of a WWW document) is recorded in a log-file together with information like time stamp, STB user Id, amount of data transferred, type of data/filename of document, etc. This allows for a detailed offline analysis of various aspects concerning the service usage.

Questions on the usability of the system, the attractiveness and the user acceptance of the new services are best investigated by methodologies like heuristic evaluation, user observation, user interviews and questionnaires. Those methodologies are applied in the AMUSE trials as well as the automatic monitoring, but this paper focuses on the presentation of analysis results of the monitoring data. For more information on the other issues the reader is referred to [2] or [3].

Due to the fact that at the time of preparation of this paper the first two AMUSE trials in Munich (May - August 1996) and Basle (October 1996 - January 1997) have been finished, the description of results is focused on a comparison of both trials to point out similarities and differences. A comprehensive summary of results of the Munich trial can be found in [2] or [4], a detailed description of all results will be found in the AMUSE deliverables (up to now reported in [5]).

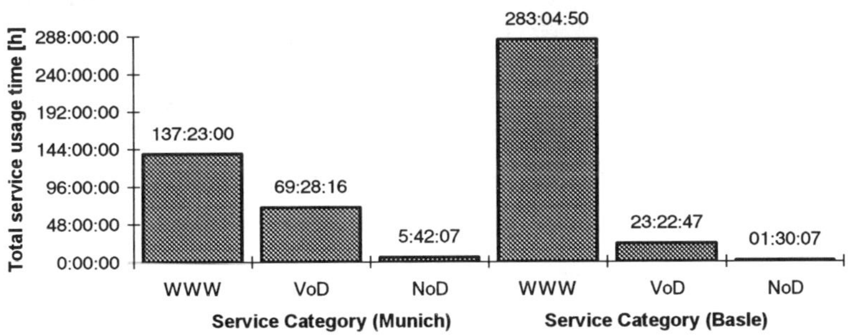

Figure 3: Total service usage for all service categories, Munich and Basle

Figure 3 shows the total usage time (in hours) for each of the three service categories provided (WWW, VoD and NoD) in the trials of Munich and Basle. Although the numbers give the total time for different trial period length and a different number of residential set top boxes (Munich: 98 days, 11 users; Basle: 105 days, 10 users), a breakdown of numbers to an average per day and single user allows to derive the same conclusions. A big large difference in terms of usage time spent on each service category shows that in both trials, the user interes focuses on particular services: The WWW service clearly is the most interesting service for the residential users. VoD is on second rank, followed by NoD usage, again with considerable difference. It is an interesting fact that the usage of WWW in Basle is much higher than in Munich, but on the other hand the usage of VoD and NoD in Basle is much lower than in Munich.

To compare the profile of interest of the individual users in a particular service, the following figures present the average usage time per day for each residential user of Munich and Basle trial (identified by their set top box STB Id.). The bars labelled as „M“ and „B“ show the average time per day over all STBs in Munich and Basle respectively.

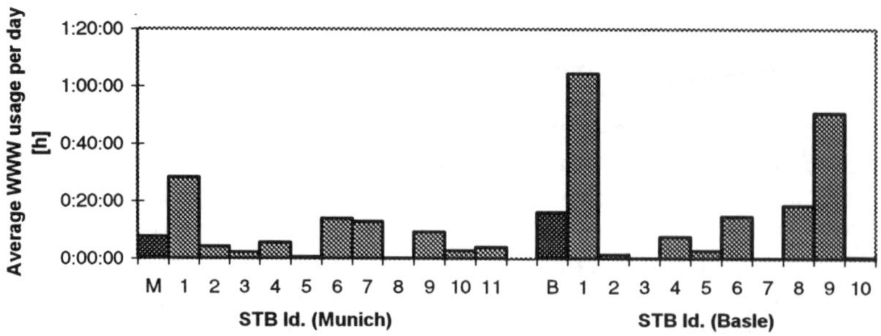

Figure 4: Average WWW usage per day for all Munich + Basle STBs

Figure 4 presents the profile for the WWW service. The users can be classified into three groups with distinct differences: There are a few top users with an average usage time high above the corresponding trial average (like users M1, B1 and B9), getting close to or even above 1 hour of usage per day. There are some more users ranging close to the trial average, like M6, M7, M9 or B6, B8). The rest (more than half of all users) have a rather small service usage time indicating little interest in the service (just to point at M8, B3 and B10, who even used the service less than one hour in total).

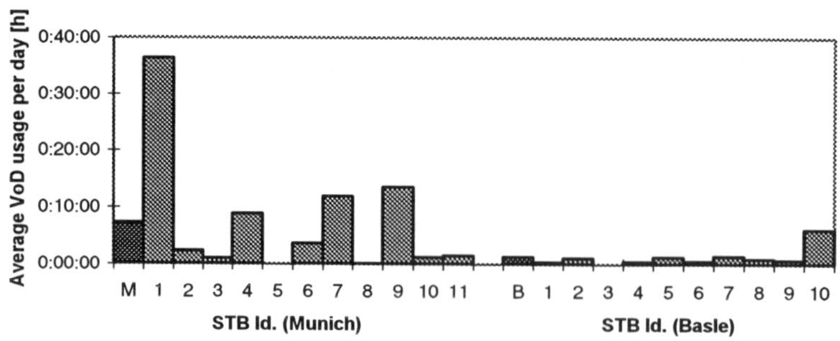

Figure 5: Average VoD usage per day for all Munich + Basle STBs

Figure 5 shows the profile for the VoD service, where the difference of general interest in this service between Munich and Basle becomes clearly visible again. Among the Munich participants, there is one top user (M1) and with considerable difference some medium users (M4, M7, and M9), who all range above the trial average. As with the WWW profile, the third and biggest group comprises of users with rather low usage (amongst those M5 and M8 with a total trial usage of less than one hour). In terms of absolute numbers, Basle has one top user (B10, who would be a medium user compared to Munich), all others are very similar to each other (even very low compared to Munich) and one user (B3) shows very little interest in the VoD service.

A figure of user profile for the NoD service is omitted because it is very similar to the one of the VoD service, but it stresses the above statements. To give a summarising interpretation of the figures for all three service categories, the differences in the usage profiles between Munich and Basle have to be pointed out: In Munich the user of STB 1 was the top user in all service categories WWW, VoD and NoD with a large distance to all other users. It has to be mentioned that the user of STB 1 in fact was a student who already had experience in using the Internet, and possibly was more open to experiment with new services than the others. In Basle there is no overall top user, in contrast to this the top user for VoD and NoD (STB 10) was amongst the least interested users for the WWW service and the WWW top users of STB 1 and STB 9 were in the third group for VoD and NoD.

The distinct gap in the usage volume between WWW and the video services (VoD, NoD) emphasises the problem of content provisioning and/or creation, which of course has considerable impact on the attractiveness of the service. Within providing the Internet access to the World Wide Web, a nearly unlimited source of contents is offered to the user, even in rather small technical trials like in the AMUSE project. But the provision of attractive and competitive contents for the video on demand service depends on the collaboration with commercial content providers. For the upcoming trials in other countries it is expected to have a broader choice of up-to-date movies and even a rather high content refresh rate. Thus, the study of the service usage will allow to investigate the impact of the content variety on the amount and frequency of service usage.

For the WWW service, which was used extensively, it was worth going into more detail for analysing the stochastic properties of the WWW traffic caused by the residential users and observed by the automatic monitoring. The results presented in the following subsection are the first step towards a detailed characterisation of the WWW traffic that future networks have to cope with.

4.2. WWW Traffic Analysis

The logging data of the field trials in Munich and Basle have been used to describe and visualise the characteristics of the produced WWW-traffic. The data allowed an evaluation of three different types of WWW-traffic: requests for data on the local AMUSE-WWW-Server, requests for data that was found in the local cache of the Proxy server and requests for data that had to be externally retrieved from the internet.

To understand the stochastic behaviour of the WWW data streams the probability distributions of the request sizes and of the request interarrival times have been investigated.

Request Sizes

The evaluation of the request sizes for the three types of WWW-traffic mentioned above showed that requests for local data have been of a very limited size. There were no large http data packets to be retrieved from the AMUSE-server. The local cache served data requests up to a size of 200 kbyte. The smaller the request size, the higher was the probability that it could be served by the local cache. While most externally served requests were quite small, there was a considerable number of large requests.

The complementary distribution functions for the request sizes of all http requests (regardless of how they were served) of the 12 STBs of the Basle trial are depicted in Figure 6 (left side). The curves show the probability of a request being of equal or greater size than drawn on the x-axis.

The curves can be categorised into two classes. Some STBs have never issued requests exceeding a request size of about 200000 bytes, whereas others requested documents with a size of more than 1 Megabyte. Again the STBs with the small request sizes are the ones that were rarely used and the busy STBs present the distribution functions including large request sizes. So the latter ones may be more representative of future users that really make use of the WWW. The evaluation of each STB of the Munich trial led to similar curves.

The thicker lines indicate the distribution function average of all STBs (all request sizes regardless of the issuing STB were taken to calculate the distribution function) and look very similar for both trials. The shape of this line has an unfortunate characteristic, since it doesn't indicate any upper bound for request sizes. It is difficult to define a mathematical function of a similar shape for the modelling of the WWW-traffic.

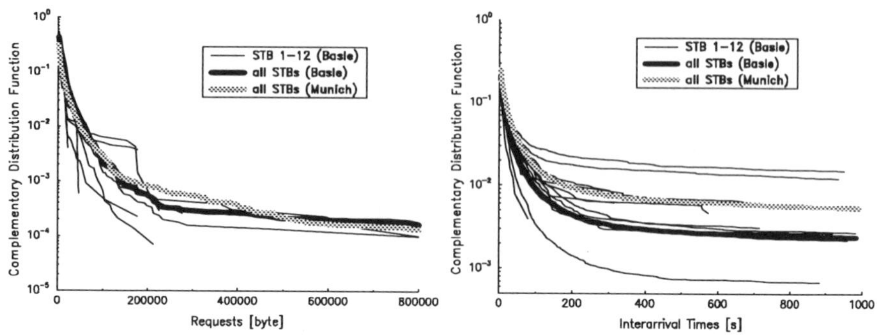

Figure 6: Complementary Distribution Functions for request sizes (left) and for request interarrival times (right), all http request types.

Request Interarrival Times

The evaluation of the request interarrival times for the three http request types showed similar results for both trials. There was a comparatively high probability for a local request to have a large interarrival time. This may be due to the fact that at the beginning of a session the local server had to be addressed to get the home page. After this, the local server was rarely used during the following session.

For all STBs Figure 6 (right side) shows the probability of a request interarrival time being equal or greater than drawn on the x-axis. Like in the Munich evaluation, the curves show two regions. First the curves drop quickly for small interarrival times, but then almost stagnate at a certain probability level. This behaviour results from the fact that the logging data contains the interarrival times within a WWW session as well as the time spans between the sessions.

Like the request size distributions, the distribution function of the interarrival times proves that some STBs were rarely used whereas others were used intensively. The thicker lines denote the average interarrival times of all STBs of the Basle trial and the Munich trial. They stagnate at a probability for very large interarrival times of about 0.23% and 0.5% respectively.

4.3. Summary

The results achieved and the experiences gained by using an automatic monitoring mechanism for the services usage have to be separated into technical and service related aspects. On the technical point of view, the tools for monitoring the service usage worked very well for all services, taking into account that the VoD/NoD and WWW services are provided by different control processes on the Multimedia server, respectively generating monitoring data separately. Altogether, the concept and methodology of automatic monitoring as being set up for the trials within the AMUSE project seems to be well suited for future trials with a larger number of users involved.

The issue of limited number of users is the main drawback for giving a solid interpretation of the results from the service point of view, gaining information on a typical usage profile and getting answers on the acceptance of the services. In addition, the behaviour of the user heavily depends on further issues like the quality of the contents provided within the service and the tariffs used.

It is interesting that the WWW traffic generated by users in Munich and Basle produced almost identical characteristics. The uncommonly heavy tails of request size distribution functions can be found in both trials. Also the two phases of the interarrival time distribution functions that can be attributed to the time spans between requests within a WWW session and the time spans between the sessions shows up in Munich and Basle.

The collection of usage data by automatic monitoring will continue for all future trials of the AMUSE project. This will increase the data basis available for the analysis and we will be prepared to do a comparison of results from the different trials which together may allow for a more sophisticated interpretation.

5. Conclusions

Network architecture and first results from the interactive Multimedia trials carried out within AMUSE were presented. The architecture based on ATM technology allows flexible utilisation and interconnection of the equipment and it can deliver broadband network access to residential homes via access networks based on the HFC or ADSL technology. More trials are planned with improved technology, enhanced services and more users.

The collection of usage information by automatic monitoring provided the opportunity to present results on the service usage of the residential users in the Munich and Basle trial. The results vary widely between the users, but in general indicate that all users at least showed a curious experimenting character in using the new services. The ultimate goal of deriving a profile of user behaviour and indicating the acceptance of the services was not achievable due to the very small number of users. Nevertheless, the experiences with establishing the monitoring mechanism and performing the evaluation showed that the concept of monitoring within the AMUSE trials is feasible also for use in trials with a larger number of users and in such an environment will hopefully allow for drawing more general conclusions from the evaluation.

The stochastic analysis of WWW traffic shows common properties of request size and interarrival time distributions observed in the AMUSE environment. The amount of data available for evaluation will grow with more AMUSE trials being monitored and evaluated. Our aim is to produce a stochastic model from the observed data by distinguishing between inter- and intra-session interarrival times and use this model for stochastic simulation and performance evaluation considering a larger user population, so that dimensioning criteria for core and access network as well as for the servers can be derived.

References

[1] DAVIC 1.0 Specification Part 02: System Reference Models and Scenarios, January 1996

[2] H. Elbert, J. Enssle, M. Frank, H. Hessenmüller, E.T. Hvannberg, T. Theimer, Multimedia Service Provisioning in the Munich Trial of the AMUSE Project. ECMAST'97, 21-23 May 1997, Milano, Italy.

[3] AMUSE Deliverable AC011/ITL/WP21/DS/P/003/b3, February, 1996

[4] H. Hessenmüller, R. Bordewisch, H. Elbert, J. Enssle, M. Frank, U. Gremmelmaier, M. Gruber, E. Lehmann, H. Martin, U. Quecke, G. Seifert, T. Theimer. Das eropäische Projekt AMUSE - Der HFC-Feldversuch in München. *Der Fernmeldeingenieur,* 1+2 1997, Verlag Georg Heidecker GmbH, Erlangen, Germany.

[5] AMUSE Deliverable AC011/GPT/WP21/DR/P/008/b1, January, 1997

Core and ATM Networks
D.W. Faulkner and A.L. Harmer (Eds.)
IOS Press 1997 © AKM

Guidelines for the evolution of Core Network capabilities for the support of advanced multimedia services

C. D. Anagnostakis, Th. B. Zahariadis, I. S. Venieris
National Technical University of Athens
Department of Electrical Engineering and Computer Science, Athens, Greece
Email: venieris@cs. ntua. gr
Fax: +30 1 772 2534, Phone: +30 1 772 2551

Abstract: The Broadband Integrated Services Digital Network (B-ISDN) opens new ways to the introduction of advanced multimedia services. However, the actual way of commercial implementation strongly depends on the relative capabilities of the Core Network. This paper aims to systematically assemble the various technical approaches, for the evolution of the Core Network in relation to the support of advanced multimedia services. The requirements of the various multimedia services in terms of signalling, intelligence and mobility demonstrate the importance of enhancements and combinations of these features as a key issue to the evolution of the Core Network. This paper outlines possible migration paths towards the support of advanced multimedia services in the context of B-ISDN, as those are followed by ongoing research in various European projects.

1. Introduction

Current advances in communications and signal processing open new ways to the introduction of advanced multimedia services to both residential and business users. Enabling the large variety of multimedia services to become progressively available across an increasing variety of networks, demands that telecommunication engineers take into account all existing and emerging systems. The ideal future network architecture should cover some basic objectives such as: ability to cope with a diversity of services to be supported, service provision by a large diversity of network operators and service providers (ranging from global carriers to local specialized service providers) and commonality of service functionality in different environments and by means of different terminals.

The Broadband Integrated Services Digital Network (B-ISDN) environment is the best candidate for supporting the provisioning of multimedia services under a common framework. To facilitate the evolution of today's infrastructure towards providing multimedia capability, it is necessary to plan a course of economically viable evolutionary steps. This paper intends to address the necessary evolution of Core Networks, so that advanced multimedia services can be supported. Section 2 describes a reference configuration considered for the provision of interactive multimedia services to residential users, focusing on the Core Network part. Section 3 looks into some basic concepts involved in the provision of advanced multimedia services likely to be introduced in the foreseeable future, and attempts to classify such services in terms of necessary signalling and IN capabilities, in order to demonstrate the increased needs of future services compared to today's ones. Possible evolutionary paths are identified in this section and discussed in more detail in section 4. Finally in section 5 we summarize the results of this study.

2. Reference Configuration

For the provision of interactive multimedia services to residential users, a reference configuration like the one shown in Figure 1, in line with the System Reference Model suggested by the Digital Audio Visual Council (DAVIC) [1] may be considered. At the user side, digital set top boxes (STB), based on the Asynchronous Transfer Mode (ATM), coupled with TV screens, PCs and network computers (NCs), are expected as Customer Premises Equipment (CPE). This equipment is connected through a cost-effective delivery system with the service provider system (SPS). SPSs will consist of various commercial multimedia servers capable of supporting simultaneously thousands of active independent multimedia sessions and storing enormous multimedia content, which may consist of any combination of video, audio, graphics, text and data. Finally, a delivery system will connect the SPS with the content provider system (CPS). Reference points of Figure 1 are in line with DAVIC 1.0 Specifications.

The specific protocols running over the reference points of Figure 1 depend on the particular approach. A variety of alternatives, regarding the type and operation of control and signalling protocols for supporting this configuration, are currently considered within the research community. Several approaches may be considered here, with the trade-off being, of course, among the fast introduction of multimedia services, the robustness of the selected solutions and the advanced service handling capabilities to be established. The basic candidates are the proposed User Network Interface (UNI) and Network Node Interface (NNI) B-ISDN signalling system (i.e. Q.2931 [2], ATM Forum v3.1 [3], B-ISUP etc.), the DSM-CC ISO standard [4] and Intelligent Network (IN) protocols (Capability Set 1) [5]. Each one of these approaches can provide the platform required for the support of multimedia services in a synergetic manner.

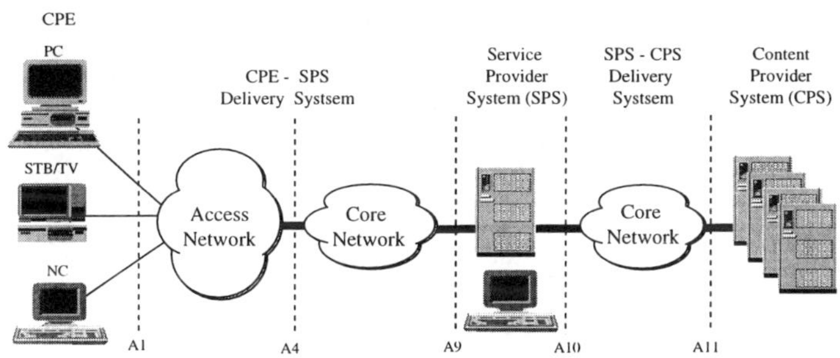

Figure 1: Reference Configuration

The Core Network provides switched connections from and to Content Providers, Service Providers (servers and brokers) and, via an Access Network, to end users. It can be as small as one switch or may extend to a worldwide network. In any case, the Core Network is expected to provide reliable transfer of information between entities like Content Providers, Service Providers and Access Networks, switching functionality to provide connections between entities, network related control functions for addressing, establishment and release of connections and network management functions for network configurations, performance and fault monitoring, billing and accounting purposes. It is, of course, assumed that the multiplexing and switching technique deployed in the Core Network is ATM, which allows the establishment of connections of virtually any bit rate up to the maximum capacity of the transmission links and does not impose any restrictions on the topology of the Core Network. Signalling conforms to the standard call and bearer control capabilities, while it should be able to support advanced service-related features. A potential solution could be the introduction of sophisticated servers realising service-related functionality transparently to the underlying network signalling system.

The Core Network and its evolution is a key aspect when advanced multimedia services are considered. The Core Network can be based on existing and future network elements offering features like advanced routing of calls based on IN service logic, support of mobility management and home network features when moving into other networks. The Core Network could be an integrated IN / B-ISDN, the current Global System for Mobile Telecommunication (GSM) network enhanced to support IN capabilities, the combination of N-ISDN with IN, or, even, an ATM / B-ISDN transport network using Telecommunications Information Networking Architecture (TINA) network elements (section 4 will address the first two alternatives, while the other two are outside the scope of this paper). It is thus clear that different combinations of network elements can be foreseen and, consequently, various approaches, allowing for an evolution from several, different, starting points can be followed.

3. Services to be supported - Concepts and Classification

In this section, new foreseeable multimedia services which require extended network capabilities are examined, with respect to the evolution of the Core Network. In particular, the evolution of call control and network functions is analyzed, so that multimedia services can be supported.

In order to highlight the necessity for the evolution of the Core Network, a set of possible target interactive multimedia services, listed in both Tables 2 and 3, is used as a basis for discussion. Most of these applications have already been available during the lifetime of some ACTS projects [6][7], and are by now under evaluation even by real residential users [7]. However, specification of the Core Network characteristics should focus on providing the ability to support, in a cost effective and easily implementable manner, all these services, and even future, more advanced ones.

The concepts of "session" and "connection", as originally introduced by the DSM-CC standard [5], have been widely accepted in the modelling of multimedia delivery systems. The following definitions have also been adopted in this paper: A *Connection* is the transport link that provides the capability to transfer information between two or more endpoints. A *Session* is an association between two or more users, providing the capability to group together the resources needed for a specific Service instance. The term *Service* denotes a collection of communicating elements that provide functions and interfaces in support of one or more applications. The distinction of a Service from other objects is that end-user access to it is controlled by a service gateway. The term *Network* identifies a collection of communication elements that provide connections and may also provide session control and / or connection control to *Users*, which, in turn, are the end-systems that are connected to a Network and can transmit information to, or receive information from, other such systems through the Network. A User may be the provider or the consumer of a Service (Server / Client respectively), or both. Based on these definitions a service may be characterized as *Multiparty* if more than two users / parties are involved, and *Multiconnection* if more than one transport links are required for the transfer of information. Although pure point-to-point connections are sufficient for most multimedia services, multipoint-to-multipoint (and point-to-multipoint) connections can help to reduce traffic in conference applications and are considered essential for broadcasting-based applications.

A first attempt to classify multimedia services on a basis of common requirements in terms of signalling and IN capabilities is presented in Tables 2 and 3 (where columns labeled S1 to S22 and IN1 to IN18 are named after Table 1). The requirements of the various multimedia services in terms of signalling and IN capabilities, as described in these tables, demonstrate the importance of enhancements and combinations of these capabilties as a key issue to the evolution of the Core Network. Discussion of such an evolution should also take into account mobility features, since they are considered an aspect that will offer a substantial added value to future multimedia services. The following section attempts to identify alternative evolutionary paths for the Core Network, so that it can efficiently support the provision of advanced multimedia services both to business and residential users.

4. Evolutionary Paths for the support of advanced multimedia services

4.1 Enhancements of Call Control capabilities

As shown in Table 3, many advanced multimedia services require multi-connection sessions. For management reasons, it is essential to associate connections to the appropriate sessions. However, Q.2931 [2] has no explicit means of allowing a connection to be associated to the session inside which it has been set up. At the network side, the entities who manage ATM signalling cannot associate connections to sessions, and cannot therefore support billing or monitoring functionality on a session basis. Multiparty and multiconnection sessions may be supported either by enhanced signalling or by IN features.

Four methods are specified in [4] by which the DSM-CC UN protocol can be employed in conjunction with Q.2931 in B-ISDN networks: the *Session Method* , the *Network Method* with AddResource messages exchanged between the Server and the Session and Resource Manager (SRM), the *Network Method* with no AddResource messages exchanged between the Server and the SRM, and the *Integrated Method.* For each of them a description of the sequence of messages exchanged among the Server, Client and SRM are given for all the message procedures required during a session (client / server session set-up, client / server add resource request / deletion, session tear-down). The first three methods use both DSM-CC and Q.2931 protocols, while the last one allows the integration of session and call / connection control by using a properly upgraded Q.2931 protocol. More details on these methods can be found in [4]. In its specifications (Part 12), DAVIC has limited the options offered by the DSM-CC UN protocol, choosing the Network Method with AddResource messages between the Server and the SRM with some variations and constraints.

4.2 Evolutionary Approach to the integration of IN and B-ISDN

The provision of new advanced multimedia services quickly and economically, as well as the ability to add, as fast as possible, new features in already settled services, in a competitive multi-provider environment, have forced the employment of IN features in Core Networks. The deployment of IN requires a fully digital network and the widespread adoption of CCITT No 7 signalling with suitable enhancements. A comprehensive discussion of the operation of IN, including its history and background, may be found in [8]. Table 1 indicates some of the advanced IN capabilities that will be required by specific multimedia services.

A single IN-based multimedia service can be composed of different B-ISDN calls. The need to correlate several different calls belonging to the same service instance demonstrates the importance of an external entity controlling the session itself. The required signalling enhancements can be achieved by either *concentrating intelligence in the network edge*, or *integrating IN and B-ISDN inside the Core Network.*

The first approach is currently recommended by DAVIC and adopted by the AMUSE project [7]. The core of this approach is the introduction of session control functionality which follows a client-server philosophy. A DAVIC-compliant session is only known to end systems (which may be terminals or session control devices). The Core Network is session independent and is not aware of the fact that different calls / connections are related to the same service and session.

The second solution is an objective of CS3-IN and B-ISDN and is currently followed by the INSIGNIA project [6]. In this case, appropriate signalling capabilities are deployed in the terminals in order to correlate different calls, while service sessions are managed by the IN service logic, that orders the establishment and tear-down of the calls / connections needed by the service. This approach assumes that network elements are fully aware of the existence of a service session concept, which is recognised by the Broadband Service Switching Points (B-SSPs) and directly controlled by the IN elements (Broadband Service Control Points / B-SCPs and B-IPs), where the service logic resides. This model has adopted a very general and powerful “session” concept, in order to coordinate two or more

B-ISDN calls. Such a concept is regarded as mandatory when IN is integrated into a B-ISDN network implementing CS1 signalling capabilities and allowing only point-to-point single connection calls. When more advanced features (e.g. point-to-multipoint calls) are incorporated into the B-ISDN network, the "session" concept could slightly reduce its role, as more complex call topologies will be handled at the B-ISDN level. Another very promising aspect is that the session concept and the IN functionalities could allow a smoother transition towards future signalling capability sets.

In the DAVIC-AMUSE case, the evolution of the service is driven by enhancements of the session control protocol and CS1 signalling is considered adequate. In the integrated IN / B-ISDN signalling case, the service logic and the signalling capabilities can evolve independently in order to respectively enhance the service features and the control functionality of the Core Network and terminals. Table 4 summarises the results of the comparison between the two approaches.

4.3 Evolutionary Approach to the integration of IN and Mobility

Mobility is considered as one of the mandatory features of future multimedia services. Mobility functions are among the most serious candidates to be covered by an extended IN service logic, since the trend in the future IN specification is to extend basic IN features to cover a wider scope of functionality. The key point in the evolution of existing IN CS1 standards towards the future IN Capability Sets when mobility issues are considered, is that the relation of networks, systems and intelligence to each other changes with time, and the role of intelligence, as known by IN, is gradually increasing to include the system functions of user and terminal mobility, with intermediate steps covering supplementary services, non-call related mobility functions, call related mobility functions and long term IN.

With respect to the support of terminal and personal mobility various issues should be addressed. The use of out-of-call signalling for location management in B-ISDN and the implementation of the enhancements necessary for the introduction of additional messages for this function should be studied. Furthermore, the impact of call handling on the B-ISDN and the implementation of the necessary enhancements should also be taken into account. B-ISDN state machines (Basic Call State Model etc.) and protocols for handover should be upgraded, and if necessary, a separate Handover State Machine should be defined, developed and implemented. In addition to that, the impact of Digital European Cordless Telecommunications (DECT) Connection Handover on the B-ISDN Bearer and Call Control has to be investigated. Finally the impact of mobility on Connection Admission Control (CAC) and routing of calls and the feasibility of a signalling load reduction by means of multiplexing several signalling links in the same ATM cell should be examined and necessary enhancements should be implemented.

Smooth transition paths from diverse fixed and second generation mobile systems towards Universal Mobile Telecommunication System (UMTS), focusing on the evolution of Core Network capabilities, are under way. The evolution of the Core Network towards UMTS, requires focus on the Control Plane of the Core Network and on Intelligent Network Application Protocol (INAP) enhancements. This includes enhancements of B-ISDN signalling, deployment of distributed databases and distributed processing capabilities, along with relevant protocols. UMTS addresses the convergence of a number of heterogeneous telecommunication systems and networks. Thus a number of different paths can be envisaged facing the evolution from second generation mobile systems (e.g. DECT and GSM). The key issue is to achieve a smooth transition through the re-use and the enhancement of the existing technologies, in order to protect the significant investment value already in the field. Initial platforms, based on IN CS1 and providing personal and terminal mobility for voice and data services, in the public, private and residential domains, have already been built. The evolution of these platforms - in particular of switches and Service Control Points (SCPs) - towards UMTS needs to be addressed, in order for the impact on the Core Network, the INAP and the UNI to be determined. It is, however, quite obvious that new functionality will have to be specified and implemented. Evolution scenarios should, among other aspects, consider Basic Call Handling, Location Management, Intra- and Inter-switch Connection Handover and ATM traffic management (e.g. CAC) in the Core Network. In order for personal and terminal mobility to be supported in a broadband environment, the following tasks should be considered: identification of INAP enhancements in order to support personal and terminal mobility, definition of the impact of mobility aspects on the data access protocol at the Service Control Function (SCF) / Service Data Function (SDF) interface, definition of the impact of procedures such as location management and authentication at the Service Switching Function (SSF) / SCF interface, and, finally, implementation of the necessary enhancements.

The UMTS evolution concept is envisaged as a cost efficient way to provide speech, data and mobile services up to high-bit-rate multimedia services (namely, up to 2 Mb/s) as well as service integration. Since UMTS will be part of a rapidly changing telecommunications environment, network technology, services and regulatory issues will all be affected. As a consequence, a number of possible scenarios that approach the introduction of UMTS are considered. Exemplary evolutionary tracks include a *GSM-based evolution scenario* and a *Cordless Terminal Mobility (CTM)-based evolution scenario*. In the first case further evolution comprises the introduction of a new (UMTS as well) radio interface, increased bit rates and enhancements of the Mobility Application Protocol (MAP) to fulfil UMTS system requirements (total coverage, global roaming, service availability), with the introduction of new base station system technologies also being a possibility. The second case foresees the introduction of ATM / B-ISDN and IN capabilities with mobility support and a radio access sub-system to the fixed network. These approaches are currently under study in the framework of the DOLMEN [9] and EXODUS [10] projects.

5. Conclusions

In this paper, the importance of enhancements and combinations of signalling and IN capabilities and mobility features, when the evolution of the Core Network is considered, has been demonstrated. Target multimedia services have been classified in terms of signalling and intelligence requirements; possible migration paths have been presented and alternative scenarios for the integration of IN / B-ISDN and IN / mobility have been discussed in more detail.

6. Acknowledgements

This paper is based on work undertaken in the framework of the ACTS project EURORIM, and has covered aspects of the ACTS projects INSIGNIA, AMUSE, EXODUS and DOLMEN.

7. References

[1] DAVIC, "System Reference Models and Scenarios", Specification 1. 0 Part 2, Rev. 4. 1, 1995.

[2] ITU-T, "Recommendation Q. 2931 - B-ISDN Digital Subscriber Signalling System n. 2 (DSS2). UNI layer 3 Specification for basic call/connection control", Feb. 1995.

[3] ATM Forum, "ATM User-Network Interface Specification", V3. 1, 1995.

[4] ISO, "DSM-CC - Digital Storage Media Command & Control", ISO/IEC 13818-6, Jun. 1995.

[5] ITU-T, "Intelligent Network Capability Set 1", Q. 12xx Series of Recommendations.

[6] AC068 - INSIGNIA, "IN and B-ISDN Signalling Integration on ATM Platforms", http://www.fokus.gmd.de/nthp/insignia.

[7] AC011 - AMUSE, "Advanced Multimedia Services for Residential Users", http://www.italtel.it/drsc/amuse/amuse.htm.

[8] J. Thorner, "Intelligent Networks", Artech House, 1994.

[9] AC068 - DOLMEN, "Service Machine Development for an Open Long-term Mobile and Fixed Network Environment", http://www.telecom.ntua.gr/activities/projects/DOLMEN/index.html.

[10] AC013 - EXODUS, "EXperiments On the Deployment of UMTS", http://www.italtel.it/drsc/exodus/exodus.htm.

8. Appendix

Table 1: B-ISDN Signalling and IN Capabilities

S1	Call with multiple connections (sequential establishment)	**IN1**	Authentication (User authentication, Terminal authentication)
S2	Simultaneous establishment of connection	**IN2**	Authorisation
S3	Negotiation of traffic characteristics during lifetime of call	**IN3**	Billing/accounting policies (e.g. premium rate, split-charging, reverse charging)
S4	Renegotiation of traffic characteristics during lifetime of call	**IN4**	Call forwarding
S5	Two parties call establishment without look ahead	**IN5**	Call screening
S6	Two parties call establishment with look ahead	**IN6**	Call transfer
S7	Multiparty call establishment (for point to multipoint only)	**IN7**	Follow-me diversion
S8	Addition of one or more connection of type 1 and attachment to existing parties under control of the initiating party	**IN8**	Mobility features (Handover, Terminal location registration, Terminal attach / detach, Terminal paging)
S9	Release, by the initiating party, of one or more connections associated with the call	**IN9**	Navigation aids
S10	Addition, by the initiating party, of one or more new parties to an existing call with attachment to an existing connection	**IN10**	Network-initiated call / connection setup
S11	Release, by the initiating party, of one or more parties from the call	**IN11**	Profiles (for user, service or terminal)
S12	Non-initiating parties can release themselves from call	**IN12**	Routing algorithms (Call distribution, Origin dependent routing, Time-dependent routing)
S13	Call release by the initiating party	**IN13**	Service broking
S14	When a non initiating party unilaterally releases / detaches itself from a call / connection , initiating party is notified	**IN14**	Session management (complex call configurations)
S15	Third party call setup	**IN15**	Virtual Private Network features (Private numbering plan, Off-net access, Off-net calling)
S16	Proxy signalling	**IN16**	B-ISDN Multiple connections point-to-point
S17	Multipoint-to-multipoint configurations	**IN17**	B-ISDN Multi-casting
S18	Addition of one or more party in a multipoint to multipoint configuration	**IN18**	B-ISDN Conferencing
S19	Release of one or more party in a multipoint to multipoint configuration		
S20	Explicit call transfer		
S21	Explicit bearer transfer		
S22	Third party Call Control		

Table 2: Classification of multimedia services in terms of IN capabilities

Service	IN1	IN2	IN3	IN4	IN5	IN6	IN7	IN8	IN9	IN10	IN11	IN12	IN13	IN14	IN15	IN16	IN17	IN18
MoD	X	X	X	-	X	-	X	-	X	-	X	X	X	X	X	X	-	-
Teleshopping	X	X	X	-	X	X	X	X	X	-	X	X	X	X	X	X	-	-
Broadcast TV	X	X	X	-	X	-	-	-	X	-	X	-	-	X	X	-	X	-
Near VoD	X	X	X	-	X	-	-	-	X	-	X	-	-	X	X	-	X	-
Delayed Broadcast	X	X	X	-	X	-	X	-	X	-	X	X	X	X	X	-	X	-
Games	X	X	X	X	X	X	X	-	X	-	X	X	X	X	X	X	X	X
Telework (CSCW)	X	X	X	X	X	X	X	X	X	X	X	X	X	X	X	X	X	X
Home Banking	X	X	X	-	X	-	X	X	-	-	X	X	-	X	X	X	-	-
News on Demand	X	X	X	X	X	-	X	X	X	-	X	X	X	X	X	-	-	-
Distant Learning	X	X	X	X	X	X	X	X	X	X	X	X	X	X	X	X	X	-
Karaoke on Demand	X	X	X	-	X	-	X	-	X	-	X	X	X	X	X	-	-	-
Video Conference	X	X	X	X	X	X	X	X	X	X	X	X	X	X	X	X	X	X
Internet Access	X	X	X	X	X	X	X	X	X	X	X	X	X	X	X	X	-	-
Virtual CD-ROM	X	X	X	-	X	-	X	-	X	-	X	X	X	X	X	-	-	-
Audio on Demand	X	X	X	-	X	-	X	-	X	-	X	X	X	X	X	-	X	-
Telemedicine	X	X	X	X	X	X	X	X	X	X	X	X	X	X	X	X	X	X
Pay per View	X	X	X	-	X	X	X	-	X	-	X	X	X	X	X	X	X	-
Information Base Retrieval	X	X	X	X	X	X	X	X	X	-	X	X	X	X	X	X	-	-
X: Required or preferred										**-: Not required**								

Table 3: Classification of multimedia services in terms of connection configurations, traffic types, multiconnection / multiparty requirements & signalling capabilities

Service	Connection Configuration	Traffic Type	Multiconnection	Multiparty	S1	S2	S3	S4	S5	S6	S7	S8	S9	S10	S11	S12	S13	S14	S15	S16	S17	S18	S19	S20	S21	S22
MoD	PTP	CBR	X	-	-	-	-	-	X	X	-	-	-	-	-	X	X	X	-	-	-	-	-	X	X	X
Teleshopping	PTP	VBR	-	-	-	-	-	-	X	X	-	-	-	-	-	X	X	X	-	-	-	-	-	-	-	X
Broadcast TV	PTM	CBR	-	X	-	-	-	-	X	X	X	X	X	X	X	X	X	-	-	-	-	-	-	-	-	-
Near VoD	PTM	CBR	-	X	-	-	-	-	X	-	X	X	X	X	X	X	X	X	-	-	-	-	-	-	-	-
Delayed Broadcast	PTP / PTM	CBR	-	X	-	-	-	-	X	X	X	X	X	X	X	X	X	X	-	-	-	-	-	X	X	-
Games	MTM	VBR	X	X	X	X	X	X	X	X	X	X	X	X	X	X	X	X	-	-	X	X	X	X	X	-
Telework (CSCW)	MTM	VBR	X	X	X	X	X	X	X	X	X	X	X	X	X	X	X	X	X	-	X	X	X	X	X	X
Home Banking	PTP	CBR	-	-	-	-	-	-	X	-	-	-	-	-	X	X	X	X	-	-	-	-	-	-	-	-
News on Demand	PTP	CBR	X	-	-	-	X	X	X	X	-	-	-	-	X	X	X	X	-	-	-	-	-	-	-	-
Distant Learning	PTP / PTM	VBR	X	X	-	-	X	X	X	X	X	X	X	X	X	X	X	X	-	-	-	-	-	X	X	X
Karaoke on Demand	PTP	CBR	X	-	-	-	X	X	X	-	-	-	-	-	X	X	X	X	-	-	-	-	-	-	-	-
Video Conference	MTM	VBR / User	X	X	X	X	X	X	X	X	X	X	X	X	X	X	X	X	X	-	X	X	X	X	X	X
Internet Access	PTP	VBR	X	-	X	X	X	X	X	-	-	-	-	-	X	X	X	X	-	X	-	-	-	-	-	X
Virtual CD ROM	PTP	VBR	X	-	-	-	X	X	X	-	-	-	-	-	X	X	X	X	-	-	-	-	-	-	-	-
Audio on Demand	PTP	CBR	X	-	-	-	X	X	X	-	-	-	-	-	X	X	X	X	-	-	-	-	-	-	-	-
Telemedicine	MTM	VBR	X	X	X	X	X	X	X	X	X	X	X	X	X	X	X	X	X	-	X	X	X	X	X	X
Pay per View	PTM	CBR	X	-	-	-	X	X	X	X	-	-	-	-	X	X	X	X	-	-	-	-	-	-	-	-
Information Base Retrieval	PTP	VBR	X	-	-	-	X	-	X	-	-	-	-	-	X	X	X	X	-	-	-	-	-	X	X	X

PTP: Point-to-Point | PTM: Point-to-Multipoint | MTM: Multipoint-to-Multipoint | CBR/VBR/ABR: Constant/Variable/Available Bit Rate | X: Required or preferred | -: Not required

Table 4: IN / B-ISDN Integration

	AMUSE approach	INSIGNIA approach
Common capabilities	B-ISDN CS1	B-ISDN CS1
Call correlation	Session control	Service Switching Function
Service control	Session entity (outside the Core Network)	B-SCP / B-IP (inside the Core Network)
Service evolution has an impact on	End systems	Core Network
Signalling evolution needed	Not requested	Possible
Service evolution performed by	Session control	IN Capabilities
Network topology flexibility	Low	High
Service flexibility	Limited to client-server services and particularly focused on VoD	General

Core and ATM Networks
D.W. Faulkner and A.L. Harmer (Eds.)
IOS Press 1997 © AKM

Broadband Video Conference in an Integrated IN/B-ISDN Architecture

Hartmut Brandt, Carsten Tittel, Petia Todorova
GMD-FOKUS, Hardenbergplatz 2, D-10623 Berlin, Germany

Abstract. The current state of Broadband ISDN (B-ISDN) signalling, based on ATM technology is not sufficient for advanced multimedia applications where interaction and greater intelligence will be needed in the network. Future B-ISDN multimedia services will require additional capabilities for service selection, service provider selection, billing, to name just a few.

With the move towards the liberalisation of the telecommunication market there are increasing demands on rapid services introduction in a multivendor service provisioning environment. As a result, an important issue facing network operators is how much of the service functionality and control should reside in the network and how much will be provided in a new generation of multimedia terminals or third-party service providers.

The paper proposes an approach for the provisioning of Broadband Video Conference service in an intelligent broadband network offering the advantages of both IN and B-ISDN architectures. This approach is under development in the ACTS-INSIGNIA Project [1].

1. Introduction

The Broadband Video Conference (B-VC) service is a multiparty service providing real-time conferencing in which both audio and video can be exchanged among single individuals or groups of individuals at two or more locations via the B-ISDN. Users can be either multiple users, grouped in conference rooms with their equipment (audio and video systems, PC, etc.) or single users with multimedia workstations.

2. Reference Architecture

The reference architecture for B-VC service deployment in an integrated IN/B-ISDN network (see Figure 1) includes end system, B-ISDN switching system and IN components [2], [3].

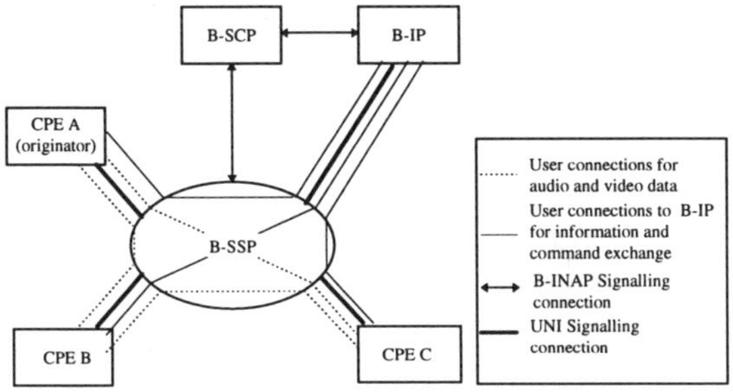

Figure 1: IN-based Broadband Video Conference

The **CPE** is a computer device serving as an endpoint for the audio and video connections. It provides network access and a graphical user interface to the video conference software. Audio and video multiplexing is done at the CPE. Every user having access to a CPE can take part in a conference as a *conferee*. He is connected to the broadband Service Switching Point (B-SSP) via UNI interface.

The **B-SSP** is an ATM switch with an added IN functionality. It provides standard UNI [4], [5] and NNI [6] interfaces to the users and to other B-SSPs and IN interface (B-INAP) to the intelligent node called broadband Service Control Point (B-SCP).

The **B-SCP** hosts the Service Logic able to guide the service provisioning process by giving instructions to the B-SSP. Additionally it implements logic for Low Level Conference Management including:

- establishing of connections;
- computing conference information (the list of connections to be established, etc.);
- managing requests presented by users;
- communicating with the Broadband Intelligent Peripheral (B-IP) to exchange user raleted data.

The **B-IP** is employed in order to handle user interactions. It implements logic which is in charge of a High Level Conference Management. This includes:

- management of static aspects of conferences;
- performing advanced communication with users;
- acting as a relay for communication between B-SCP and users.

A connection between B-IP and B-SSP is needed in order to allow the transfer of user choices performed on the B-IP intelligent node.

3. Broadband Video Conference Scenarios

A conference is defined by its *conference state*, *conferees* and all established/to be established *connection sets* between each pair of conferees: user data connections for audio, video and data transfer and connections between each Customer Premises Equipments (CPEs) and the B-IP used to run a conference management protocol (see Figure 2).

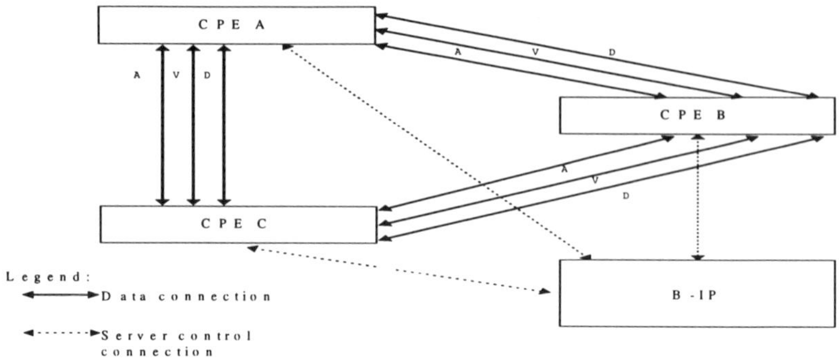

Figure 2: B-VC with three Conferees

A conference may be in one of two basic states: *inactive* or *established.*

Before the B-VC can be started, it has to be defined. The user at a CPE can invoke the service *Conference Management* [7] according to a set of scenarios defined later on for B-VC creating, establishing, etc. The B-SCP initiates a bearer connection between the user and the B-IP for information exchange. The conference description (topological definition, attributes and state of a certain conference, of the conferees and the connection sets) is stored in the Conference Information Record (CIR) within the Conference Information Data Base (CIDB), which is located in the B-IP and the B-SCP. Following scenarios are provided:

Conference Creation. The B-VC can be started by any user having this right by calling a previously specified number to invoke the scenario Conference Creation. The B-SCP instructs the B-IP to communicate with the user and create a CIR. After the creation of the conference the B-SCP updates the CIDB by including the CIR. The conference is defined and is in an inactive state.

Conference Establishing. The B-VC can be established (activated) by any user having this right by initiating a basic call and informing the B-SCP about invoking the service Conference Establishing. The B-SCP creates a list with all connections required by the defined conference. All actions necessary for audio and video negotiation will be performed and QoS information's are added to the list.

Static conference management. In this scenario the B-IP is responsible for interacting with the originating user to modify conference. The interaction uses a application level protocol to modify or delete a conference by means of a CIR. During conference management the user can request information, for example lists of conferees. The user is allowed to abort the conference modification at any time during the user interaction.

Adding user to conference. In this scenario the conference is already running. The B-IP is responsible for correlating the incoming request from user A to the running conference. The B-SCP will update the CIR within the CIDB according to the information provided by the user.

Invite conferee. The invitation of users or inactive conferees is a required service. The functionality is needed during conference establishment and whenever a conferee wants new users to participate. The B-SCP is responsible for checking of the conferee *permission* to invite a conferee and to instruct the B-SSP to establish all connections between the new participant and the existing ones. The B-IP only relays messages towards the B-SCP.

Join conferee. The operation of joining a conference is applicable to conferees which could not participate from the beginning, or for users which where raised to the conferees role by a conference management operation. In difference to an invitation process, the process may be initiated by a user currently unknown to the conference management, who asks about a change of the conference configuration.

Leave conferee. The scenario gives conferees the possibility to end their participation in a B-VC.After completition of this operation the user remains a conferee to this conference as long as he is not removed from the conference description by the service Conference Management.

Check invitation. In this scenario an inactive conferee is able to check for stored invitations.

Disconnect conferee. The scenario enables a conferee being authorised to perform this action to disconnect another active conferee from a running conference.

Closing down. The video conference can be closed by any user having the right by informing the B-SCP about invoking of the Close Conference service. The B-SCP instructs the B-IP to communicate with the user and receives as result a list of connections to be released. Upon entering the inactive state all established connection sets are released.An inactive conference might be removed upon request of an authorised conferee.

4. Broadband Video Conference Provisioning

The provisioning of B-VC on an integrated IN/B-ISDN architectures is shown by taking as an example the *Conference Creation* scenario description (see Figure 3) as follows:

1 User A dials the IN number of the video conference service 'Conference Creation'.

2 The Service Logic is activated. The Service Logic retrieves the calling party information and checks the permission of the user to create conferences. If the permission is denied, the B-SCP instructs the B-SSP to release the connection to user A and sends a error indication to User A.

3 The B-SCP instructs the B-SSP to connect user A to the B-IP.

4 A bearer connection from user A to B-IP is established.

5 The B-IP asks the B-SCP for instructions.

6 The B-SCP instructs the B-IP to interact with the user for function 'Conference Creation' and to report all changes within the CIDB.

6' The B-IP interacts with the user. The result of the interaction may be a new CIR including the corresponding Conference Identifier (CID).

6'' If a new CIR is created, the B-IP will lock the corresponding CID and block all other management requests until the conference creation has ended.

7 The B-IP reports the result of the interaction to the B-SCP. The result may include a new CIR.

7' The B-SCP updates the CIDB according to the new CIR.

8 The B-SCP instructs the B-IP to unlock the created conference and to report the result.

9 The B-IP unlocks the conference and reports the result to the B-SCP.

10 The B-SCP instructs the B-SSP to release the connection to the B-IP.

11 The B-SCP instructs the B-SSP to release the connection to originating user (user A).

12 The B-SSP releases the connection to user A.

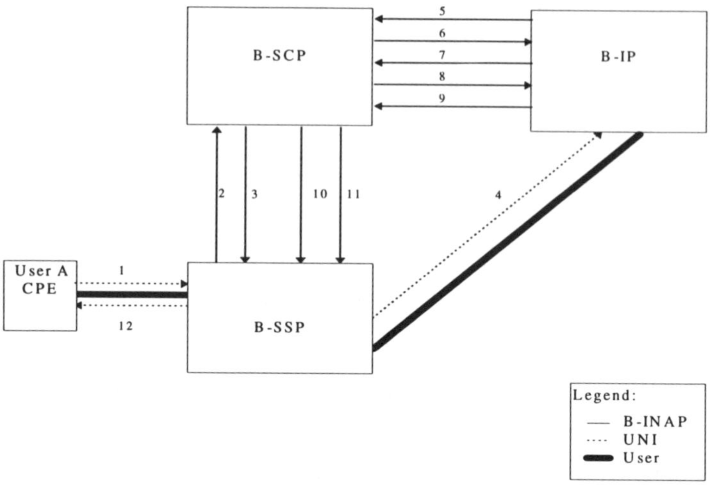

Figure 3: Conference Creation

5. CPE Requirements

The minimum requirements for running the B-VC application at the End System are listed below:

- Sun SparcStation 10 with Sun Solaris 2.5
- Parallax PowerVideo board
- FORE SBA 200 ATM Interface board
- Sun Microphone
- Video Camera with Parallax XVideo Software for OpenWindows 3.5

Using more powerful workstations generally allows more conference participants to a conference without degrading the overall system performance especially when mixing audio streams. The use of FlexCam cameras and headsets is for reasons of convenience and to reduce echo effects in a noisy environment.

References

[1] INSIGNIA Project Description (URL: http://www.fokus.gmd.de/nthp/insignia/entry.html)

[2] ITU-T Rec. Q.1214, Distributed Functional Plane for Intelligent Networks CS-1.

[3] G.v.d. Straten, J.Totzke, R. Zygan-Maus, Realization of B-IN Services in a Multinational ATM Network, ICIN(4th International Conference on Intelligence in Networks), Bordeaux, Dec.1996.

[4] ITU-T Rec. Q.2931, B-ISDN-DSS2-User Network Interface Layer 3 Specification for Basic Call/Connection control

[5] ATM Forum UNI Specification, Version 3.1

[6] ITU-T Recs. Q.762, Q.763, Q.764

[7] H. Hussmann, F.J. Herrera Galvez, R. Pasquali, P. Todorova, V. Venieris, F. Zizza, A transnational IN/B-ISDN integrated network for the provision of multimedia services, submitted to ECMATS'97 2nd European Conference on Multimedia Applications, Services and Techniques, Milan-Italy, May 1997

Experiments on Integrated Traffic Control for ATM in the ACTS Project EXPERT

MARTIN LORANG, THOMAS RENGER
University of Stuttgart, Inst. of Communication Networks and Computer Engineering, Pfaffenwaldring 47, 70569 Stuttgart, Germany

HARALD PETTERSEN, EGIL AARSTAD
Telenor Research & Development, P.O. Box 83, 2007 Kjeller, Norway

LAURENT JAUSSI
Swiss Federal Institute of Technology (EPFL), Telecom Lab, 1015 Lausanne, Switzerland

MARTIN POTTS
Association Swiss PTT/ASCOM (ASPA), Morgenstrasse 129, 3018 Bern, Switzerland

Abstract. In this paper first an overview of the ACTS project EXPERT is given by briefly describing the major project goals and the related technical work items, the sophisticated testbed infrastructure and the various collaborations EXPERT has with other R&D projects. Focus is then given to an important area of work within EXPERT, namely the definition and investigation of a traffic control architecture which allows different ATM service categories to share common resources and thereby to integrate them into one ATM network. For switches which employ delay priorities among service categories, a CAC method is presented which takes into account the specific requirements of each category as well as the underlying buffer architecture. This concept is validated for rt-VBR and nrt-VBR type of traffic. Extensive experiments and measurements have been made to investigate the performance of TCP over UBR with and without packet discard schemes dependent on many parameters of the involved protocol layers.

1 Introduction

EXPERT is one of more than 150 projects in the European Union 4th Framework Programme ACTS: "Advanced Communications Technologies and Services". It operates probably the largest broadband test environment in Europe. One objective of EXPERT is to provide a well-supported broadband platform for its own experimentation and trials, and for use by other projects within and outside the ACTS Programme. This platform also serves as an ACTS showcase for visitors with varied interests in ATM aspects and in EU Telecommunications R&D Programmes in general.

In addition, EXPERT is investigating several technical areas, including broadband signalling, service-related control, resource management, user Quality of Service (QoS), traffic control in a mixed traffic environment, and the development of an ATM network performance evaluation tool.

EXPERT provides a unique broadband (ATM) test environment comprising varied and sophisticated equipment both in Basel, Switzerland, and Leidschendam, The Netherlands. In association with these sites, advanced application trials are being performed to validate and demonstrate the capabilities and advantages of new broadband services and to show the benefits of the underlying ATM technology. The testbeds in Basel and Leidschendam currently comprise 13 (commercial and prototype) ATM switches, various terminals and test equipment, broadband signalling, network interworking units and broadband applications. The introduction of the Available Bit Rate (ABR) service is being prepared.

Due to the comprehensive facilities at its disposal, EXPERT has a major role within the ACTS Programme for the identification and validation of a framework of traffic control functions which enable an ATM network to support the partly conflicting QoS demands of the five service categories (CBR, rt-VBR, nrt-VBR, ABR, UBR) as defined by the ATM Forum [1]. These functions should operate efficiently for each individual category. However, in order to enable an ATM network to simultaneously support traffic from different service categories contending for common resources, the set of control functions has to be integrated into a robust and consistent framework. Partners in EXPERT are not only working on the theoretical definition of such a framework, but are also implementing, validating and comparing the proposed solutions through extensive experiments and trials. It is especially important that solutions for guaranteeing that user-requested QoS for all types of applications can be demonstrated, before network operators can offer commercial services.

2 The EXPERT Project

2.1 Work Items

The work being undertaken in the EXPERT project covers a broad scope of ATM-oriented broadband communication activities, centred on the comprehensive, supported test environment available to the partners in the project, and to collaborating users or projects. The platform is being continually extended and upgraded with new equipment, functionalities and applications, through developments within the project, co-operation agreements with manufacturers, or purchase.

Developments in the project include an integrated services switch (for attaching both terminals with ATM-25 interfaces and N-ISDN equipment) and a performance evaluation tool for the design and planning of ATM networks. Additional items provided by partners in the project include a new ring-based gigabit campus network, and an ATM Passive Optical Network (APON) with associated VB5.1 interface functionalities. Partners are also implementing the latest ATM Forum and ITU-T signalling capabilities, and dynamic routing and resource management algorithms.

However, and of particular concern for this paper, a substantial proportion of the effort in the project is concentrated on the definition, implementation and validation of an integrated traffic control framework to enable an ATM network to support a broad range of QoS requirements. This work continues from earlier individual investigations of Connection Admission Control (CAC) and Usage Parameter Control (UPC), whilst incorporating aspects related with the newly specified ABR and Unspecified Bit Rate (UBR) services. The recent availability of long-distance (intercontinental) ATM links adds a further dimension to these topics.

2.2 Infrastructure

The EXPERT platform is illustrated in Fig. 1 below. The collection of broadband facilities offered by EXPERT (applications, terminal equipment, 13 different ATM switches, signalling capabilities and network interworking units) is unique in Europe. Terminal types include PCs, SUN workstations, TV/video equipment and N-ISDN terminals, all capable of being attached to the ATM network using terminal adapters. Most of the ATM switches have different internal architectures, but have sufficient common external interfaces to allow them to be interconnected as required for particular trials. Applications include CSCW (Computer-Supported Collaborative

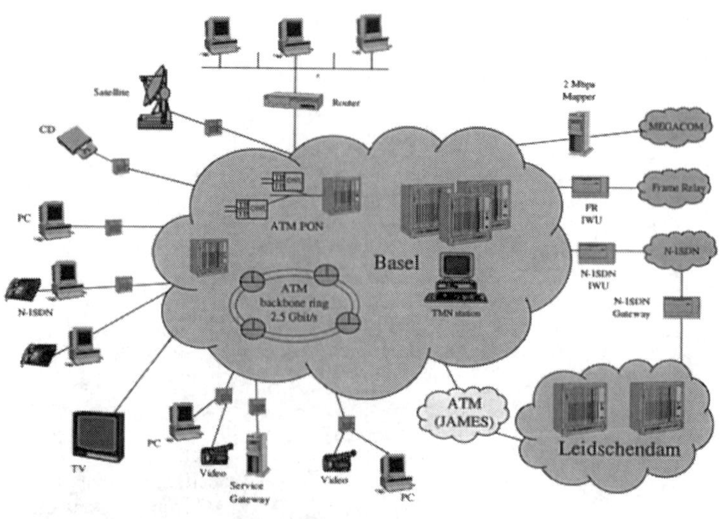

Fig. 1: The EXPERT platform

Working), standard TV/video, 3-D video, CD-audio and multimedia conferencing. Network interworking is provided to/from the Frame Relay and N-ISDN networks. Permanent access is available to the European ATM network (operated by the EU project "JAMES"). Via this network, part of Deutsche Telekom Berkom´s commercial network and the trans-atlantic fibre "CANTAT-3" (operated by Teleglobe), similar testbed sites in Canada can also be reached.

2.3 Relationship to other Projects

In consideration of the facilities available to the project, EXPERT is committed to supporting other projects both within and outside the ACTS Programme. For example, in the field of UMTS (in a local hospital environment) EXPERT is providing the location for the ATM/UMTS interworking equipment. EXPERT will also collaborate with the Finnish Multimedia Programme for trialling VoD, and is a platform for ATM accounting and signalling experiments, tele-education trials and TMN (availability) aspects. The EXPERT project also has very close associations with the National Hosts of Switzerland and The Netherlands (direct physical connections), Denmark, Norway, Finland and the UK. Due to its reputation, EXPERT is also involved with collaborations from outside of the ACTS community. A particularly interesting co-operation is currently taking place with the Canadian BADLAB Testbed (from the CANARIE Programme) whereby in the framework of the G7/GIBN (Global Interoperability of Broadband Networks) Programme regular "Virtual Classroom" sessions are held between schools in Ottawa and Basel.

3 Integrated Traffic Control for ATM

ATM Service Categories (ASCs) have been defined in [1] to offer optimal support for the divergent QoS requirements of existing and future broadband applications (such as voice, video and data) and to allow for a high bandwidth utilisation. Individual ASCs are designed such that their properties complement each other with respect to criteria such as time-relation (real-time vs. non-real-time), QoS model (guaranteed vs. best effort) and control methodology (preventive vs. reactive). Currently there exist five ASCs: CBR, rt/nrt-VBR, ABR and UBR. Appropriate traffic control functions have to be associated with each ASC enabling it to offer a cost-effective service, i.e. ensure negotiated QoS with minimal network resources effort. The diversity in the properties of individual ASCs is reflected in the type and design of the required control functions. The major goal of integrated traffic control is to integrate all ASC-related control functions into a common traffic management framework which allows to share network resources efficiently among connections belonging to different ASCs.

The set of control functions together with the underlying switch buffer management is referred to as the integrated traffic control architecture. Among others, the following main criteria to assess and compare different control architectures should be considered: QoS protection, efficient use of network resources, robust operation of the control system and implementation complexity.

The basis for ASC integration is the underlying buffer management. The traditional FIFO service strategy is not applicable in integrated services scenarios and some kind of priority scheme (e.g., delay priorities, round robin or weighted fair queueing based) is mandatory. For an overview of suitable service strategies for ATM it is referred to [2, 3]. Furthermore, the buffer capacity is usually partitioned into a number of logical queues (LQ) making it possible to assign each connection (per-VC), group of connections (per-connection-group) or each ASC (per-ASC) to a separate LQ. The service sequence between LQs is controlled by the service discipline. Certainly there is a trade-off between the partitioning principle applied and the related implementation effort. While per-VC queueing offers the highest control flexibility and QoS protection, it requires to set-up and manage for each connection (VC) a separate LQ which may become unfeasible in public network scenarios where hundreds or even thousands of connections share the same link.

CAC plays a key role within the control framework since it directly controls the network load by deciding if enough resources are available to accept a new connection. Since the resource allocation for a new connection is strongly dependent on the requested ASC, it is not possible to apply a common CAC algorithm for every ASC. Hence, for each ASC a suitable CAC method has to be identified and it has to be ensured that these ASC-specific functions can co-operate in the resource allocation (bandwidth and buffer capacity). The CAC framework is closely related to the buffer management implemented in the switch. In the next section a CAC concept based on a buffer architecture employing delay priorities between ASCs will be presented.

UPC is needed to enforce the negotiated traffic parameters. The Peak Cell Rate (PCR) and additionally the Sustainable Cell Rate (SCR) with Burst Tolerance (BT) are controlled for CBR and rt/nrt-VBR connections, respectively. ABR requires a dynamic UPC to control the varying value of the allowed cell rate while UBR in principle does not need any policing.

Other important control functions are traffic shaping and packet discard schemes (e.g., Early Packet Discard (EPD), Partial Packet Discard (PPD)). While the former helps to generate a traffic pattern which allows the network to allocate less resources and, therefore, to offer a cheaper service, the latter reduces overload and avoids waste of bandwidth by discarding cells which belong to corrupted AAL-packets. In section 5 the impacts of EPD and PPD as part of the UBR control framework on the performance of TCP will be shown and discussed.

In EXPERT different solutions for integrated traffic control architectures have been defined and investigated. So far, one proposal based on delay priorities between LQs for different ASCs and a rate-based approach with explicit reservation of bandwidth and buffer capacity have been identified [4]. While the first one requires only one LQ for each considered ASC, the latter is more flexible since it can employ any grouping principle from per-VC to per-ASC queueing.

Investigations on the control mechanisms for the ABR service category and their inclusion into the integrated traffic control architecture are also performed in EXPERT [4, 5, 6]. Further studies on the interactions between the control functions and higher layer protocols are currently being extended to include ABR.

4 CAC in a Delay Priority System

Experiments have been performed within the EXPERT testbed to identify a CAC method appropriate for systems applying delay priorities between rt-VBR and nrt-VBR traffic [4, 5]. All experiments so far have been based on a Fore ASX200 switch implementing delay priority queues. The system has been loaded using synthesised traffic generators which are capable of generating artificial test traffic according to a Markov model [7]. The buffer configuration is shown in Fig. 2. The service discipline is such that the low priority queue (non-real-time traffic) is served only when the high priority queue (real-time traffic) is empty.

Through measurements a CAC boundary has been determined that can be compared with CAC boundaries calculated using different CAC algorithms. N_{rt} on/off sources are multiplexed into the rt-VBR queue and N_{nrt} on/off sources into the nrt-VBR queue. All sources have a peak bit rate of 31.1 Mbit/s and exponentially distributed on- and off-periods with a mean length equal to 20 ms and 80 ms, respectively. The rt-VBR buffer was set to 48 cells, the nrt-VBR buffer to 4000 cells.

Fig. 3 shows the measured CAC boundary giving the maximum number of rt-VBR and nrt-VBR sources that can be multiplexed with a target cell loss ratio equal to 10^{-4}. The boundaries corresponding to a convolution based CAC (bufferless model) and a linear CAC (effective bandwidth) are displayed together with the measured results. The measurements reveal the increased utilisation that may be achieved using a priority scheme between the rt-VBR and nrt-VBR classes, thereby allowing to utilise larger buffers without affecting the delay performance for the real-time traffic.

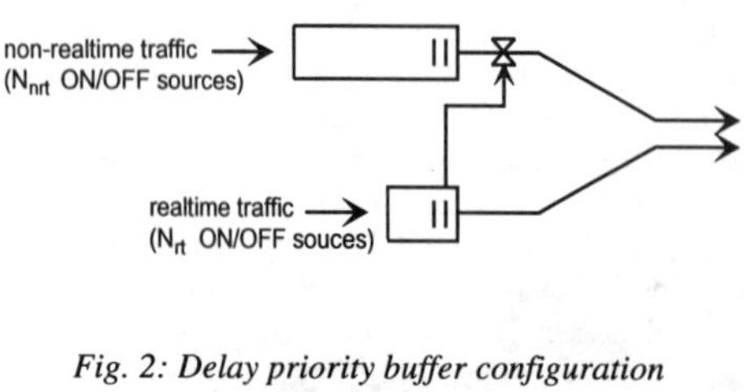

Fig. 2: Delay priority buffer configuration

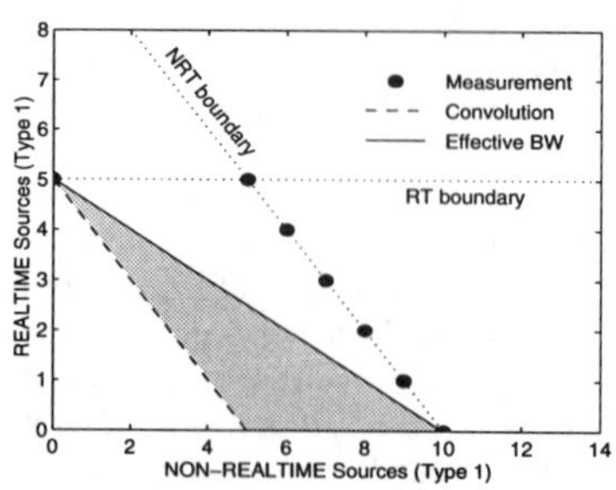

Fig. 3: CAC boundaries for rt-VBR and nrt-VBR traffic

The convolution based CAC gives low utilisation since a bufferless model cannot benefit from the large buffers available. The shaded area in Fig. 3 shows the increased gain that could be obtained using a linear CAC based on the effective bandwidth for each class. The effective bandwidth for the rt-VBR traffic is determined using the rt-buffer capacity, while the calculation of the effective bandwidth for the nrt-VBR traffic is based on the nrt-buffer capacity.

However, as can be seen from Fig. 3, a significantly better link utilisation can be achieved by limiting the CAC acceptance region with a boundary (denoted 'RT boundary') given by rt-VBR CAC convolution calculations together with a boundary (denoted 'NRT boundary') obtained by computing effective bandwidths as if all connections are of the nrt-VBR type and could use the large nrt-buffer. By the latter it is meant to allocate resources according to:

$$N_{rt} \cdot EB_{rt} + N_{nrt} \cdot EB_{nrt}$$

where EB_{rt} is the effective bandwidth for the rt-VBR and EB_{nrt} is the corresponding effective bandwidth for the nrt-VBR traffic, both calculated using the nrt-buffer capacity. Fig. 3 shows that this combined approach gives an acceptance region close to the one determined by the measurements.

The following overall CAC strategy may be applied when ASCs are separated by delay priorities:

- CBR and rt-VBR: Convolution based CAC calculation allowing statistical multiplexing gain
- nrt-VBR: Allocation of capacity to nrt-VBR connections based on effective rates including rt-VBR connections in the calculation as if they are nrt-VBR connections
- ABR: Linear allocation of capacity to individual ABR connections based on Minimum Cell Rate (MCR) taking into account the effective rates used for the nrt-VBR allocation
- UBR: No allocation of capacity to individual UBR connections

To ensure the robustness of this CAC solution, it is required that the calculated effective rates define a conservative CAC boundary. One way to obtain the effective bandwidths could be to estimate them off-line using a homogeneous fluid flow model. This approach should be compared to a CAC scheme based on worst-case traffic assumptions. For the proposal of such a scheme and for more information about the use of fluid flow models for architectures based on delay priorities it is referred to [5].

5 Performance of TCP over UBR

The UBR service is intended for nrt-applications that do not require guaranteed QoS commitments. With additional, relatively inexpensive control functions such as packet discard schemes, UBR could become a cost-effective alternative for the transmission of data traffic, offering a flexible solution as opposed to nrt-VBR or ABR with its sophisticated and complex rate-control protocol. Therefore, it is essential to evaluate the TCP over UBR performance, to identify how UBR can be integrated efficiently with other ASCs, and, in a final step, to compare its performance with ABR. This section surveys some key results obtained from a comprehensive set of experiments with TCP over UBR comprising measurements taken on different protocol layers.

5.1 Configuration for the TCP over UBR Experiments

Four workstations and the Fore switch build the local ATM network based on 100 Mbit/s TAXI interfaces as depicted in Fig. 4. The workstations are equipped with Fore SBA200 adapters which encapsulate IP packets in AAL5. The user application accesses the TCP/IP protocol stack and finally the ATM network interface through the BSD based socket API. The benchmarking tool *ttcp* measures the TCP memory-to-memory throughput. During the measurement, the source *ttcp* application is backlogged at any time instant. By means of VP shaping, the bottleneck rate is set to 37.44 Mbit/s, which is a rate both pairs of workstations can exploit completely.

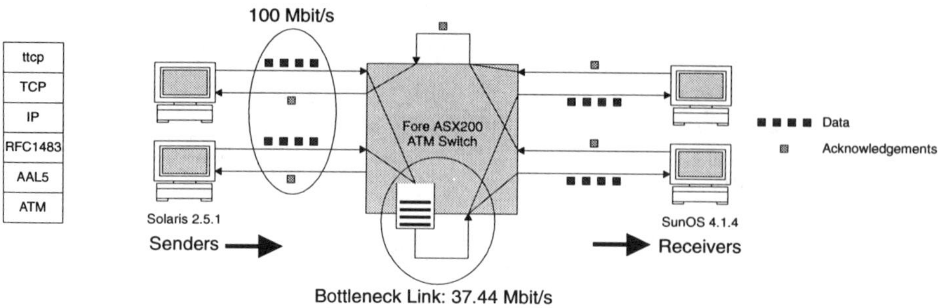

Fig. 4: Configuration for the TCP over UBR experiments

5.2 TCP over Plain UBR

TCP provides end-to-end window based flow control by means of several congestion control algorithms such as Slow Start, Congestion Avoidance, Fast Retransmit and Recovery. All these algorithms assume that packet loss indicates congestion in the network which has to be resolved by an appropriate rate decrease. Packet loss is either detected by a retransmission timeout or duplicate acknowledgements. Waiting for timeouts or duplicate acknowledgements can cause TCP to stop the transmission, and therefore, the switch buffer for UBR has to be dimensioned appropriately to ensure that packet loss can be kept within reasonable bounds.

TCP achieves the maximum throughput when no packets are lost, which means that the buffer at the multiplexing port has to be equal or larger than the sum of the receiver windows of all TCP connections involved. In this case, the TCP connections share the available bandwidth almost perfectly, as can be seen in Fig. 5 and 6 with a switch buffer size of 2176 cells. These figures also show goodput, Cell and Frame Loss Ratio (CLR and FLR) for a switch buffer size between one and two TCP windows and for different Message Transfer Units (MTU).

As expected, the goodput increases and the CLR decreases with increasing buffer size up to the point where the buffer is large enough to prevent any cell loss. As opposed to simulation results [8], the goodput grows with a larger MTU. This is caused by the relatively high processing overhead for the transmission of smaller packets. This processing overhead compensates the fact that the loss of a single cell has a more negative impact on the goodput for larger packets. Fig. 6 shows as well that a larger switch buffer is required to ensure zero loss for smaller packets. This is because the effective maximum TCP window size depends on MTU [5]. As expected, the FLR is significantly larger than the CLR because the loss of a single cell inside a frame inevitably leads to the discard of the whole AAL5 frame at the destination. Only if all cells were dropped at the switch, the CLR would equal the FLR.

5.3 TCP over UBR with EPD

In its implementation of EPD, the Fore switch defines an EPD threshold (*EPDTh*) for the shared switch buffer. If the fill level of the shared buffer exceeds this threshold when the first cell of an

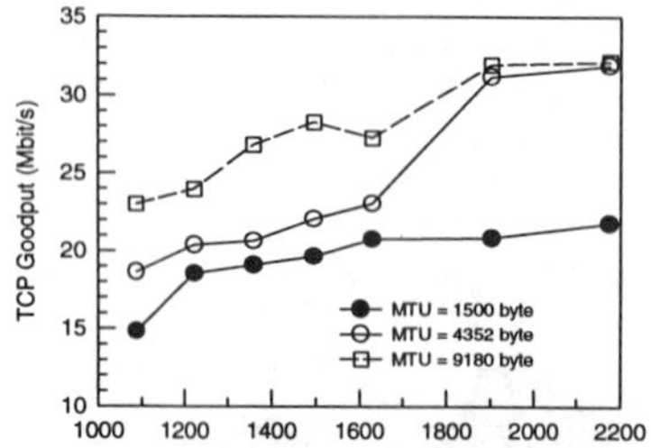

Fig. 5: TCP goodput as a function of the buffer size

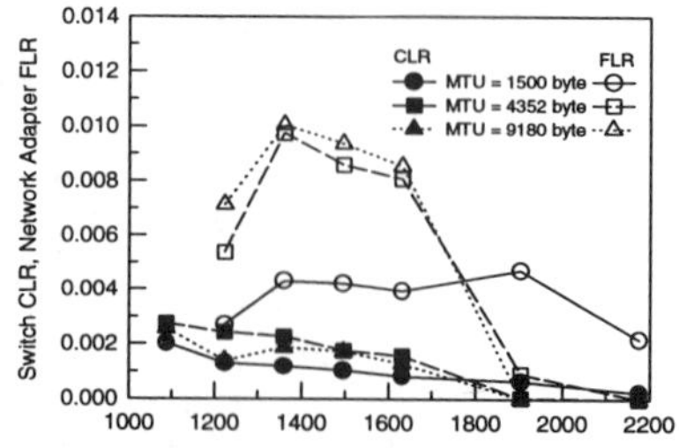

Fig. 6: FLR and CLR as a function of the buffer size

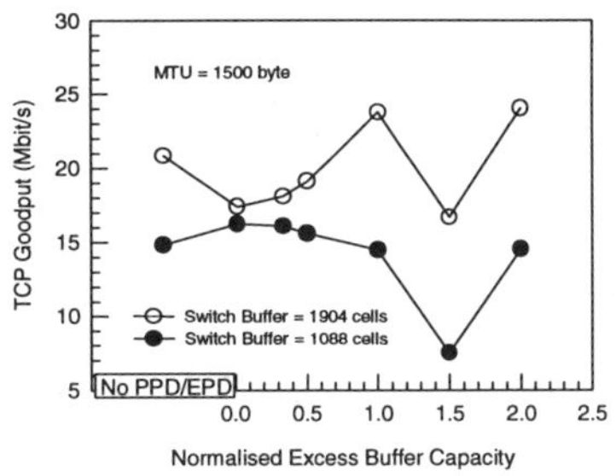

Fig. 7: TCP goodput as a function of the normalised excess capacity b

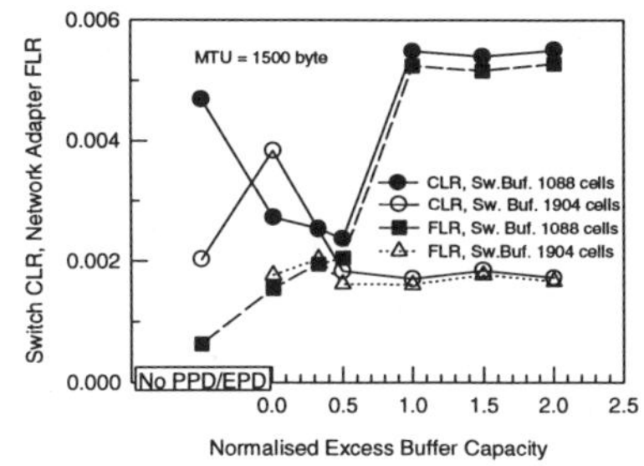

Fig. 8: FLR and CLR as a function of the normalised excess capacity b

AAL5 frame arrives, all incoming cells of this frame will be discarded except for the last one which allows to delimit the frames. If a cell in the middle of a frame is lost, the later incoming cells are discarded as well (Partial Packet Discard, PPD). Further variants of EPD are described in [8, 9].

In the following, the normalised excess capacity b (defined by the relation $EPDTh = SwitchBuffer - b \cdot MTU$) is applied to compare the performance of EPD for different threshold values with a fixed switch buffer. PPD is realised by setting b to zero.

Fig. 7 and 8 show the total goodput, the CLR and FLR for two TCP sessions, different excess capacities and without EPD for comparison. With small buffers (1088 cells), the best goodput is achieved with PPD and not with EPD, as one might expect. For a larger buffer (1904 cells), EPD improves the goodput best when the excess capacity is an integer multiple of MTU, since the excess capacity then allows to store an entire frame thus avoiding that frames are transmitted partially. As can be seen, the CLR and FLR differs for plain UBR and PPD whereas they become approximately equal when the normalised excess capacity is larger than 0.5. With a larger excess capacity, the cell loss process is driven more and more by the EPD mechanism which discards a whole frame. For larger MTU values [5], EPD usually does not improve or even reduces the goodput achieved by PPD. The buffer space allocated as the excess capacity is seen rather as a buffer reduction than as a means to avoid the transmission of corrupted frames. For a more detailed discussion refer to [5]. The experiments have shown that the dimensioning of the EPD threshold and the excess capacity are separate issues and that the most important issue for TCP over UBR is the dimensioning of the threshold, i.e. the buffer which can be used without any restriction.

5.4 The Impact of Delayed Acknowledgements

The default SunOS TCP refrains from sending acknowledgements until the receiver window has slid more than two maximum packets or 35 % of the maximum TCP window (delayed acknowledgement, DACK). The comparison of TCP with and without DACK both for PPD and EPD in Fig. 9 and 10 reveals that TCP without DACK suffers from relatively low goodput and a significantly lower fairness index (definition taken from [10]). Obviously, TCP without DACK allows a dominant session to conserve its resource share in the network. Conversely, DACK inserts randomness to the transmission of packets. Since the ACK for every second packet is deferred,

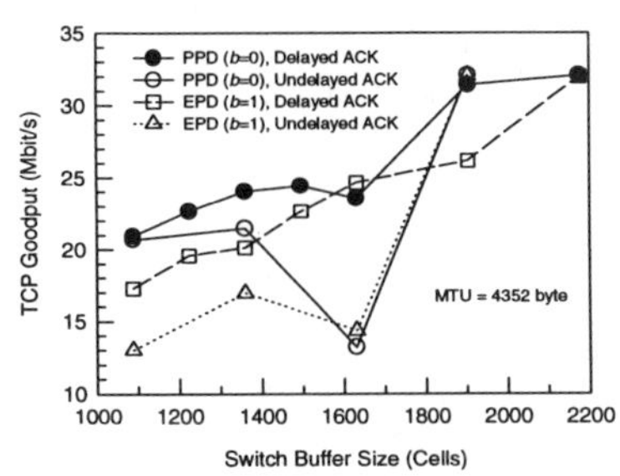

Fig. 9: TCP goodput for delayed/undelayed ACK

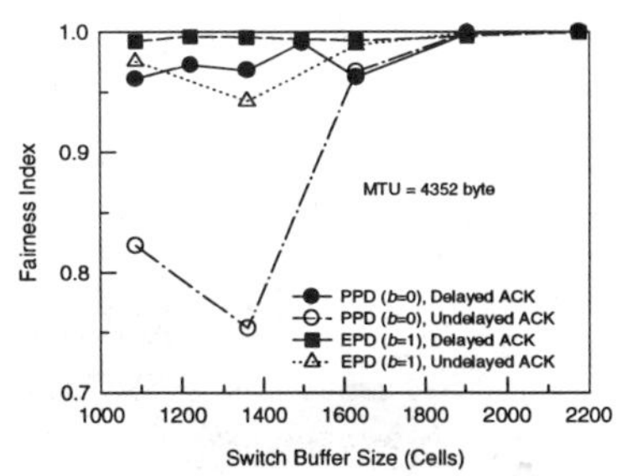

Fig. 10: TCP fairness for delayed/undelayed ACK

DACK prevents the synchronous sequence: „packet-ACK-new packet". If the buffer space exceeds the size of the TCP window, a connection starting first may occupy buffer corresponding to its window size thus ensuring that its TCP congestion control works very efficiently compensating the low goodput of the other connection. TCP without DACK tends to maintain synchronous, deterministic periods. Acceptable overall goodput despite unfairness is possible, as can be seen for a switch buffer of 1360 cells. Without DACK, TCP's congestion control works inefficiently when the buffer is increased further to 1632 cells and the second connection comes into play. The higher fairness for EPD suggests that EPD adds randomness to the system similar to DACK.

6 Conclusions

The broadband test environment of the EXPERT project with its extensive collaborations with other R&D projects has been described focusing on the important area of ATM traffic control. Effective control functions are required to enable an ATM network to efficiently support different ATM service categories. These functions have to co-operate efficiently with the buffer management used in the switches to achieve a robust integrated control system able to satisfy the divergent QoS requirements.

Architectures based on delay priorities between different service categories and a rate-based approach with explicit reservation of bandwidth and buffer capacity either per connection or per group of connections have been considered. A CAC strategy for delay priority architectures has been developed and shown to achieve increased resource utilisation without affecting the real-time delay performance.

Packet discard schemes proposed to enhance the UBR performance have been investigated in extensive TCP over UBR experiments. Larger goodput is observed for larger values of MTU when using plain UBR. In the configuration used, the TCP sessions benefit only little from PPD or EPD. This is contrary to what has been shown in several simulation studies and demonstrates the importance of experiments with real systems. Future work will focus on the integration of the ABR service into the traffic control framework and the related experimental activities.

Acknowledgements: The achievements being made within the project EXPERT and the comprehensive development work being undertaken is only possible with the conscientious co-operation of all partners especially of WorkPackage 4.2, together with the support of the Project Officer and funding from the Commission of the European Union, the Swiss Bundesamt für Bildung und Wissenschaft and other national bodies. The contribution of all these players is therefore hereby gratefully acknowledged.

References

[1] ATM Forum: *Traffic Management Specification Version 4.0.* AF-TM-0056, ATM Forum Technical Committee, April 96.

[2] EXPERT WP4.2: *Initial Specification of Bandwidth Management (taking into account particularly routeing strategies and the use of ABR).* ACTS 094 Deliverable 4, March 1996.

[3] H. Zhang: *Service Disciplines for Guaranteed Performance Service in Packet Switching Networks.* Proceedings of the IEEE, Vol. 83, N-10, Oct. 1995.

[4] EXPERT WP4.2: *Specification of Integrated Traffic Control Architecture.* ACTS 094 Deliverable 6, Sept. 1996.

[5] EXPERT WP4.2: *First Results from Trials of Optimised Traffic Control Features.* ACTS 094 Deliverable 10, March 1997.

[6] C. Blondia, O. Casals: *Throughput Analysis of the Explicit Rate Congestion Control Mechanism in ATM Networks.* Proceedings of the 10th ITC Specialist Seminar, Lund, September 1996.

[7] B. Helvik, O. Melteig, L. Morland: *The Synthesized Traffic Generator: Objectives, Design and Capabilities.* Proceedings of the IBCN&S, Copenhagen, April 20-23, 1993.

[8] A. Romanov, S. Floyd: *Dynamics of TCP over ATM Networks.* IEEE JSAC, vol. 13, No 4, pp.633-641, May 1995.

[9] H. Li et al: *A Simulation Study of TCP Performance in ATM Networks with ABR and UBR Services in ATM LANs.* IEICE Trans. Comm., Vol. E79-B, No 5, pp.658-667, May 1996.

[10] R. Goyal et al.: *Further results on UBR+: Effect of Fast Retransmit and Recovery.* ATM Forum Contribution 96-1761, Dec. 96.

Core and ATM Networks
D.W. Faulkner and A.L. Harmer (Eds.)
IOS Press 1997 © AKM

Integrated Dynamic Routing, Load Balancing and Survivability in ATM-based IBCN

Y. T'Joens[1], P. Georgatsos[2], S. Sartzetakis[3], D. Ranc[4]

Abstract. Network intelligence in ATM-based IBCN of the future should provide cost effective network performance and availability under both normal and fault situations. The system described in this paper integrates dynamic routing, intelligent load balancing and fast and reliable ATM network layer survivability functions in order to reach these objectives. The proposed system incorporates both control and management planes of network operation, and is designed and implemented adopting the emerging TINA framework and using a CORBA based distributed processing environment. The paper focuses on the functional specifications and architectural aspects of the proposed system. Demonstration, validation and assessment issues are also discussed. The system will be tested through international ATM field trials, on the National Host networks of Switzerland, the Netherlands, Norway and Greece.

1. Introduction

The use of the Asynchronous Transfer Mode (ATM) in an Integrated Broadband Communication Network (IBCN) on top of high capacity fibre networks allows for a wide range of services to be multiplexed on a single network infrastructure. Combining this with intelligent resource management techniques will lead to a cost-effective service provisioning.

However, the use of high capacity fibre networks implies that greater volumes of traffic will be concentrated on sparser network topologies, thereby augmenting the vulnerability of service support to network failures [1]. The adoption of B-ISDN has pushed telecom manufacturers and operators to strengthen their investigation into intelligent resource management and survivability techniques at the ATM network layer.

Recognizing the need for combining the functional capabilities of the control and management planes for the benefit of network operation, this paper introduces the REFORM (REsource Failure and restORation Management in ATM -based IBCN) system with the purpose to ensure network performance and availability within acceptable levels under normal and fault conditions. The specifications of this system both from the functional and architectural aspects is the main theme of this paper.

The proposed system covers both the control and management planes of network operation, with emphasis on fault, performance, and configuration management. The system incorporates rapid and reliable ATM network layer self-healing mechanisms,

[1] *Alcatel Telecom Research, F. Wellesplein 1, 2000 Antwerp Belgium, e-m: tjoensy@rc.bel.alcatel.be*
[2] *Algosystems S.A., 4, Sardeon str., 171 21 N.Smyrni, Athens, Greece, E-mail:pgeorgat@algo.com.gr*
[3] *Institute of Computer Science, Foundation of Research and Technology - Hellas, Heraklion, GR71110, Crete, Greece, e-m : stelios@ics.forth.gr*
[4] *Alcatel Alsthom Recherche, Route de Nozay, Marcoussis 91460, France, e-m : Daniel.Ranc@aar.alcatel-alsthom.fr*

intelligent load balancing, dynamic routing and resource management functions all interworking together with the overall goal to ensure cost-effective network performance and availability under normal and fault situations. Restoration mechanisms in the ATM network layer are integrated with the control and management plane functionality, aiming at providing an integral and network-wide treatment to the problem of fault recovery. The REFORM system is one of the first systems to provide such integrated solutions.

The REFORM system comprises two distinct parts: the REFORM control part (embedded in the NEs) hosts the fast responding restoration capabilities of the system; and the REFORM management part, hosts the required management functionality. Routing and resource management functionality is spread across these two parts through a hierarchical model, operating in different levels of time-scale and abstraction. The control part of the REFORM system will be realised with appropriate means to guarantee the time-critical requirements of its desired operation, while management part realisation will be based on the emerging TINA management framework and by using a CORBA-based distributed processing environment.

In the following section we provide briefly some background information we consider helpful in understanding the scope of our system. Section 3, presents the scope and approach taken to obtain the REFORM system. Section 4 presents the REFORM system specifications, and section 5 presents the REFORM architecture. Following that, we provide some information concerning the future validation of our system. Finally Section 7 summarises the paper and describes aspects of future work.

2. Background Information - State of the Art

2.1 Resource and Routing Management

ATM as being a multi-service, multi-access system, guaranteeing a certain service level to its individual users, should certainly address the problem of how access to the resources is granted. Resource management refers to the necessary means and functions for resolving resource contention problems. Resource management functions work within the limits of cost-effectiveness criteria and constraints of general performance requirements. ATM networks, by their asynchronous multiplexing nature and the use of logical layers such as the VPC layer, offer several degrees of freedom in traffic multiplexing. Although this flexibility eases the task of traffic admission and multiplexing it may have a disastrous effect on network performance if it is not managed properly.

Very closely related to the problem of resource management is the routing issue. Routes in ATM networks are defined in terms of VPCs and VPCs are defined and installed by the network operator in order to support the routing of switched VCs, the basic service we assume in the REFORM system.

Routing functionality is generally spread over both control and management planes of the network and network elements. More specifically route definition functionality is generally accepted to belong to the network management plane, while route selection is part of the call and connection control functions present at the individual network elements.

Route definition functionality relies heavily on estimates concerning both service characteristics and service load upon the network. Since user behaviour changes dynamically over different time scales, it is very likely that the network may become inefficient when the bandwidth allocated to VPCs on the existing routes is not in accordance with the quantity of traffic that is required to be routed over them. This may lead both to under-usage of certain parts of the network, as well as increased degrees of call/connection blocking in other parts of the network.

Routing mechanisms and algorithms should therefore cater for traffic changes, both temporarily as well as in longer-term and should adapt gracefully to topological changes due to network failures, avoiding routing through damaged areas of the network. Moreover routing functions should harmoniously coexist with the resource management functions in all layers of network operation with the overall goal of optimizing network performance. Different routing mechanisms may be defined, differing in the placement of the associated functionality, the type of algorithms they use, the degree of adaptivity they offer and the type of information they utilize [2].

Load balancing is widely accepted as an objective of the routing function, as to allow a better utilization of free bandwidth over the network, improve adaptivity to variations in the offered load, minimize the overall network possibility of call/connection blocking and to decrease service interruption during the network failure situation [3].

2.2 Fault management

The ATM-based network operation and maintenance is organized in a layered fashion. Five hierarchical OAM layers and the relative OAM information flows have been defined (F1-F5). The first three layers are at the physical layer, the latter two (F4 and F5) are at the ATM layer, VP and VC respectively [4].

The OAM functions at the ATM layer are independent of the underlying transmission system. F4 and F5 OAM cells must be multiplexed into the user cell stream. They offer a unique possibility to evaluate networks performance on the actual connections in use. The in-service performance measurements offer realistic information on active connections while the OAM flow on a connection can account for as low as 0.1% of the user load.

The importance of integration between the management system and OAM functionality at the ATM layer has been identified by the ITU-T. Recommendation I.610 [4] proposes an interaction scheme between the TMN and OAM. This scheme involves both X & Q interfaces for the entire path starting from the customer premises, passing the access local exchange up to the core network.

Successful fault management requires that information is gathered from network elements in a standardized way. Practical experience on existing networks shows that some fault scenarios involve alarm burst phenomena (e.g. uncontrolled pouring-in of numerous redundant alarm notifications) which tend to overload the Operation System, in a situation where availability and synthetic information are both required in order to take management decisions as efficiently and accurately as possible.

2.3 Self Healing

Self healing refers to the capability of the network to restore services in a distributed way, as opposed to service restoration in a centralised way through interactions with operations systems. According to the various network topologies on which they are defined, a distinction can be made between ring based topology restoration [5] and mesh based topology restoration [9]. This paper will concentrate on the use of VPC layer meshed ATM network topology restoration.

When relying on VPC restoration, a number of different procedures can be considered in order to install within the network the protection VPCs. These will be discussed next.

2.3.1 Static Restoration schemes

Static schemes refer to the pre-installation of a backup VP prior to the occurrence of a failure. Various authors have described the used of static schemes and their performance. Kawamura et al. describe in various papers in detail the use of preassigned VPs [6, 7, 8, 9].

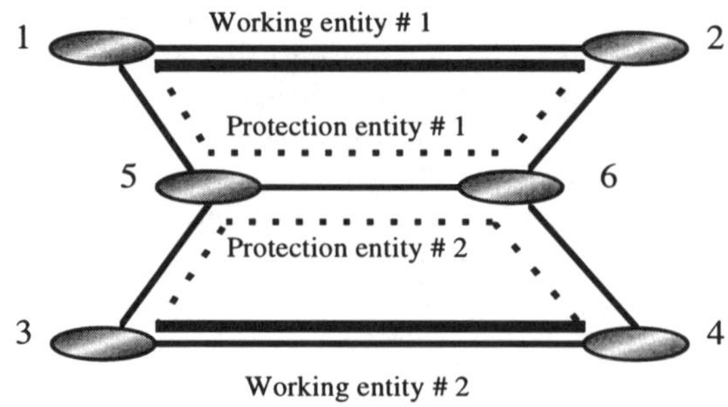

Figure 1: Static protection scheme

Static schemes, although simple to implement in the control plane and very fast inherent by their nature, require the installation of very effective and complex planning and network fault management functions. A further drawback of static schemes is their consumption of VP identifiers (which should also be regarded as a valuable network resources) and the fact that the resources are pre-planned, and should explicitly be monitored to adopt to changes in the state of the network.

ATM has the inherent advantage over physical layer protection schemes, that route and bandwidth occupation are both independent logical entities. Following the independent assignment of routes and bandwidth, the following classification of static schemes can be considered.

In a dedicated static scheme, also known as an Automatic Protection Scheme (APS), a pre-defined bandwidth is reserved for the backup VPC. Although this scheme offers very fast and up to 100% restoration from multiple link failures (given that the multiple failures do not hit working and protection entity in parallel), the scheme is very demanding in terms of bandwidth occupation, and is therefore not cost-effective. When considering Figure 1, the restoration resources allocated on link 5-6 should occupy the sum of working entities #1 and #2.

On the other hand, in a shared static scheme, use is made of the possibility to assign routes and bandwidth independently, and so install zero bandwidth VPCs in the network. In this case, restoration resources are shared among a number of backup VPCs, that protect working VPCs on disjoint physical paths. The capacity is shared in links between multiple backups which are potentially activated due to different failures. Depending on the network planning, a pre-defined level of restoration capability can be introduced within the network. The restoration resources depicted in Figure 1 would only occupy *max*(working entity #1, working entity #2) in stead of the sum of both when considering the dedicated static scheme.

Although the shared scheme is slightly slower and more complex than its dedicated counterpart (because of the restoration resource control protocol needed to activate the backup VPC), it is known as very cost-effective since the restoration bandwidth is shared. It is for this reason that this scheme is included in the REFORM system.

2.3.2 Dynamic Restoration schemes

As already indicated, static schemes do not allow for easy adoption to the present usage of the network, and introduce the need for complex planning tools and network fault management functions. When using dynamic schemes, no explicit pre-planning phase is needed, and no occupation of identifier resources is necessary prior to the failure.

Flooding schemes, also known as route search schemes, make use of flooding messages to obtain resources between sender and chooser nodes. The routes are not pre-established, but the scheme relies on the availability of spare capacity floating around the network. As to the specific details of the different flooding schemes, a lot of literature can be found. An overview of this literature was presented in [2], a specific description of a flooding scheme can be found in [10]. Flooding has the inherent disadvantage that it introduces a lot of messaging within a network. Given the fact that a large number of working VPCs can be impacted by a single network failure, this may give rise to an enormous explosion of messages.

3. REFORM system scope and approach

Adopting a network operator's viewpoint and by taking into account the broadband and multi-service nature of the IBC network environment, the scope of the REFORM system is to provide the necessary means and functions for ensuring network performance and availability within acceptable levels under both normal and fault situations. Ensuring network performance addresses the ability of maintaining the performance of existing connections within acceptable levels, while ensuring network availability refers to the ability of the network to accept new connections guaranteeing their performance requirements.

In order to achieve this, the REFORM system offers fast fault detection, alarm indication and report functions, rapid and reliable network self healing mechanisms, efficient dynamic and/or static resource and routing management schemes and algorithms with inherent load balancing functionality, coping with fault conditions, and efficient resource migration algorithms. These functions will co-exist and co-operate with the overall goal to ensure both cost effective network performance and availability in normal conditions and cost-effective, reliable and robust network recovery in the performance of existing connections as well as the availability for new connections during fault situations.

We build upon the results of the RACE II projects ICM [12], IMMUNE [10] and TRIBUNE. Integration and augmentation of the functionality studied within these individual projects has lead to the following functional decomposition of the system.

4. REFORM system functional specification

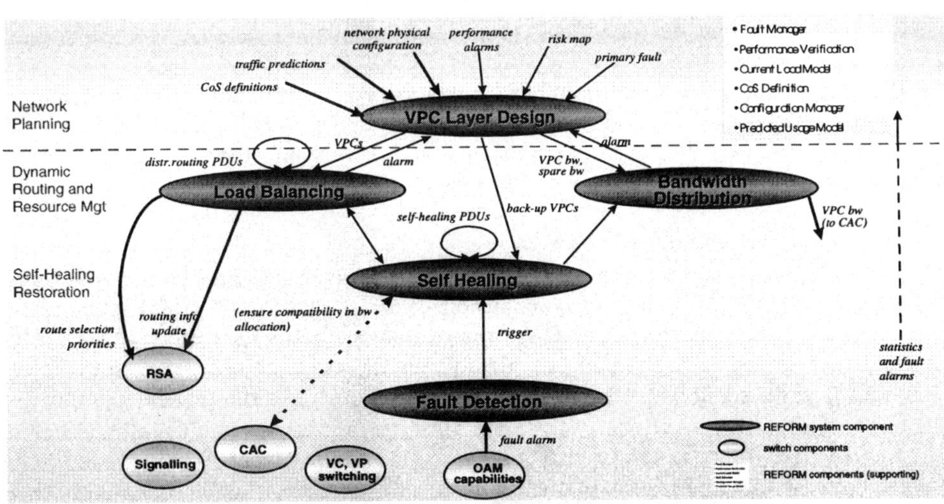

Figure 2 : The REFORM system functional overview

Figure 2 presents the functional decomposition of the system. At the highest level, the VPC Layer Design takes into account the traffic predictions and network physical configuration among other information to calculate and install the meshed VPC layer, this is referred to as the network planning activity. Given the installed VPC layer, it is the specific task of the Load Balancing function, to influence routing decisions within the network from a short-term perspective. The Load Balancing function will accomplish this by updating the Route Selection Parameters within the individual network elements. It has been designed as a distributed function, both to improve on the time response to network topology changes and to augment overall network robustness. The distributed routing protocol that is associated to the Load Balancing function is based on the emerging PNNI Routing Protocol [11].

During normal operation of the network, the Route Selection Algorithms (RSA) select appropriate routes for new connections according to their algorithms [3]. The Load Balancing functions manage the route selection criteria used by the RSAs to balance the load over the network as evenly as possible by altering the selection parameters associated with each route selection entry.

Given that actual resource usage of the network may deviate from the predictions that lead to the VPC layer designed, it is the task of the Bandwidth Distribution function, to accommodate for these changes by redistributing the available bandwidth over the present VPCs, while respecting the agreed availability and survivability constraints.

The Self Healing function has as specific task, to accept failures indicated by the Fault Detection function, and to both seize the restoration resources along the backup VPC, and cater for the actual switch over of active traffic between failed and backup VPC.

The Fault Detection function is concerned with failures caused in or escalated to the ATM network layer. For example, when a physical link failure has occurred and cannot be recovered by the restoration mechanisms in the physical layer(s), the fault is escalated to the ATM layer. This type of failure will cause a large number of alarms, as many as the VPLs defined on this link. REFORM therefore will provide for alarm correlation functionality in order to determine the original link failure. The failures will be identified through the standardised OAM alarms. Restoration mechanisms and escalation procedures in the network levels below the ATM layer are not addressed.

For the previously identified components, cost-effective solutions are pursued by means of considering suitable objective functions in the optimisation problems corresponding to the route definition, restoration resource allocation and management algorithms. The detailed description of a functional model decomposing the above functionality into individual components and specifying the information exchanges between them is beyond the scope of this paper.

5. REFORM System Architecture

As already mentioned in previous sections, the REFORM system comprises both a control plane part and a management plane part.

The key aspect of the REFORM architectural approach is the adoption of the emerging TINA framework with the purpose of evaluating its suitability and assessing its potential benefits when applied to network management, while comparing it as well to established technologies (OSI-based TMN, Internet SNMP). We use a commercial CORBA-based platform for providing the Distributed Processing Environment (DPE) for designing and implementing the management plane functionality. Existing TINA-C architectures in the area of Resource Management are taken into account.

TMN architectures for configuration and performance management, such as the ICM Virtual Path Connection and Routing Management [12, 13], constitute the starting point for the management plane architecture. These are enhanced and modified in the light of the TINA-C developments and ported on the CORBA-based DPE. They are complemented by new functionality, not yet addressed by TINA and not available in existing TMN architectures.

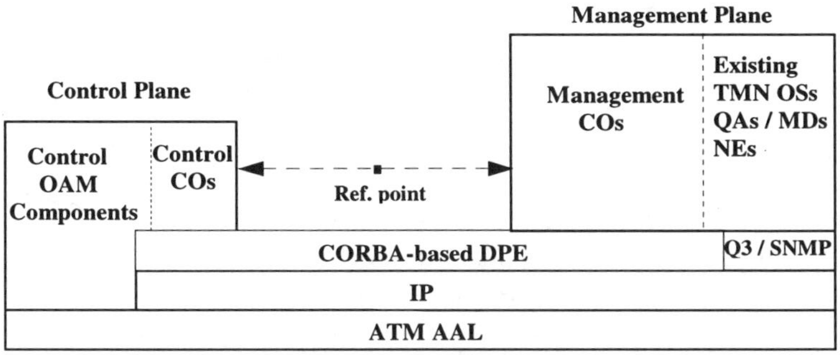

Figure 3 : Overall REFORM system architecture

The control plane functionality will be supported by relevant control logic and OAM transport protocols operating directly over the ATM adaptation layer. The state of the art in terms of projects such as IMMUNE and TRIBUNE together with the relevant ongoing ITU-T work is the starting point. The integration between the control and management plane domains will be accomplished through the interaction of computational objects over the DPE. As such, the control plane domain includes computational objects for the purpose of interfacing to the management plane. Such interactions are expected to be bi-directional.

All functions and operational components of our system in the leftmost box in Figure 3 are regarded as local inside a node, and will be implemented according to the particular capabilities of the network elements. We take into account the low level interfaces that provide connection with the hardware subsystem. Thus, direct message exchange will be feasible.

6. Experiments

Our work is centered around a number of trials and experiments to validate the system and assess its performance in a realistic environment. The field trials are based on European National Host (NH) platforms (ACTS [14]). The NHs are consolidated advanced platforms comprising communications infrastructure, services and generic applications. The NHs are made available to projects interested in performing operational trials or experiments of leading edge applications, services and management that involve real networks, services and users.

The field trials are planned in a phased approach, the first phase to be held in November 1997 will consist of a limited subset of functionality, in order to get experience on the environment, and to test some basic functionality (self healing and load balancing functions). The second trial to be held is scheduled for November 1998, and will accommodate the full REFORM system functionality.

7. Conclusions

We introduced the preliminary investigation in an integrated framework for ensuring network performance and availability in multi-class ATM networks under normal and fault conditions. We integrate control and management plane functionality, covering the complete failure management cycle (prior, post, normalisation), and combine in one system rapid and reliable network self-healing mechanisms, spread across the control and management layers of the network operation; intelligent routing coping with fault conditions, taking into account the multi-service environment; intelligent load balancing functionality with inherent self-healing capabilities, and cost-effective restoration resource allocation and appropriate management schemes.

Since we adopted the emerging TINA framework based on a commercial CORBA-type platform, we will evaluate, validate and demonstrate through experimentation in real network environment the suitability of such approaches to network management.

We plan to carry out field trials promoting in parallel the use of advanced networks and services as they are provided by NH environment incorporating real user communities in Europe. This way we demonstrate how the functional capabilities of the developed system can improve performance and reliability of end-to-end connections.

Much of the above work is still on-going, and experimentation is planed for this and the next years. REFORM is a research project (July 96 - June 99) of the ACTS programme partially funded by the Commission of the European Union. The authors wish to thank all their project colleagues who contributed in many ways to the formation of the ideas described here. Without them this work and our co-operation for these three years would never have started. Up to date information on the REFORM project and its results is widely available on-line [15].

8. References

[1] Wu T-H., "Emerging Technologies for Fiber Network Survivability", IEEE Communications Magazine, February 1995

[2] "REFORM system requirements and analysis", AC208-REFORM, Deliverable 1, January 1997

[3] Georgatsos P., Griffin D., "A management System for Load Balancing through Adaptive Routing in Multi-Service ATM Networks", IEEE INFOCOM'96, Vol.2, pp.863-869, 1996

[4] ITU-T Recommendation I.610, "B-ISDN operation and maintenance principles and functions", Geneva 1995

[5] Imai K., Honda T., Kasahara H., Ito T., "ATMR: Ring Architecture for Broadband Networks", Proceedings of IEEE GLOBECOM'90, December 1990.

[6] Kawamura R., Sato K-I, Tokizawa I., "Self-Healing ATM Network Techniques utilizing Virtual Paths", Networks '92, Kobe , Japan.

[7] Kawamura R., Sato K-I., Tokizawa I., "Self-Healing ATM Networks Based on Virtual Path Concepts", IEEE JSAC, vol. 12, no. 1, January 1994

[8] Kawamura R., Hadama H., Tokizawa I., "Implementation of Self-healing Function in ATM Networks Based on Virtual Path Concept", Proceedings Infocom 95 (?), 3b.1 , p303-311

[9] Kawamura R., Tokizawa I., "Self-Healing Virtual Path Architecture in ATM Networks", IEEE Communications Magazine, September 1995.

[10] Nederlof L., Struyve K., O'Shea C., Misser H., Du Y., Tamayo B. , "End-to-end Survivable Broadband Networks" IEEE Communications Magazine, September 1995

[11] "Private Network-Network Interface Specification Version 1.0 (PNNI v1.0)", af-pnni-0055.000, The ATM Forum, March 1996

[12] "Integrated Communications Management of Broadband Networks", ed. Griffin D., Crete University Press, Heraklion, Greece, ISBN 960 524 006 8

[13] Griffin D., Georgatsos P. , "A TMN system for VPC and routing management in ATM Networks", Integrated Network Management IV, Proc. Of 4th ISINM 1995, ed. A.S. Sethi, et al., Chapman & Hall, UK, 1995

[14] ACTS , On line Information: http://www.uk.infowin.org/ACTS

[15] REFORM, On line information : http://www.algo.com.gr/acts/reform

Core and ATM Networks
D.W. Faulkner and A.L. Harmer (Eds.)
IOS Press 1997 © AKM

Experimental Results on Distributed Restoration in an ATM laboratory Testbed (RACE IMMUNE)

Leo NEDERLOF, Hans VANDERSTRAETEN,
Dominique CHANTRAIN, Luc VAN HAUWERMEIREN
Alcatel Telecom, Research Division, F. Wellesplein 1, B-2018 Antwerp, Belgium
E-mail: LNed@rc.bel.alcatel.be

Abstract: This paper reports on the experiments and the results obtained with distributed restoration on an ATM laboratory testbed. These experiments were performed within the framework of the RACE project IMMUNE, the objective of which was the development of a common strategic view on end-to-end network survivability. Next to two testbed implementations and a comparison of survivability techniques this project encompassed the assessment of issues and requirements and the definition of user and service classes, the development of evaluation methods, and a study of network management issues and escalation between network layers. The focus of this paper is on the ATM testbed that was set up for experimenting and demonstration of distributed restoration based on a message flooding algorithm. After the generation of a physical failure, the restoration mechanism is able to restore the traffic obtained from a video source within three to five seconds.

1. Introduction

During 1994 and 1995, a consortium of six European companies, universities and operators[1] joined forces to work on the area of end-to-end broadband network survivability. Until then, some concepts and technologies had been proposed or developed, such as protection switching, self-healing rings and self-healing networks, but an integrated view on application of those technologies in complex networks still lacked.

This paper reports on the results and conclusions learned from the implementation of a distributed restoration algorithm (DRA) on a laboratory testbed and observations thereof. First, in section 2, an overview is given of the IMMUNE project as a framework for the experiments. Section 3 describes the components and the configuration of the testbed, and section 4 the operating principles of the DRA that was developed and implemented.

The results of the various experiments are given in section 5, followed by the conclusions.

---

[1] The IMMUNE consortium consisted of Alcatel Bell (B, leading partner), BT Labs (GB), IMEC/University of Gent (B), KPN Research (NL), Alcatel SESA (E) and Philips Kommunikationsindustrie (D)

2. Overview of the IMMUNE project

The IMMUNE project has set the objective to develop a common strategic view on end-to-end network survivability. Based on this view, it has defined and demonstrated restoration methods for future public and corporate Integrated Broadband Communication (IBC) networks that can ensure end-to-end survivability in a seamless and cost-effective way.

The first focus point was to understand what the issues are and who the key players are in broadband network survivability. Users and service providers are placing greater demands on network operators to provide a high grade telecommunications network. The additional commercial impact of telecommunications together with the advent of new technologies and services and the relaxation of government-owned monopolies has placed the issue of improving network survivability high on the list of priorities for network operators. The IMMUNE project has described and analysed the needs in terms of survivability of the users, service providers and network operators, resulting in a matrix of user and services classes [1].

Before the actual deployment of a survivability strategy in a network, an assessment of the performance versus cost of the strategy must be made. Hence, the IMMUNE project developed a framework to evaluate a survivability strategy under realistic networking conditions [2]. The main concerns of this part of the project were the performance evaluation of restoration techniques through simulation, and the planning of survivable networks and the dimensioning of spare capacity.

The IMMUNE project has reviewed possible survivable network architectures and technologies for each layer of the B-ISDN transport network: SDH, ATM and optical. For each of these three layers, Protection Switching, Self-Healing Rings and mesh networks can be implemented. Several techniques have been compared by means of simulation on a reference network [3].

Finally, a major area of interest of the IMMUNE project was survivability in layered and partitioned networks and the management thereof. If different parts or layers of a network are protected by different restoration mechanisms, the interaction or escalation between those mechanisms requires careful study. Introducing distributed restoration in a centrally managed network brings along the necessity of a changed attitude towards traditional network management paradigms. Several possible scenarios for escalation and interaction between restoration mechanisms and the network management system have been studied, and the critical issues have been identified [4].

A full report on these studies and the exploitation of the developed strategies can be found in [5] and a summary of the project and the main achievements was published in [6]. In addition to these studies, two lab demos were part of the IMMUNE project. PKI in Aachen developed and demonstrated an ATM multi-ring network of duplex transceiver ADMs, with a restoration mechanism based on loopback and bypass rerouting strategies [7]. The testbed that Alcatel set up in Antwerp will be described in the remainder of this paper.

3. Testbed description

In the research laboratory of Alcatel in Antwerp, a testbed was set up for experimenting with a distributed restoration algorithm (DRA). As shown in Figure 1, this testbed consists of a number of ATM switches interconncected by optical fibre STM1 interfaces. Each network element (NE) consists of an 8x8 ATM switch connected to a control workstation (CS), and was configured to be used as a VP cross-connect. The DRA was added as a software module on the control workstation.

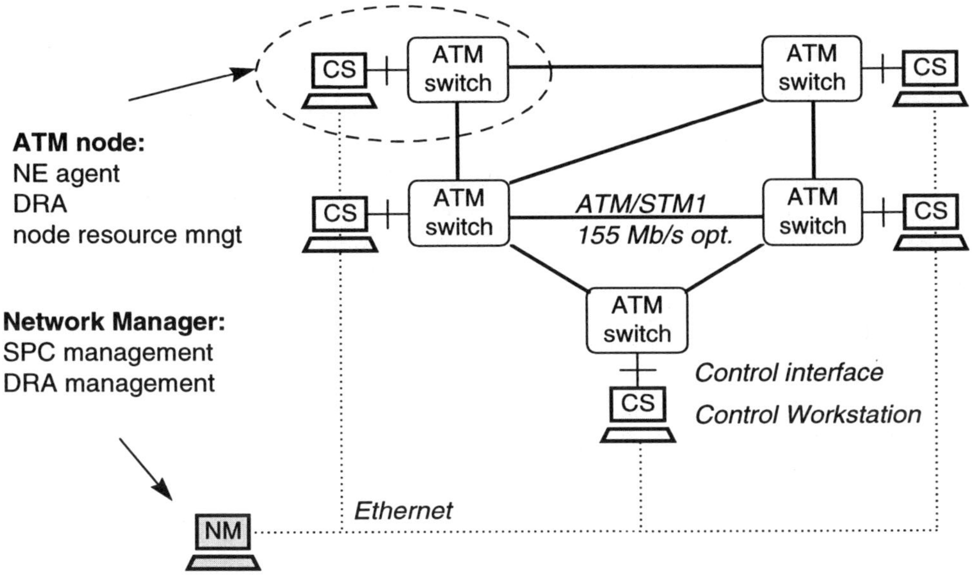

Figure 1 The testbed for experimenting with restoration, consisting of five 8x8 ATM switches, managed by a central network management application

Considering the nature and the elementary features of a DRA, the basic requirements to the testbed were:

- a transparent channel between DRA instances in adjacent nodes for the exchange of DRA protocol messages
- a (software) interface between the DRA and the ATM switch to perform the basic tasks of setting up connections and receiving information about active connections, links' statuses and alarms.

These requirements were realised by 1) an AAL5 process that was added on the on-board controller (OBC) of the ATM switch and 2) adaptation of the NE agent to provide the DRA with the necessary information and functionalities.

In order to facilitate experimentation and demonstrations, a network management (NM) system was added to the testbed. In principle, the NM could communicate with the NEs via a data communications network carried by the ATM transport infrastructure, as it would in a real network, but in this testbed the NM communicates with the NEs via the ethernet network present in the laboratory. This does not reduce the value of the experiments, since the NM only plays a passive role of monitoring during the restoration. This monitoring is exploited in the testbed by an on-line graphical representation of the restoration process for demonstration purposes, which would be hampered if the NM communicated over the ATM links that are subject to cuts.

For demonstrations, an interactive Video-on-Demand (VOD) application was available. The VOD server and user were connected to different nodes in the testbed, and a VP connection was set up from the management station. Alternatively, an ATM traffic generator and analyser (ATGA) was available that could generate, terminate and monitor a number of ATM traffic streams in parallel.

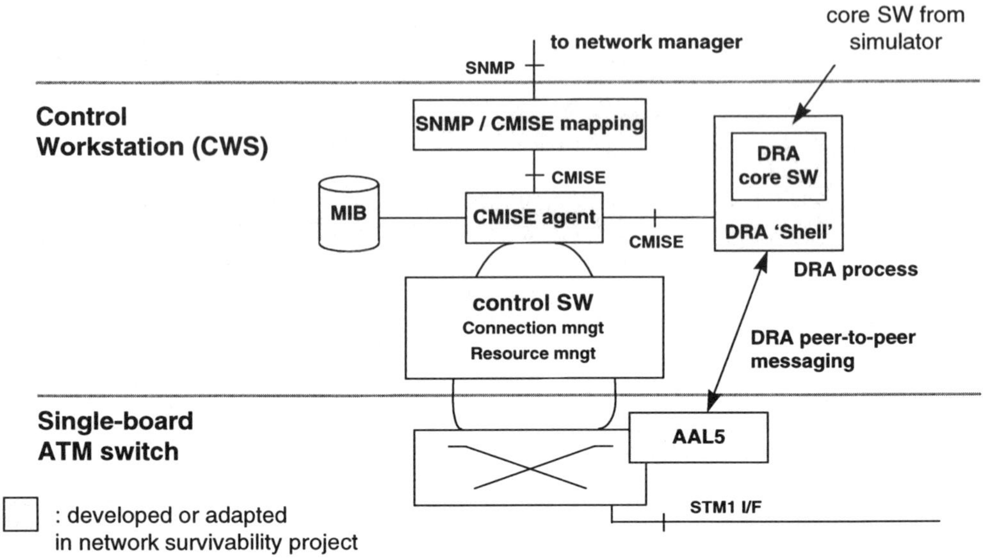

Figure 2 Software structure of the network elements in the testbed

3.1 Software architecture

The general structure of the software of the network elements is given in Figure 2. This figure illustrates some necessary changes that were made for the integration and the management of the DRA in the testbed, as well as the AAL5 segmentation and reassembly process that was added to the on-board controller of the ATM switch to accommodate DRA messaging between neighbouring nodes. The software module that contains the core code of the algorithm was directly imported from the simulation platform on which the DRA was developed and validated. This, as a side effect, offered valuable opportunities to study and compare the behaviour of the algorithm in a simulated and a real environment.

3.2 Network management applications

Although a DRA can operate autonomously without interaction with the network manager, applying a management platform in the testbed still proved to be very useful. Some applications were developed for managing the DRA and some extensions were made to the management information base (MIB). In the initial implementation in the Alcatel research lab, this management interaction, was not a focus of the experiments, but was just meant to facilitate the experiments and adding value to the demonstrations by a graphical display of events during the restoration process. Furthermore, all events relevant to the restoration process are stored in logs by the nodes locally, to be retrieved by the network manager afterwards for analysis of the restoration process.

4. Algorithm

The algorithm that forms the core of the testbed demonstrations is a distributed restoration algorithm based on a message flooding principle after detection of a failure [8]. A significant increase in the speed of restoration is attained by initiating the message flooding from both ends of the failure simultaneously. This 'two-prong' method was devised by Nicholson [9],

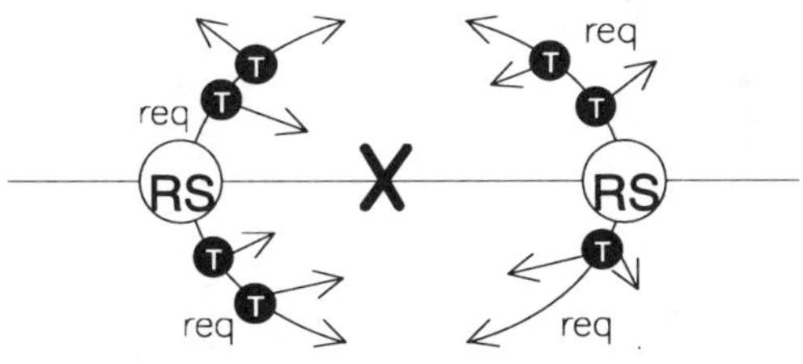

Figure 3 Request message flooding

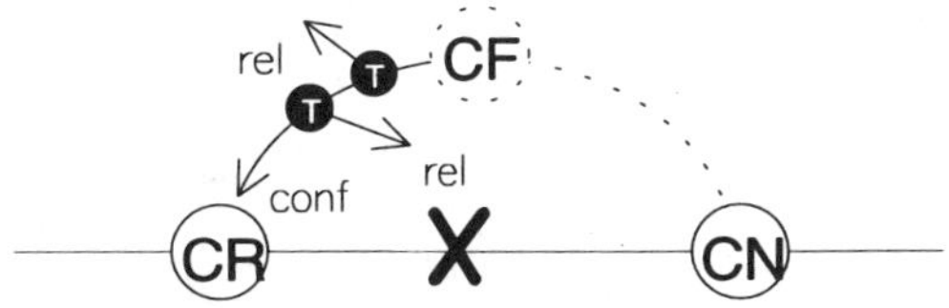

Figure 4 Confirmation

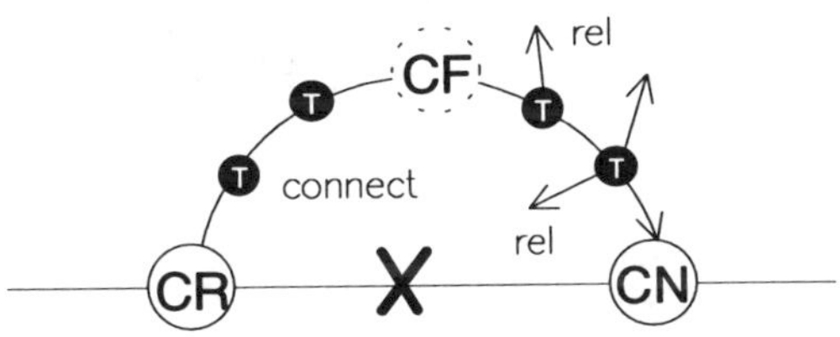

Figure 5 Reconfiguration

and first applied in a restoration algorithm by Chow [10]. The progress of the restoration algorithm is schematically illustrated by the following figures.

The route search process is based on a flooding of *Request messages*, originating from the *Request Source* (RS) nodes adjacent to the failure as in Figure 3. Intermediate nodes, referred to as *Tandem* (T) nodes, store the request, and forward updated copies of the Request message. As these messages propagate between the nodes, spare capacity that is requested will be explicitly reserved for each message until this capacity is captured as part of an actual alternative route, or released to make place for a different alternative route. A limit is set to the flooding procedure by specifying the maximum number of tandem nodes by which a request message can be remote from the original RS node. This parameter is usually referred to as the *hop count.*

When a route is found, i.e., when request messages meet somewhere between the two RS nodes, the Tandem node where the branches meet, now labeled *Confirm* (CF) node, designates the RS node with lowest ID as *Chooser* (CR) and the other as *Chosen* (CN). The Confirm node sends *Confirm messages* down the candidate alternative route towards the Chooser as illustrated by Figure 4. Any surplus bandwidth reservations are cancelled implicitly, or by sending *Release messages* on obsolete request branches.

Figure 5 shows the Chooser node that makes a selection of the VPs that can be allocated to candidate alternative routes. Information about the rerouted connections is sent along the route to the Tandem nodes and the Chosen node in *Connect messages*. Each Tandem node between the CF node and the CN node now releases obsolete request branches.

To evaluate the performance of this algorithm, extensive simulations have been performed, which have shown that this algorithm outperformes most alternative algorithms when simulated under identical simulations. These simulations are reported a.o. in [11]. The relative performances of the different algorithms are given in [3].

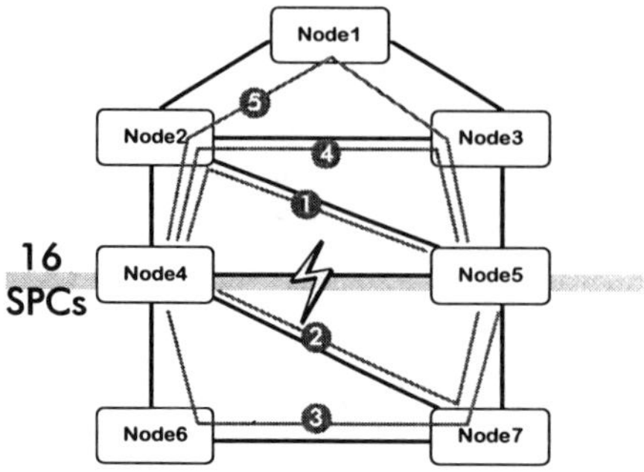

Figure 6 Testbed re-routing

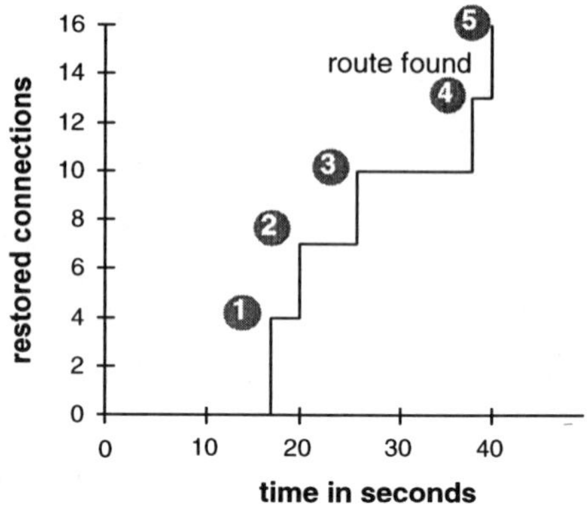

Figure 7 Time results of restoration in lab testbed

5. Testbed results

Before giving details on the performance results obtained with the testbed experiments, it is important to stress that the main objective of the testbed was to demonstrate the viability of distributed restoration by making it work and showing its robustness to cope with an exhaustive set of failure situations. No real efforts have been spent on improving the performance of the testbed. As a result, the performance figures given here may seem far from the 2-second objective, but at least we have identified the main bottlenecks from which guidelines for future products can be formulated.

Most of the demonstrations have been given with one single VP SPC carrying a video-on-demand application, with a VOD server and a VOD user connected to different nodes in a 5 node network. After simulating a link failure by disconnecting a fibre, the video image is usually restored 3 to 5 seconds later. More extensive tests have been performed on a 7 node network as in Figure 6, with a traffic generator generating and receiving 16 VP SPCs. The link between Node4 and Node5 was cut, and the spare capacity in the network was dimensioned such that the DRA was forced to restore the connections on 5 different alternative routes, as shown in Figure 6.

The resulting restoration times are given by the graph in Figure 7. It shows the time needed to find the five different alternative routes after the failure has occurred. The total time needed for restoration can be split in three parts:

- The alarm detection time: This is the time that passes between the cutting of the fibre and the activation of the DRA, usually 3 to 5 seconds.
- The DRA execution time (indicated in Figure 7): This is the time that the DRA takes to find alternative routes with spare capacity, and to allocate failed connections to those routes.
- The time needed to set up the connections in all nodes: This is the time needed to complete the creation of the connections in all nodes that are involved in the re-routing.

The execution time of the DRA is for the largest part a result of the message transfer delays. The bottleneck here is the AAL5 process that runs on the OBC of the TLK, and handles both incoming and outgoing messages. The resulting message transfer delay depends on the current load of the OBC and on the length of the DRA message, but rapidly runs into the seconds timescale. A major increase in performance can be achieved by improving the

AAL5 handling process, for instance by connecting the CS with an ATM link to the node. The message transfer delay should be no more than several milliseconds then. Simulations that have been done with these values have shown that this phase of the restoration will then take about 0,5 to 0,8 seconds.

A second significantly large delay is the time that is needed to set up the connections, once the DRA has determined the new routes. This is a direct result of the limited functionalities that the cross-connect software offered. The only way for the DRA to set up a connection is to pass through the local element manager using a CMISE request, each call taking several seconds to complete. Besides, the nodes that terminate the bypass – nodes 4 and 5 in Figure 6– first have to delete the original connection explicitly, which requires another two calls to the element manager. It is evident that the connection set-up time can be greatly reduced by giving the DRA direct access to the routing tables.

6. Conclusions and further work

A prototype implementation of a distributed restoration algorithm has successfully been made on a testbed in the Alcatel Telecom research laboratory in Antwerp, as part of the RACE project IMMUNE. The message flooding algorithm that forms the core of this prototype was developed and validated by means of simulations, which have shown a competitive performance when compared to alternative algorithms. Even though the implementation of the DRA is bound to a number of restrictions as a consequence of the prototype nature of the testbed, experiments have shown the ability of a DRA to perform restoration within seconds after a failure.

The value of these experiments are twofold. First, the mere fact of being able to demonstrate what is otherwise just proven to work by computer simulations has shown to have a significant impact. In this, the network management console display on which the network, the failure and the restoration process are represented graphically have even further added value the demonstrations. The second benefit to be taken from these experiments lies in the lessons learnt from going through the process of integrating the DRA in the control and management plane of a network of ATM cross-connects. Some performance bottlenecks have been clearly revealed by this prototype implementation so that requirements and recommendations can be formulated towards the commercialisation of distributed restoration.

Now that distributed restoration by means of a flooding algorithm has been shown to work, a further challenge lies in optimising the operation of an integrated restoration strategy. In a multi-layered network different restoration mechanisms will be applicable to different parts of the network or to different service classes of the traffic carried. The interaction between these mechanisms and the escalation strategy is a subject of further study in this area.

Experiments like these can be a firm basis for contributing to standardisation. Activities are currently going on in the different standardisation bodies on the area of protection switching and survivability in general. It is questionable whether one single DRA can or should be standardised, but at least a framework of (software) interfaces and messaging channels could be defined in which a DRA can be plugged in.

References

[1] RACE II IMMUNE, Deliverable D2, "Requirements and Reference Configurations for Survivability", June 29, 1994.

[2] RACE II IMMUNE, Deliverable, D6, Final Report on Performance and Resource Allocation Evaluation Methods, June 30, 1995.

[3] RACE II IMMUNE, Deliverable D7, "Performance Report on Network Restoration Algorithms", September 22, 1995.

[4] RACE II IMMUNE, Deliverable D8, "Final Report on Escalation Systems and TMN Interaction", September 30, 1995.

[5] RACE II IMMUNE, Deliverable D12, "Report on IBC Network Survivability Strategies and their Exploitation", December 31, 1995.

[6] L. Nederlof, K. Struyve, C. O'Shea, H. Misser, Y. Du and B. Tamayo, End-to-end Survivable Broadband Networks, IEEE Communcation Magazine, vol. 33, no. 9, pp. 63-70, September 1995

[7] K.P. May, P. Semal, Y. Du and C. Hermann, A fast restoration system for ATM-ring-based LANs, IEEE Communcation Magazine, vol. 33, no. 9, pp. 90-98, September 1995

[8] L. Nederlof, H. Vanderstraeten, P. Vankwikelberge (Alcatel Belgium), A new Distributed Restoration Algorithm to protect ATM Meshed Networks against Link and Node Failures, International Switching Symposium (ISS '95), pp. 398-402, April 24-28 1995, Berlin, Germany

[9] T.A. Nicholson, Finding the shortest route between two points in a network, Computer J., vol 9., pp. 276-280, 1966.

[10] C.E. Chow, J. Bicknell, S. McCaughey, A Fast Distributed Network Restoration Algorithm, International Phoenix Conference on Computer and Communications, March 22-26, 1993, Tempe, Az.

[11] G. Ocakoglu, L. Nederlof, L. Van Hauwermeiren, ATM Meshed Networks, Protected by Distributed Restoration Mechanisms, Thirteenth Annual Conference on European Fibre Optic Communications and Networks (EFOC&N'95), pp. 102-105, June 27-30 1995, Brighton, England

Core and ATM Networks
D.W. Faulkner and A.L. Harmer (Eds.)
IOS Press 1997 © AKM

PANEL - Protection Across NEtwork Layers

P. Demeester, M. Gryseels, K. Struyve K. Vandoorselaere — *IMEC-University of Gent, Belgium*
A. Autenrieth — *Technical University Munich, Germany*
C. Brianza — *Italtel, Italy*
G. Signorelli — *SIRTI, Italy*
R. Clemente, M. Ravera, — *CSELT, Italy*
A. Jajszczyk, M. Roszkiewicz — *ITTI, Poland*
G. Kalbe — *Belgacom, Belgium*
Y. Harada, T. Yada — *NTT, Japan*
A.G. Rhissa — *INT, France*

Correspondence : P. Demeester
IMEC-University of Gent
Department of Information Technology
Sint-Pietersnieuwstraat, 41
B-9000, Gent, Belgium
tel.: +32-9-264 3334, fax : +32-9-264 3593
e-mail : demeester@intec.rug.ac.be

Abstract

The integration of different transport network technologies requires a better coordination between network layers such as ATM, SDH and WDM. This paper addresses specifically the problem of interaction mechanisms between multiple layers in order to improve the overall network reliability. Various parameters defined by the operators have to be taken into account such as : survivability, availability, serveability. In order to develop new interaction strategies, a detailed knowledge is required on the fault propagation and on the recovery mechanisms available in the individual layers. From this, interaction strategies can be developed which have to be supported by detailed simulations in order to obtain a better understanding of their operation and performance. The goal of the interaction between the different layers is not only the improvement of the overall network reliability but as well the reduction of the overall network cost. This requires however also an integrated multilayer approach for the network planning.

These aspects are addressed by the ACTS project PANEL (Protection Across NEtwork Layers) and are discussed in this paper. They are split in 4 activities covering the work on operator requirements, interworking concepts and protocols, simulation of the recovery mechanism and multilayer network planning.

1. Problem formulation

Figure 1 gives an example of a multilayer network composed of a WDM (Wavelength Division Multiplexing) layer, an SDH (Synchronous Digital Hierarchy) layer and an ATM (Asynchronous Transfer Mode) layer. If a cable breaks, it is possible to recover the failure in the WDM layer if enough spare capacity has been planned in this layer (which would normally be the case). Eventually a second cable failure can occur which can not be restored in the WDM layer. In that case it might be possible to recover part of the failed SDH link connections (carried by the failed WDM network connections), if spare capacity is still available in the SDH layer. The same applies when propagating the failure to the ATM layer. From this simple example, a number of conclusions can be drawn.

- What do we really require ? A full recovery of all failure possibilities would result in an unrealistic high network cost. The requirements have to be defined by network operators who have to satisfy the customer needs at a reasonable cost.
- It is clear that by having an interworking between the different network layers, it is possible to improve the overall network reliability. Indeed, if no recovery is possible in the SDH layer, nothing could be done about the second cable break. New interworking strategies have to be developed and tested.
- A cable cut will result in a number of failures detected in the WDM, SDH and ATM layers. A detailed understanding of the propagation of the alarms and the fault detection mechanisms is required.
- The spare capacity of the network has to be planned carefully in order to enable recovery of a large number of failure scenario's at a reasonable cost.
- An overall view of the network status has to be maintained during the multilayer recovery and if necessary a centralised management needs to take action. This will be based on the use of TMN.

The European ACTS project PANEL is addressing the above raised issues. The strategy will be described in more detail below.

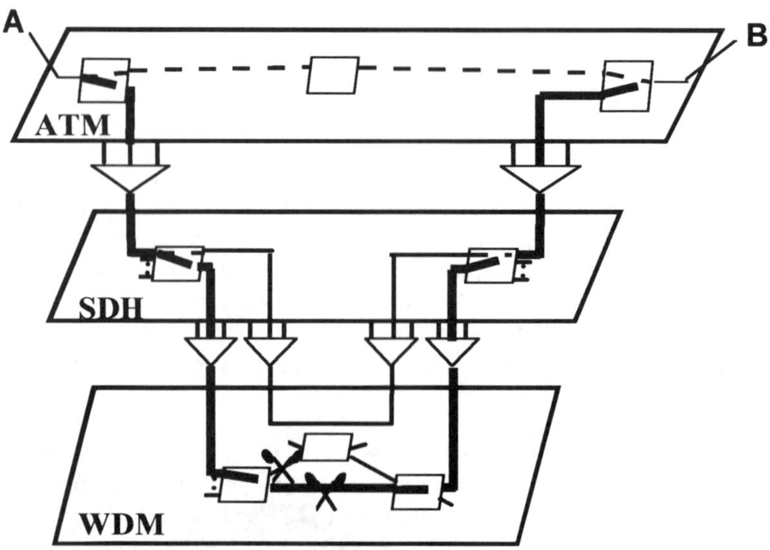

Figure 1. Example of a multilayer network (ATM/SDH/WDM)

2. Operator requirements

The information described in this part has been collected from the telecommunication operators experience and reflect their targets concerning the network performances and service quality. This data is essential for the following activities in PANEL giving input to the development of the recovery mechanisms, their subsequent simulation and the planning of the model network. In general, the results obtained here help to define a model network based on realistic scenarios and will serve as threshold level the developed recovery mechanisms will have at least to satisfy or improve.

The results obtained can be grouped in three classes: a) the operator requirements expressed in terms of network performance, b) the impacts of network failures or connection interruptions on the networks and services c) the parameters needed for the simulation, planning and evaluation activities of this project.

From the different existing possibilities it was decided to measure the network performances by the availability, serveability and survivability. The values obtained were taken from the operators experience of the day-to-day operation of the networks. Were available, they are compared against targets specified by standards. An example is shown in figure 2.

The impacts of service interruption on the service performance were analyzed as a function of interruption duration and service class. This was illustrated by an example of a mayor natural disaster, e.g. the heavy Hanshin earthquake in Japan (in 1995). This example also showed the impacts on the network itself.

Besides the relation between interruption duration and service performance a second timing aspect is studied: the undetected fault time. The knowledge of this time for the different network layers is essential for the interworking of the recovery mechanisms of the different individual layers. In a number of cases it is important that a failure can be restored before the higher layer is detecting the fault and taking any parallel action.

As input to subsequent activities in this project the values of MTBF (Mean Time Between Failure), MTTR (Mean Time To Repair) and availability for fibre optics cable and ATM, SDH and WDM equipment was calculated. For the same reasons a large effort was made to define the reference networks, specifying the layering structure, the topology and the node configuration. These networks are based on real examples from the involved network operators. In addition the traffic distribution as a function of service class was defined.

Network Layer	AVAILABILITY [%]
PSTN	99
ATM VC	99.6
ATM VP	99.6
SDH VC-12	99.7
SDH VC-4	99.9
WDM	99.99

Figure 2. Availability objectives for different network layers

3. Interworking concepts and protocols

The basic scope of this activity is the study and definition of suitable interworking strategies between recovery mechanisms in multi-layered networks. These will represent the input to the simulation activity in charge of validating them and proving their efficiency compared to the traditional single layer approach to provide network survivability. In order to fulfill the activity goals the following topics have been studied in detail, their knowledge being necessary to elaborate sound and concrete survivability strategies.

First of all a deep and careful analysis of the relevant aspects of the technologies (WDM, SDH and ATM) involved in the reference scenarios addressed by the project has been performed. This has included the study of the main characteristics of the technology, the format and meaning of the information overhead, the functionality, reference architectures and typical configurations of the equipment and the typical network topologies for the deployment of such technologies.

The study of the events happening in a multi-layered network as a consequence of failures has been recognised in fact as a precondition to study the interworking among recovery mechanisms, being their activation directly related to failure occurrence. Thus after a brief survey of the fault types, their possible causes and their implications, in terms of defects detected at the different network layers, we have concentrated on the analysis of the fault detection and indication mechanisms within the different network layers. Such mechanisms are based on the monitoring and generation of particular maintenance signals propagated in each layer as bi-directional OAM flow. This has implied to analyse in detail the OAM functions supporting the fault management in the different network layers. Such survey has been based not only on the standards but also (especially for the optical layers) by capturing information from other sources like the existent technical literature.

As further step we have applied such concepts to investigate the reactions of the equipment to the network failures in a multi-layered network scenario like the one shown in the figure 3 below.

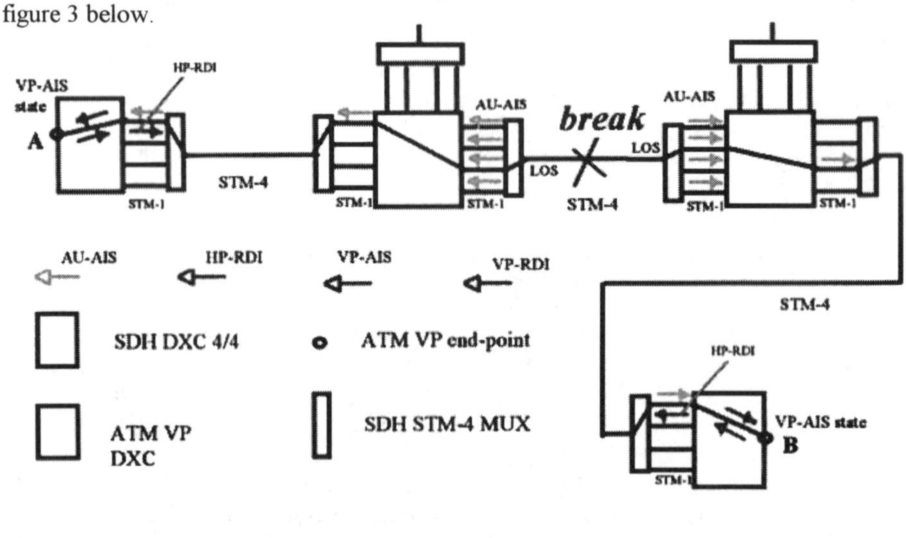

Figure 3. fault scenario in a sample hybrid ATM/SDH network

Before treating the interworking issues, we have completed this preliminary research activity by making a survey of the most relevant recovery mechanisms for the ATM, SDH and optical layers. Also in this case we have looked not only at the standards but at the most reliable proposals made in literature as well. The description of the recovery mechanisms has been performed using a critical approach, i.e. identifying both their merits and weak points. In our survey we have put major emphasis to the distributed protection/restoration mechanisms, because, being directly activated by the fault detection mechanisms implemented in the equipment and not requiring the intervention of the management system, they are able to better satisfy the operator's need to squeeze as soon as possible the recovery time.

The detailed analysis of the existing protection/restoration mechanisms has put in evidence that an optimum mechanism representing the best solution to provide survivability to the network, whatever are its topology, its layering structure and its target costs, does not exist. Each mechanism has a proper application field, some advantages and inevitably some disadvantages. There is a trade-off of excess capacity versus complexity and associated time-to-restore. It is expected that network operators will adopt to protect their multi-layered networks with hybrid strategies, implying the adoption of more than one recovery mechanism. This again proves the necessity to study the inter-working between different recovery mechanisms operating in different layers.

Recovery mechanism	*Centralised restoration*	*Distributed restoration*	*Protection switching*
Control mechanism	Centralised	Distributed	Distributed
Routes Calculation	real-time/ pre-calculated	real-time/ pre-calculated	pre-calculated
Bandwidth assignement	real-time	real-time	pre-assigned
ATM Layer	link/path based restoration	flooding/back-up Vps based	back-up Vps based
SDH Layer	link/path based restoration	flooding based	SHRs, SNC prot., MS linear prot.
Optical Layer	link/path based restoration		link/path based protection, WDM SHR

Figure 4. Overview of classification of single layer recovery mechanisms studied

4. Simulation and Evaluation

Since complex, distributed time dependent systems cannot be analysed with analytical methods and a network trial is economically unfeasible, a network simulator is required to assess the performance of recovery mechanisms and multilayer interworking strategies. Given the network architecture from planning, i.e. the topology and dimensioning, it permits one to identify areas for improvement in the recovery procedure and, finally, to verify its correct operation. The signalling and timing of the layered network must be modelled to enable performance simulations and to achieve the data for the evaluation of the investigated interworking strategy. The restoration ratio as a function of time and the global network availability are examples of such evaluation parameters. The development of a single multi-layer simulation environment using state of the art tools, and its use for the evaluation of the investigated strategy therefore forms both an important project activity and result.

The network model and simulation strategy will be described using the below figure. It gives an overview of SELANE (Simulation Environment for Layered Networks), the simulation environment developed in PANEL. Depicted are the main elements as well as the interfaces and relations to other PANEL activities and to the graphical user interface for the visualisation. The network simulator in SELANE is implemented using SDL (Specification and Description Language, ITU Z.100), a formal description language especially suited to model telecommunication applications. Represented in the figure is the system level of the SDL description. The SDL system models the transport, control and management functions of a network based on recommendations (G.805, G.783, I.732). Details about specific aspects of the equipment like the transit delay or the fault propagation of SDH DXCs are provided by the equipment manufacturers within the project. In the nodes the recovery algorithms and interworking strategies proposed and developed by the project are implemented. Additionally, the relevant procedures for the TMN interactions are implemented in the Central Manager (CM). The network topology and dimensioning data for the configuration of the simulator are obtained from the planning activity. A visualisation of the network with a graphical user interface (GUI) is a very useful feature to interactively investigate and evaluate different survivability strategies. At the same time, the GUI will be the user interface for the final demonstration of the PANEL project.

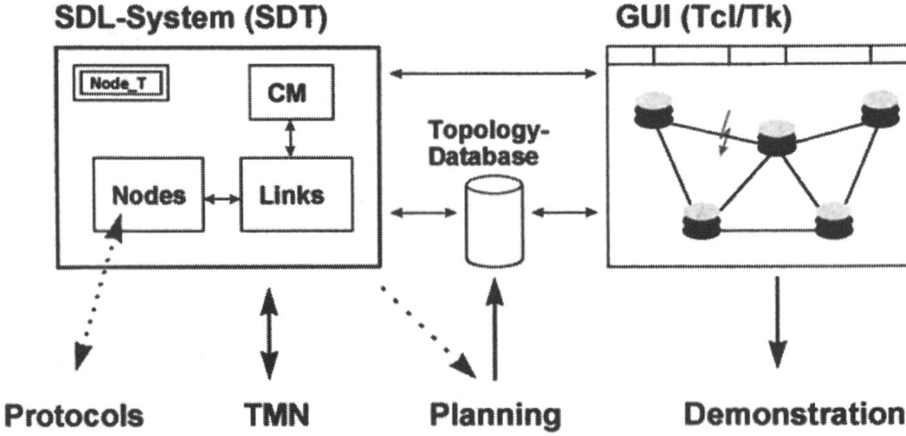

Figure 5. General simulation environment based on the use of SDL

5. Multilayer Network Planning

The planning activity in PANEL has the following objectives :

- provide the means to compare the network capacity required under different multi-layer recovery schemes and to carry out simulation activities;
- define a global methodology for the planning of multi-layer networks, taking into account the interaction of recovery mechanisms at different layers.

For this purpose the following solutions for multilayer network recovery are taken into consideration: each layer recovers its own failures; all failures are recovered in the highest layer; higher layers partially recover lower layer failures. Particularly, it will be investigated how interworking between recovery mechanism of different layers allows to enhance the availability of the network without deployment of additional spare capacity and to save capacity (in the server) layer without a change of the whole set of the network's survivability requirements.

So far, a number of reference network scenarios have been defined. (e.g. ATM VP over SDH HOP). The focus is put on sample networks composed by a single layered survivability domain and involving a single recovery mechanism at each layer. Two main planning approaches based on network layering have been identified by the project: multi single-layer planning and single multi-layer planning. The second approach is based on an iterative optimisation of the overall network and, due to the lack of suitable tools, will not be applied in PANEL. The first approach, on the other hand, permits the maximum reuse of the existing knowledge and available planning software and will be adopted in PANEL mainly aiming at a sequential planning of the single layers. The main inputs for the sequential planning are the connection demand matrix for each layer and the physical network sites and links. This multi single-layer sequential approach can be implemented in two ways: top-down approach and bottom-up approach. The top-down approach consists of : a topology planning problem at the client layer; a recovery planning for a set of expected failures of the client layer and a traditional single layer planning of the server layer. In contrast, the bottom-up approach starts with : a topology planning of the server layer (using a 'translation' of upper layer matrices) and continues with a recovery planning taking into account a set of expected failures of the server layer ending with a traditional single layer planning of the client layer taking into account a set of expected failures of the client layer. With both approaches, spare resources can be allocated in each layer to recover from failures in the layer itself. After planning the network with one of both approaches, the benefits of interworking can be estimated through the simulation of the availability performances and by comparing the required total spare capacity.

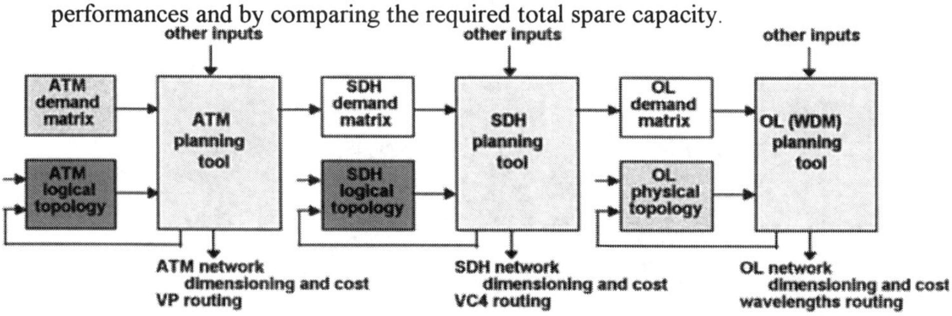

Figure 6. Multilayer top-down planning approach

6. Example

An example of a multilayer recovery strategy is described in this last paragraph. The network considered is an ATM VP network on top of an SDH HOP network. The single layer recovery mechanisms are : back-up VP at the ATM layer and distributed link restoration at the SDH layer. The interworking strategy will first recover at the lowest layer

(SDH) and then propagate to the next layer (ATM). Two propagation strategies are used : a hold-off timer (a certain time is allocated to the SDH layer to recover as much as possible and then the recovery continues at the ATM layer) or a recovery ratio (the recovery at the SDH layer will continue until a certain ratio of the disrupted connections are recovered at the SDH layer, then the ATM layer continues). The failures considered are complete node failures (both ATM and SDH nodes at the same location fail at the same time). A specific spare capacity planning strategy was used, taking into account the used single layer recovery mechanisms.

An example of the results obtained is shown in figure 7. As a function of time, we observe the average recovery ratio measured at the ATM VP layer. The different curves with time indications correspond to the hold-off timer and the curve with the 90% indication corresponds to a recovery ratio of 90% at the SDH layer. If we first consider a hold-off timer of 6 seconds, we observe that a plateau is formed (time below 6 seconds) which corresponds to the maximum recovery ratio which could be obtained if only recovery is used in the SDH layer (about 70%). After 6 seconds the recovery continues at the ATM layer and 100% recovery can be obtained. It is clear that smaller hold-off timers result in better time performances. The best result is however obtained when using a recovery ratio for the propagation decision.

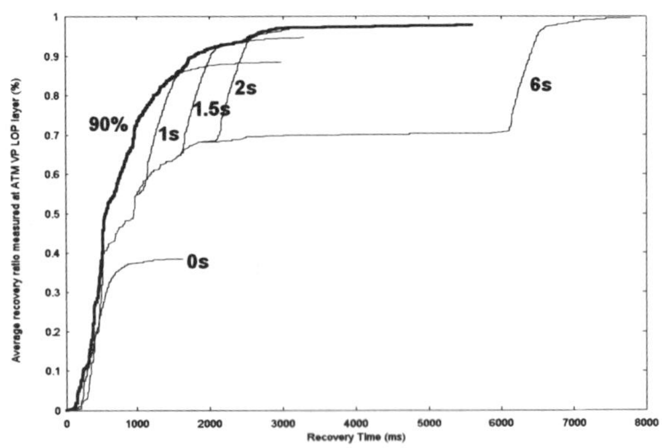

Figure 7. Example of recovery ratio as a function of time in case of a multilayer recovery strategy

Conclusions

This paper has presented a general framework for the development of multilayer recovery strategies, as studied in the ACTS project PANEL.

Acknowledgment : We would like to acknowledge the support of the European Commission through the ACTS project PANEL. Michael Gryseels thanks the FWO-V for financial support.

References

[1] L. Nederlof, K. Struyve, C. O'Shea, H. Misser, Y. Du, B. Tamayo, "End-to-end survivable Broadband Networks", IEEE Communications Magazine, September 1995

[2] T. Noh, "End-to-end Self-healing SDH/ATM Networks", Globecom'96, November 1996, London

[3] D. Johnson, "Survivability Strategies for Broadband Networks", Globecom'96, November 1996, London

[4] R. Kawamura, I. Tokizawa, "Self-healing Virtual Path Architecture in ATM Networks", IEEE Communications Magazine, September 1995

Core and ATM Networks
D.W. Faulkner and A.L. Harmer (Eds.)
IOS Press 1997 © AKM

Optimal Space Capacity Design for Various ATM Protection Switching Methods

Thomas Bauschert Thomas Frisanco

Lehrstuhl für Kommunikationsnetze
Technische Universität, München, Germany

Abstract

Todays broadband ATM networks are much more sensitive to failures of network elements since even during a short disconnection a large amount of data can be lost. Different restoration mechanisms have been proposed for preserving network performance by routing the affected traffic streams over still intact network parts. However, the appropriate application of these mechanisms requires the provision of additional capacity on the network links.

In this paper a classification of restoration mechanisms for ATM networks is given and a subset of these mechanisms – the so called protection switching methods – is considered in detail. Our objective is to determine the optimal restoration paths and the minimum amount of spare link bandwidths so that no traffic is lost in case of arbitrary single failures. Mathematical optimization models for four different ATM protection switching methods as well as a general solution procedure are presented. Finally we discuss results for a 23 node wide-area ATM example network.

1 Introduction

ATM restoration mechanisms can be classified according to different restoration control principles and different instants of restoration path determination and bandwidth reservation. If the restoration control is performed in a centralized fashion by the network management, the restoration mechanism is called **Rerouting**. Moreover, Rerouting is a dynamic method as the restoration path search and the assignment of spare link bandwidths to the restoration paths are carried out just at the failure instant. In contrary to that **Protection Switching**- and **Self-Healing**-mechanisms operate in a decentralized manner. Self-Healing like Rerouting works dynamically whereas with Protection Switching the restoration paths and capacities have to be predetermined. A comprehensive classification and description of ATM restoration methods can be found in [1].

In this paper, only virtual path (VP)-based ATM Protection Switching methods are considered where the VPs, which are affected by an arbitrary failure event, are switched to restoration VPs with fixed routes. It is further assumed, that the bandwidths of the restoration VPs are identical to those of the original VPs, i.e. no VP traffic splitting is allowed. The general objective is to determine the optimal routes of the restoration VPs so that the sum of the link capacities, which have to be reserved for the restoration VPs, is minimized under the constraint that no traffic is lost in case of arbitrary single failures (*minimization of redundancy*). The mathematical formulation of this optimization problem and the corresponding solution procedure is presented in section 3.

It should be pointed out that only for this class of restoration mechanisms the practical determination of optimal spare link capacity values is possible. For all other restoration mechanisms the spare capacity demand can only be estimated using the same procedure. However, a rigorous optimization is not possible as the restoration paths and bandwidths are determined at the failure instant and therefore depend on the current network state.

In most contributions concerning robust network design, restoration mechanisms with predefined restoration paths and bandwidths are considered and similar assumptions as in this work are made. Only the method of [3] allows the consideration of two simultaneous failure events whereas in [4]-[6] a Self-Healing restoration method is assumed. Moreover, there exist many different proposals for the calculation of the optimal spare link bandwidths: some authors use linear programming based solutions, others pure heuristics or a combination of both approaches, see e.g. [7]-[16]. Solutions which are based on linear flow models imply, that the VP carried traffic can be allocated in arbitrary small units to more than one restoration VP. This, however would hardly be practical in real failure situations. In contrary to that, in our approach VP traffic splitting is not allowed and the optimal routes of the restoration VPs as well as the spare link capacity values are determined simultaneously by the solution of a mixed integer linear program (MIP).

2 Classification of VP-based Protection Switching Methods

The VP-based Protection Switching mechanisms can be divided into several subclasses according to specific restrictions concerning the routes of the restoration VPs. Here we consider four different cases:

- With **Link Protection Switching** the VPs affected by a link failure are only rerouted between the adjacent nodes of the interrupted link. For all VPs the same restoration path applies and the restoration VPs are required to use all intact links of the respective original VPs.
- With **VP-Link Protection Switching** like with Link Protection Switching the affected VPs are only rerouted between the adjacent nodes of the interruped link and the restoration VPs are required to use all intact links of the original VPs. However the restoration paths may be different for each affected VP.
- With **VP-Fragment Protection Switching** all VPs affected by a link or node failure are rerouted between their terminating nodes. The restoration VPs may have links in common with the respective original VPs.
- With **VP-End-to-End Protection Switching** like with VP-Fragment Protection Switching all VPs affected by a link or node failure are rerouted between their terminating nodes. However the restoration VPs are not allowed to have links in common with the original VPs i.e. the routes of the original and the restoration VPs are link-disjoint. Here two variants are possible: either for every VP only one restoration VP is assigned independent of the location of the failure, or every VP may have different failure location dependent restoration VPs.

Note, that for practical applications one has to consider that these mechanisms not only differ in their spare capacity requirements but also show significant differences in respect of their operational simplicity. From this point of view, Link and VP-Link Protection Switching as well as the failure location independent version of VP-End-to-End Protection Switching are favourable.

3 Determination of Optimal Restoration Paths and Spare Link Bandwidths

3.1 Modeling Assumptions

Our optimization procedure is based on the following assumptions:

- only independent single link failure events are considered
- no traffic may be lost in arbitrary failure events (complete (100%) restoration)
- no VP traffic splitting is allowed
- the spare link capacities may be shared among different restoration VPs

Note, that with VP-Fragment and VP-End-to-End Protection Switching also the restoration of node failures is possible. Our optimization model can be easily extendet to this cases. However if complete restoration of node failures is assumed, the sum of additional required link capacities would be higher than the basic capacity demand during normal network operation. The same would be true if multiple simultaneous link failures are considered. In practical applications such large amounts of spare capacity would hardly be provisioned.

3.2 Problem Formulation

The optimal spare capacity dimensioning problem can be formulated as follows:

Given the physical network structure as soon as the routes and the capacities of the original VPs find:

- the routes of the restoration VPs
- the required spare link bandwidths

so that the sum of the spare link capacities is minimized under the constraint that no traffic is lost in case of arbitrary single failures.

3.3 Optimization Procedure

Because of complexity reasons is not possible to consider all feasible restoration paths when searching for the optimal solution of the above stated problem nor it is possible to perform the path search during the optimization. Therefore our solution procedure consists of two separate modules (see Figure 1): the *Generate Candidate Restoration Paths* and the *Path Selection and Spare Capacity Optimization* module.

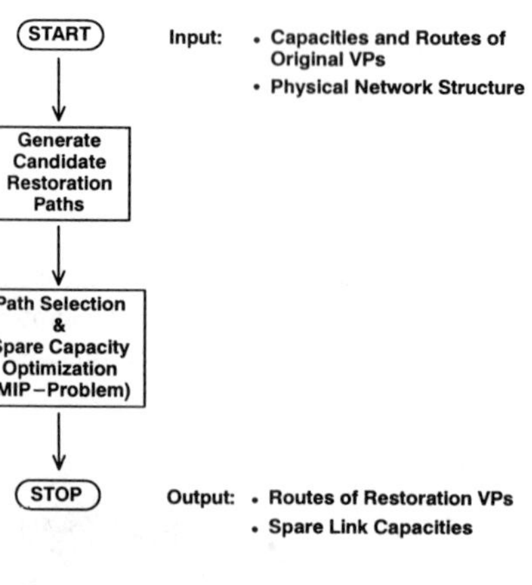

Figure 1: Optimization Procedure

In the *Generate Candidate Restoration Paths* module for each failure event a restricted set of candidate restoration paths is determined. Note, that the candidate path sets are different for different Protection Switching methods. The path search (within the physical network) is carried out by using a K-shortest-path-algorithm, see e.g. [17] or [18].

The goal of the *Path Selection and Spare Capacity Optimization* module is to select the optimal restoration paths (and thus the optimal routes of the restoration VPs) from the candidate path set for each failure event and to determine the optimal spare link capacities. This problem is formulated as a linear optimization problem with continuous variables (which represent the spare link bandwidths) and discrete binary variables (which decide for each candidate restoration path if it is selected or not). Optimization problems of this kind are called mixed integer linear programs (MIP). The numerical solution of our MIP-problem is obtained by using the MIP-solver of the commercial CPLEX [19] optimization package. With that, good results have been obtained for medium scale networks (up to approximately 30 nodes) within acceptable computation times. The mathematical formulation of the MIP again depends on the specific Protection Switching method. In the next section, mathematical models for all four cases described above are given.

3.4 Mathematical Models

The objective function of the MIP is the same for all cases:

Minimize the sum of spare capacities ΔN_l which have to be reserved for the restoration VPs on the network links:

$$\sum_l \Delta N_l \tag{1}$$

Thus, in the following, only the specific constraints are listet. Note that the number of continuous MIP-variables is always equal to the number of physical network links, whereas the number of discrete variables strongly depends on the specific Protection Switching method.

For **Link Protection Switching** the following constraints are valid:

$$\sum_{\text{RestorationPaths } r \in R_e} x_e^r = 1 \quad \forall \text{ Failures } e \tag{2}$$

$$\sum_{\substack{\text{RestorationPaths } r \in R_e \\ \text{and } r \in \text{Link } l}} x_e^r \, N_e \le \Delta N_l \quad \forall \text{ Failures } e, \forall \text{ Links } l \tag{3}$$

$$x_e^r \in \{0,1\} \quad \forall \; e, r \tag{4}$$

Here R_e denotes the set of candidate restoration routes for failure event e (i.e. failure on link $l = e$) and N_e is the bandwidth of the failure link. Optimization variables are the binary decision variables x_e^r and the continuous spare bandwidths ΔN_l. The link capacities N_l are input variables: they can be obtained by summing up the bandwidths of all original VPs running along each link.

Equation 2 enforces the selection of a single restoration path out of the set R_e for each failure event. The left side of equation 3 represents the required spare capacity on link l in case of failure event e. In order to achieve 100% restoration, the spare capacity ΔN_l must be at least as big as the maximum value of the left expression.

For **VP-Link Protection Switching** the following constraints are valid:

$$\sum_{\text{RestorationPaths } r \in R_e} x_{pe}^r = 1 \quad \forall \text{ Failures } e, \forall \text{ VPs } p \in VP_e \tag{5}$$

$$\sum_{\text{VPs } p \in VP_e} \; \sum_{\substack{\text{RestorationPaths } r \in R_e \\ \text{and } r \in \text{Link } l}} x_{pe}^r \, N_p \le \Delta N_l \quad \forall \text{ Failures } e, \forall \text{ Links } l \tag{6}$$

$$x_{pe}^r \in \{0,1\} \quad \forall \; e, p, r \tag{7}$$

With VP-Link Protection Switching the same candidate restoration route set like with Link Protection Switching applies. However, as the restoration routes may be different for each VP that is affected by failure event e, we need the VP-index p to be included in the binary decision variables x^r_{pe}. Moreover, VP_e denotes the set of VPs, which are affected by failure event e and N_p is the capacity of the original VP p (input variable).

For **VP-Fragment Protection Switching** the constraints are as follows:

$$\sum_{\text{RestorationPaths } r \in R_{pe}} x^r_{pe} = 1 \quad \forall \text{ Failures } e, \forall \text{ VPs } p \in VP_e \tag{8}$$

$$\sum_{\text{VPs } p \in VP_e} \sum_{\substack{\text{RestorationPaths } r \in R_{pe} \\ \text{and } r \in \text{Link } l}} x^r_{pe} \, N_p \leq \Delta N_l \quad \forall \text{ Failures } e, \forall \text{ Links } l \tag{9}$$

$$x^r_{pe} \in \{0,1\} \quad \forall \; e, p, r \tag{10}$$

In case of VP-Fragment Protection Switching the candidate restoration route set R_{pe} is the largest of all Protection Switching methods. For each original VP p and each failure e, which affects the respective VP, candidate restoration routes have to be supplied. Therefore two indices, p and e are necessary.

With **VP-End-to-End Protection Switching** the candidate restoration route set R_p does not depend on the failure event e but only on the respective original VP p (as the restoration routes have no links in common with the original VP). However, we have different constraints for the two variants described in section 2. If the routes of the restoration VPs are dependent on the failure location, the constraints are similar to those of VP-Fragment Protection Switching (the only difference lies in the restoration route set):

$$\sum_{\text{RestorationPaths } r \in R_p} x^r_{pe} = 1 \quad \forall \text{ Failures } e, \forall \text{ VPs } p \in VP_e \tag{11}$$

$$\sum_{\text{VPs } p \in VP_e} \sum_{\substack{\text{RestorationPaths } r \in R_p \\ \text{and } r \in \text{Link } l}} x^r_{pe} \, N_p \leq \Delta N_l \quad \forall \text{ Failures } e, \forall \text{ Links } l \tag{12}$$

$$x^r_{pe} \in \{0,1\} \quad \forall \; e, p, r \tag{13}$$

If, for every VP, only one restoration VP is assigned independently of the failure location, the failure index e has to be omitted from the binary decision variables and the contraints are as follows:

$$\sum_{\text{RestorationPaths } r \in R_p} x^r_p = 1 \quad \forall \text{ VPs } p \tag{14}$$

$$\sum_{\text{VPs } p \in VP_e} \sum_{\substack{\text{RestorationPaths } r \in R_p \\ \text{and } r \in \text{Link } l}} x^r_p \, N_p \leq \Delta N_l \quad \forall \text{ Failures } e, \forall \text{ Links } l \tag{15}$$

$$x^r_p \in \{0,1\} \quad \forall \; p, r \tag{16}$$

4 Numerical Studies and Discussion

The numerical studies in this contribution are based on an 23 node/41 link network example (see [2]) which could form the ATM backbone network for the wide-area traffic in Germany. The basic VP network design, i.e. the determination of the routes and capacities of the original VPs without taking into account any restoration mechanism, was done by the method of T. Bauschert, which is described in [2]. In the basic network design the End-to-End-traffic values are given, and the optimization objective is to minimize the overall link capacities under the constraint, that the end-to-end blocking probabilities are below a certain threshold (here:

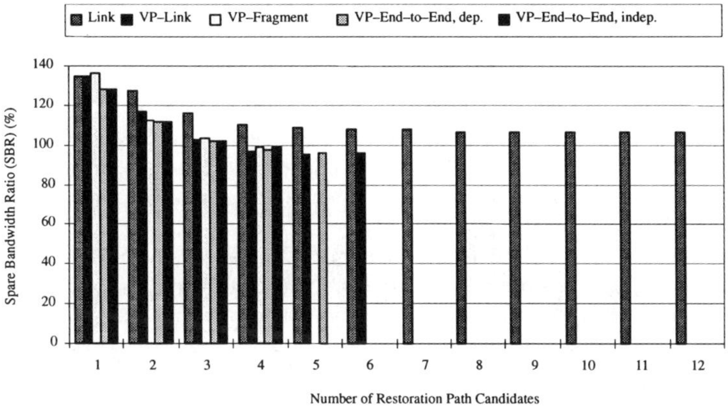

Figure 2: SBR vs Number of Restoration Path Candidates

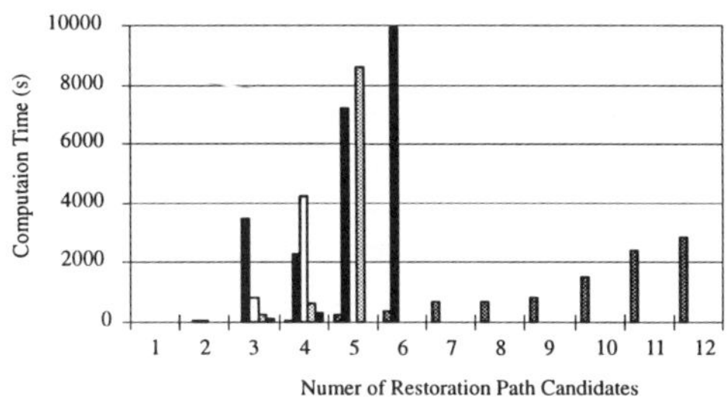

Figure 3: Computation Time vs Number of Restoration Path Candidates

1%). For the traffic load assumed in [2] an optimal basic VP network configuration with 127 VPs has been found.

First, Figure 2 shows the dependency of the spare bandwidth ratio (SBR) on the number K of restoration path candidates per failure event e and/or VP p for each Protection Switching method described in this paper. The SBR is defined as the ratio of the sum of spare link capacities (obtained through our optimization procedure) and the sum of the basic link capacities (which is known from [2]). As expected, the SBR gets smaller with increasing K values, because more restoration paths are considered in the MIP solution. However, for larger K values, only slight improvements can be observed. Note, that the decrease of the SBR happens at the expense of an increase in the mean restoration path lenght. Thus, ATM cell delays on the restoration VPs also increase.

For $K = 1$ no optimization is necessary: in this case the cell delays on the restoration paths are minimized (rather than the sum of spare link bandwidths) since always the shortest restoration path is selected. There are only small differences regarding the specific SBR values. The sum of the cell delays is smallest with VP-Fragment Protection Switching as for every VP without restrictions the shortest possible restoration path is selected, while with VP-Link Protection Switching the restoration paths include the intact links of the respective original VPs and for VP-End-to-End Protection Switching the restoration paths and the routes of the original VPs are disjoint. Note, that in the case of $K = 1$ the results for Link and VP-Link

Protection Switching are identical since all VPs which are affected by a specific link failure have to use the same restoration path. The same yields for the both variants of VP-End-to-End Protection Switching.

For $K > 1$ Link Protection Switching always requires more restoration capacity than the other methods. However, the results for these methods differ only slightly. Actually, the lowest SBR value would be expected for VP-Fragment Protection Switching as it is the most general method (without any restrictions regarding the restoration paths). But, obviously the number of restoration paths is still not large enough, so that only suboptimal results are obtained.

The computation time needed in the *Generate Candidate Restoration Paths* module of our procedure is negligible compared with the time needed for the MIP-solution. The times depicted in Figure 3 are observed on a SUN SparcStation20 workstation with 64 MB of main memory. They heavily depend on the specific Protection Switching method (because of the underlying complexity of the MIP). Note, that for Link Protection Switching the computation time is significantly smaller than for the other methods. Normally, the computation time increases strongly with increasing K value but exceptions are possible. Such exceptions may arise from situations where the MIP can be solved more easily (despite of its higher dimension) if one or more additional restoration paths are included in the path set. As can be seen from Figure 3, for some methods, K values higher than 5 or 6 already lead to prohibitively high requirements in computation time and/or main memory.

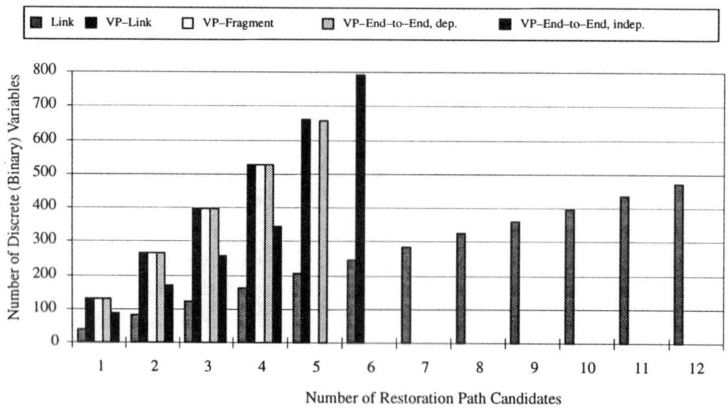

Figure 4: Number of Discrete (Binary) Variables vs Number of Restoration Path Candidates

The computation times are strongly interrelated to the complexity of the MIP, particularly to the number of MIP-variables. Figure 4 shows the dependency of the number of discrete binary variables on the number of restoration paths candidates for each Protection Switching method (Note, that the number of continuous variables is independent of the specific method). Normally the number of binary variables increases linearly with increasing K values. Exceptions are due to the fact, that for certain failure events it is not possible to find exactly K restoration paths within the physical network. For Link Protection Switching the number of binary variables is smallest compared with the other methods. Although the optimization models for VP-Link, VP-Fragment and the failure location dependent variant of VP-End-to-End Protection Switching have the same number of binary variables, there are differences in computation time. This is because the specific forms of the MIP-constraints (which are different due to the different sets of candidate restoration paths) also have an impact on the solution speed.

5 Conclusion

In this contribution a procedure for the calculation of minimum spare capacities and optimal restoration paths for VP-based Protection Switching is presented. Results are discussed for four different Protection Switching methods. As expected, Link Protection Switching requires more spare capacity (approximately 15%) than the other methods. However, the other methods differ only slightly in their spare capacity requirements. Because of that, we recommend the use of an operationally simple method for real implementation (i.e. VP-Link Protection Switching or the failure location independent variant of VP-End-to-End Protection Switching).

The computation times differ significantly for different Protection Switching methods and can take rather large values. But fortunately, as for Protection Switching the calculation of spare bandwidths and VP restoration routes can be carried out offline, we have no real time requirements. Furthermore, according to our experience, in most cases K values about 4 or 5 are sufficient for obtaining good suboptimal results within acceptable computation times.

References

[1] B. Edmaier, "Pfad-Ersatzschalteverfahren mit verteilter Steuerung für ATM-Netze", Ph.D.-Thesis, Technische Universität München, Lehrstuhl für Kommunikationsnetze, 1996.

[2] T. Bauschert, "Design of VP-based ATM Networks with Nonhierarchical Routing", 7'th International Network Planning Symposium, NETWORKS'96, 1996.

[3] I. Ouveysi, A. Wirth, "An Efficient Heuristic for the Design of a Survivable Network Architecture for Dynamic Routing", IEEE Globecom '95, 1995.

[4] K. Murakami, H. S. Kim, "Virtual Path Routing for Survivable ATM Networks", IEEE/ACM Transactions on Networking, 4, no. 1, February 1996.

[5] K. Murakami, H. S. Kim, "Near-Optimal Virtual Path routing for Survivable ATM Networks", IEEE Infocom '94, 1994.

[6] P. A. Veitch, D. G. Smith, I. Hawker, "Design and Performance Issues of Protection Virtual Path Networks", IFIP Conference on Broadband Communications '96.

[7] J. Veerasamy, S. Venkatesan, J. C. Shah, "Spare Capacity Assignment in Telecom Networks using Path Restoration", MASCOTS '95, January 1995.

[8] D. Medhi, R. Khurana, "Optimization and Performance of Restoration Schemes for Wide-Area Teletraffic Networks", Journal of Network and Systems Management, 3, no. 3, 1995.

[9] T. H. Noh, D. R. Vaman, X. Gu, "Reconfiguration for Service and Self-Healing in ATM-Networks based on Virtual Paths", ICCC '95, August 1995.

[10] J. Yamada, "A Spare Capacity Design Method for Restorable Networks", IEEE Globecom '95, 1995.

[11] K-S. Song, A. K. Somani, "Fault Tolerant ATM Backbone Network Design Considering Cell Loss Rates and End-to-End Delay Constraints", International Phoenix Conference on Computers and Communications '94, April 1994.

[12] D. Medhi, "A Unified Approach to Network Survivability for Teletraffic Networks: Models, Algorithms and Analysis", IEEE Transactions on Communications, 42, Nr. 2/3/4, February/March/April 1994.

[13] A. Gersht, A. Shulman, "Optimal Dynamic Bandwidth Allocation and Restoration in ATM Networks", IEEE Globecom '94, 1994.

[14] I. Hawker, D. Johnson, R. Cheng, "Distributed Restoration in Telecommunications Networks", IEE Conference on Telecommunications '95, March 1995.

[15] K. Murakami, H. S. Kim, "Joint Optimization of Capacity and Flow Assignment for Self-Healing ATM Networks", IEEE ICC '95, 1995.

[16] H. Lee, S-B. Kim, Y. Go, "Modeling and Algorithms for Physical Layer Network Survivability", IEEE ICC '95, 1995.

[17] J. Y. Yen, "Finding the K shortest loppless paths in a network", Management Science, 17, no. 11, July 1971.

[18] J. Y. Yen, "Shortest Path Network Problems", Mathematical Systems in Economics, 18, 1975.

[19] "CPLEX Version 4.0 Manual", CPLEX Optimization Inc., Incline Village, NV, USA, 1995.

Core and ATM Networks
D.W. Faulkner and A.L. Harmer (Eds.)
IOS Press 1997 © AKM

An Algorithmic Approach for Survivability in ATM Networks: LeapVP

Ilse Lievens, Piet Demeester
University of Gent - IMEC, Sint-Pietersnieuwstraat 41, B-9000 Gent, Belgium
ilse.lievens@intec.rug.ac.be

Abstract. In future broadband telecommunication networks Quality of Service (QoS) will be a key issue. An integrated solution for these QoS requirements is the idea of intelligent routing, which integrates the fields of survivability, congestion control and mobility. In this paper we focus on one aspect, the network survivability. We present a new distributed algorithm, LeapVP, developed for high-performant restoration of network element failures. In LeapVP the route search phase is accelerated by taking search leaps of two hops at the time, through pre-established two-hops Virtual Paths. Furthermore, an overview of the restoration area is built in one node for consistent and efficient spare resource allocation. The algorithm has been simulated in OPNET and results show that the algorithm performance is very well when compared with existing algorithms.

1. Introduction

A variety of services each with particular characteristics will be offered in future broadband telecommunication networks. Quality of Service (QoS), guaranteed to the user through negotiation contracts, will be a key issue in these networks. As there are different services with different costs and priorities, different QoS requirements will have to be fulfilled. The requirements are determined by the importance of the service to the user and thus by the cost the user is willing to pay for the quality guarantee of its service. An integrated viewpoint to these QoS requirements is the idea of intelligent routing.

The main definition and interpretation of intelligent routing is to use the available network resources as efficient as possible in terms of survivability, congestion control and support of mobility. These are several aspects which have to be taken into account to attain a required level of QoS. End-to-end reliability of user connections against network infrastructure failures needs to be guaranteed, which is dealt with in the field of survivability. Local overloads of the network should not disturb existing and new connections, referring to congestion control. Finally, as the number of mobile users keeps on growing, special attention must be paid at maintaining the quality of a connection during mobile handovers.

In this paper we focus on the network survivability aspect. The key point here is the rerouting of existing connections, as fast as possible to keep the disruption as short as possible. A novel algorithm, called LeapVP, for restoration in case of network element failures in ATM networks is presented. The algorithm can be extended towards the other aspects in intelligent routing: congestion control at connection level, which is in fact rerouting existing and new connections away from the congested area. For mobility this rerouting is also required due to movements of the users.

In the following section, some required terms and notions of network survivability, as well as an overview of the state of the art in survivability are given. The LeapVP algorithm is discussed in Section 3. The algorithm has been simulated and tested and the simulation environment as well as the results are presented in Section 4. Conclusions and future work are given in Section 5.

2. Restoration: state of the art

Several categories of mechanisms for network survivability exist. Basically they all provide an alternative way to reach the destination in case of failures. The algorithm in this paper falls under the category of restoration. This means that the spare capacity provided in the network, can be used to restore any failure at any time, which implies that spare resources are not dedicated to particular working elements.

Restoration mechanisms can be classified according to some characteristics. Two important characteristics are the moment when the alternative route is calculated, and the control under which the restoration process takes place.

The calculation of the restoration route can be done in advance - called pre-planned restoration - or at real-time. With pre-planned restoration one just has to look up the alternative route in a database in case of a failure, making it a fast technique. The routes itself are calculated in advance, taking assumptions into account on failure types and network state at the time of the failure. If however a failure scenario is not provided or the change in the network state is not expected, pre-calculated routes are not available when a failure actually occurs. In real-time restoration a restoration route is only calculated after the occurrence of a network failure. This implies a slower process, but also a more accurate technique since the actual network state at the time of the failure is taken into account in the calculation. Real-time restoration can even react on changes of the network during the route calculation making it a more robust technique.

The restoration can be performed under a central control, which means that there is one all-coordinating node in the network, which entirely controls the process. Information on node states, connectivity, spare resources, etc. is stored in that central node. All actions in the network are only done on its explicit commands, while everything that happens in the network must be reported to the central node. This makes it a slower and highly vulnerable process. The restoration can also happen under distributed control, in which every network node has some intelligence to carry out the restoration. Each node has knowledge of its local network environment and the restoration is performed by exchanging information. The nodes actually search a route by exchanging requests for spare resources. This provides a more robust technique as compared to centralised restoration and its response to network failures is also faster.

For the exchange and processing of the restoration messages, some protocol is required defining how the nodes, between which the restoration route is searched, start and finish the restoration process. In the Sender-Chooser protocol [1], the restoration routes are searched from one end (the Sender node) of the failure towards the other end (the Chooser node). The Sender starts the flooding of messages towards its neighbour nodes. Once messages reach the Chooser, candidate routes are found and the Chooser then chooses the actual restoration route. Other protocols are available, the most important one being the Two-Prong protocol [2]. Here the routes are searched from both sides of the failure, resulting in a faster, but somewhat more complex protocol.

Another distinction to be made is which part of the failed connection is replaced by a restoration route. With link restoration, an alternative route is only searched between the nodes immediately adjacent to the failure, and the undamaged parts of the connections using the failed link are kept. Another possibility is path restoration in which an entirely new route is searched between the connection's end nodes, taking slightly more time since these nodes have to be notified.

Important for the applicability of an algorithm are also the failure types it is capable to restore. The more failures it can handle, the more robust and efficient an algorithm is. The most common failures are single-link, multiple-link and node failures. Usually algorithms are tested under these failure scenarios.

Research has already been performed in the area of distributed restoration, which resulted in algorithms for SDH [1,2,3,4,5] which are currently operationally used. Algorithms for distributed restoration in ATM networks [6,7,8] are still in a phase of research and development.

Yet, ATM [9] offers several features, which can be successfully applied for a better restoration. There is the basic ATM cell which provides a natural transport mechanism to exchange the restoration messages. The bandwidth allocation policy of ATM is of great help for the efficiency of the global use of the network: restoration messages will only take bandwidth when they are required. In addition, there is also the independence between a connection's route and its bandwidth: the route can already be stored in the translation tables of the node, without any bandwidth being assigned to it. When needed, the restoration route takes its bandwidth [6].

3. Restoration with LeapVP

3.1 General features

Key designing objectives included increasing the restoration speed, without making the restoration more complex, as compared with existing techniques.

The LeapVP algorithm performs real-time, distributed restoration. The current version of the algorithm applies link restoration, chosen because of its ease to start the restoration, which results in a better restoration speed. It is designed to cope with single-link failures, multiple-link and node failures. This implies that no additional actions have to be taken in case of a multiple-link failure as compared to the restoration of a single-link failure. The mechanisms are built in the algorithm's flow. For the comfort of the reader, the working of the algorithm is explained for single-link failures. However, it will be clearly stated when the handling of more complex failures had an influence on certain design decisions.

LeapVP exchanges its messages by using the flooding mechanism. To prevent the restoration messages from useless wandering around in parts of the network which are too far away from the failure and restoration area and thus to reduce the message volume, the flooding is limited by a hop limit, which restricts the maximum length of restoration routes. The Sender-Chooser protocol was chosen to avoid a too large complexity.

The LeapVP algorithm is specifically designed and intended for restoration in ATM networks. This implies that the algorithm uses explicit ATM features, that is, it makes use of Virtual Paths to obtain the restoration.

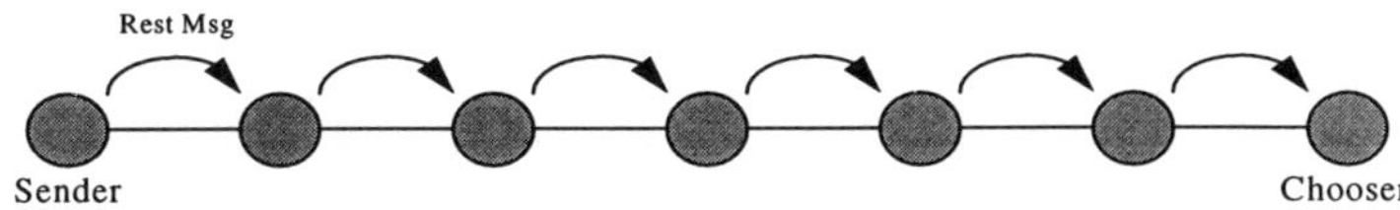

Figure 1: Normal search

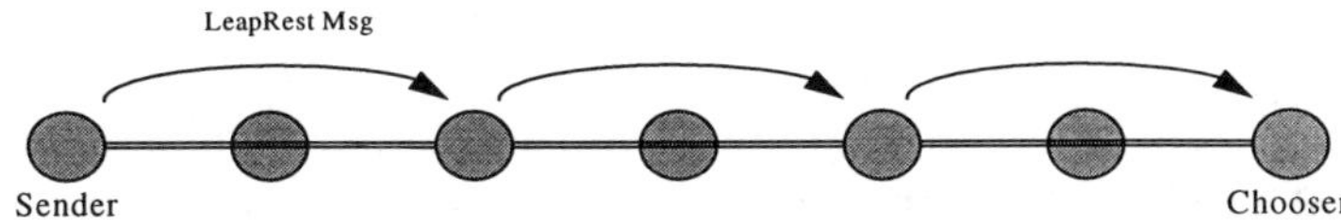

Figure 2: Leap search

3.2 Algorithm particularities

A distributed restoration algorithm has a search phase in which the algorithm searches for candidate restoration routes. This is generally done by sending restoration messages, which take one hop at the time (Figure 1). A node broadcasts a restoration message to an immediate neighbour node of his, with information on spare resources etc. On reception of this message, the neighbour node on its turn will transmit messages to its immediate neighbour nodes. So, the restoration route is searched one hop at the time. This implies that the speed of searching an alternative route is limited by n, with n the number of nodes on that alternative route.

The LeapVP algorithm accelerates this search phase by taking two hops at the time (Figure 2). It means that a node will transmit a restoration message to its second order node, in stead of an immediate neighbour. This gives a speed limit proportional with $n/2$.

To send messages at high speed to the second order neighbour nodes, the ATM Virtual Path (VP) concept is used (Figure 3). This makes a high-speed (e.g. at 155 Mbit/s) transparent transmission through the first order neighbour node possible. This intermediate node (node B) only has to switch the leap message from an inlet to the appropriate outlet, by using its VP/VC translation table. The node does not have to process the message, which is generally a time consuming action. The translation table is set up before the restoration takes place, even before a failure has occurred. This means that LeapVP needs a pre-restoration phase, in which ATM VPs are set up between every node and all its second

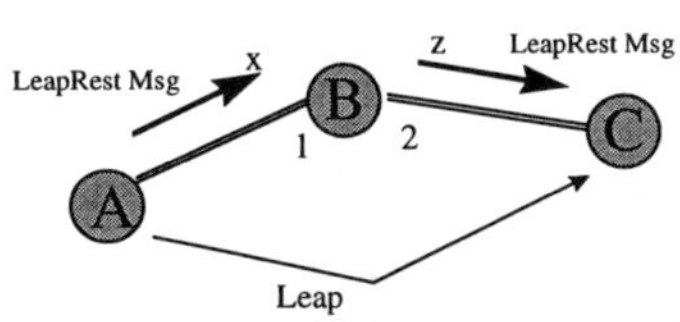

Translation table in B

In	VP_{in}	Out	VP_{out}
1	x y	2	z x

Figure 3: Use of VPs in Leap search

order neighbours reachable through the immediate neighbour nodes.

When only search leaps of two hops would be taken, the message could easily miss the Chooser node if this node is at an odd number of hops away from the Sender (Figure 4). Therefore a search leap of one hop is allowed when the Chooser node is an immediate

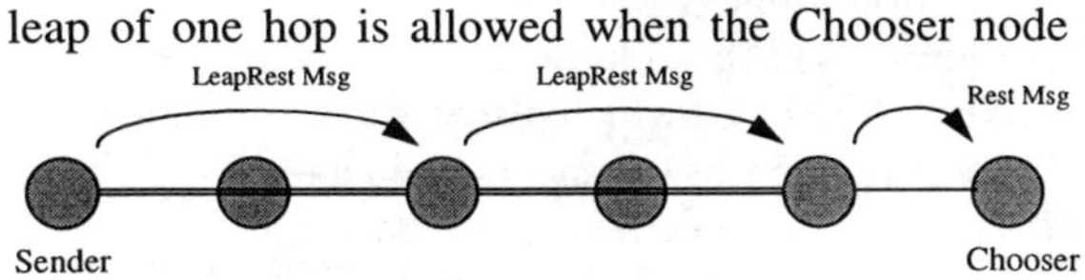

Figure 4: Chooser at odd number of hops

neighbour. None of the pre-established VPs is used then.

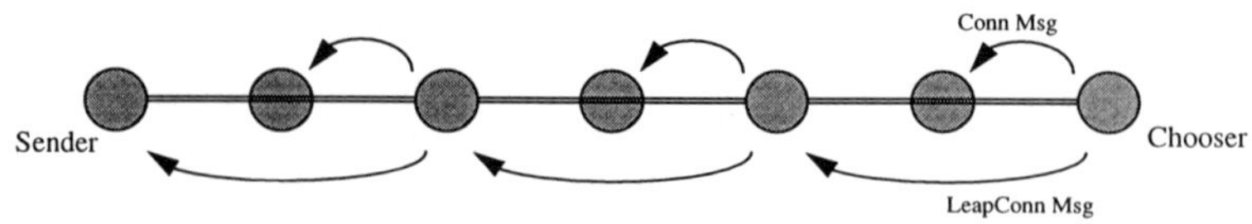

Figure 5: Connect phase in LeapVP

Once restoration messages reach the Chooser node, candidate restoration routes have been found from Sender to Chooser. An actual restoration route is not yet available since the LeapVP algorithm does not perform reservation of spare bandwidth during the search phase. A Connect phase will take place, not only to perform this reservation of spare resources, but also to notify the Sender that a route has been found. Two messages are sent in parallel along the candidate route (Figure 5). The LeapConnMsg is transmitted from the Chooser onwards to the Sender, by taking two hops at the time, with the purpose of reaching the Sender as soon as possible. Beside that, every node receiving a LeapConnMsg will also send a ConnMsg to its immediate neighbour on the route to perform the reservation on that node's outgoing link.

The allocation of the spare resources is always a complex issue in a distributed environment, since it is very easy to have inconsistencies in data, stored at several physical locations in the network, i.e. the network nodes.

Therefor the LeapVP algorithm does not make any reservation of spare resources during the actual search for routes. No complex release protocols are required to release capacity which has been unnecessarily reserved in the search phase.

For the actual reservation, an overview of the restoration area is built in the Chooser, based on the restoration messages that reach this node. These messages carry information on the links, they passed through, and the associated spare capacity. By implementing such an allocation procedure, a situation like in Figure 6 is avoided. Two restoration messages with different routes except for one link (link AB), reach the Chooser. Although these two messages could signify two different routes, these only represent one restoration route, since they use the same link, which has limited spare capacity. Therefore the Chooser not only needs to know the links that a message has passed through, but also their associated spare capacity, and can thus distinguish the real routes.

4. Simulation results

4.1 Simulation environment

In order to verify the behaviour of the LeapVP algorithm and to measure its performance, a simulation environment was built, using a commercially available discrete-

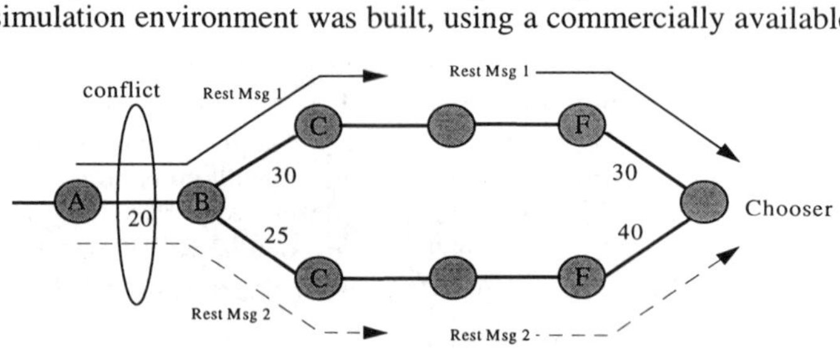

Figure 6: Risk for over-allocation

event simulator, OPNET [10]. This is actually a set of simulation and analysis tools, explicitly intended for simulating distributed systems and communication networks.

For the simulations a meshed reference network with 32 nodes and 53 links was used (Figure 7). This network was chosen to be able to compare the results with previous simulations, carried out in the context of the RACE II IMMUNE project on survivability of networks [11].

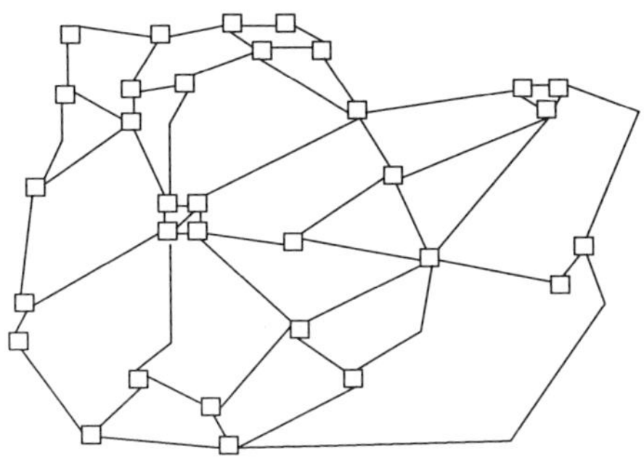

Figure 7: The reference network

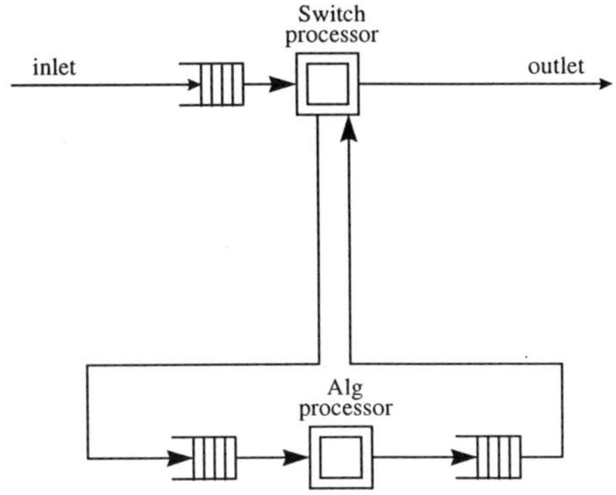

Figure 8: Node model

Each node of the network was assigned a model (Figure 8), consisting of a switch processor, which determined whether an incoming message must be switched to an outlet (according to the VP/VC translation table) or sent to the node's processor to be processed. Incoming messages are queued and serially served in this model. It takes 10 ms to read in and process a message and 10 ms to generate an outgoing message, while cross-connect times were assumed to be zero.

The load in the network was static, and assumed to be a realistic network load.

The failure scenario consisted of all single-link failures in the network. During the simulations all links were failed sequentially and before every failure the model returned to the original network state in terms of load and spare capacity. The average over all link failures has then been taken.

4.2 Simulation results

An important performance measure for restoration is the restoration speed, which has a direct impact on the total outage time due to an element failure. An outage time of 50 ms is transparent to most services [12]. When the outage time increases to 200 ms, switched voice and data can become impacted. If a failure takes up to 2s, circuit switched services can experience call dropping. Intolerable outages take more than 2s. Therefore the allowable limit for the time available to find an alternative route, is generally set at 2s.

• LeapVP, compared with other Sender-Chooser algorithms (Figure 9):

The algorithm's performance was compared with two other Sender-Chooser algorithms: the Self-Healing Network (SHN)[1] and the Fujitsu algorithm [5]. These two algorithms were simulated in the IMMUNE project with comparable parameters [11]. The results show that LeapVP has a higher restoration speed than the two other algorithms, mainly due to the accelerated search phase (by using 2 hops). While SHN and Fujitsu attain full restoration within seconds, LeapVP already has 100% restoration after 0.4 s.

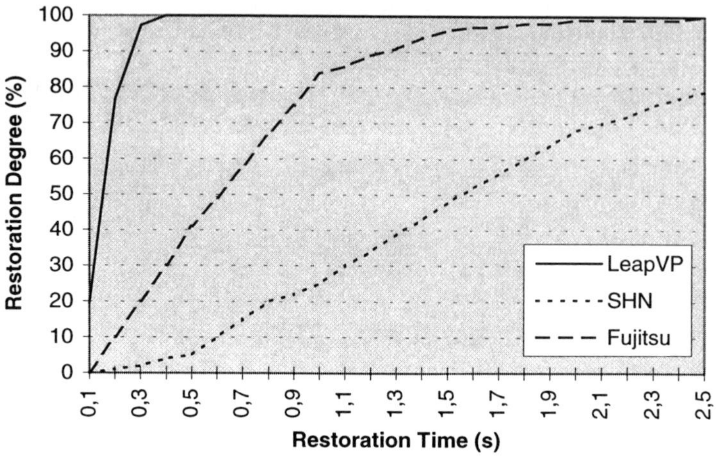

Figure 9: Comparison of LeapVP with Sender-Chooser algorithms

• LeapVP, compared with Two-Prong algorithms (Figure 10):

Another important issue is the comparison of the LeapVP algorithm with Two-Prong algorithms, which also accelerate the search phase, but by searching from both sides of the failures. The Two-Prong algorithms are LINKRES [7] and AlcDRA [8], which are both distributed link restoration algorithms. These results were also obtained in the IMMUNE project [11]. The results show that even compared to Two-Prong algorithms, the LeapVP algorithm is behaving very well.

5. Conclusions

A new distributed restoration algorithm, the LeapVP algorithm, is presented and performance results obtained by simulations in OPNET, are shown. LeapVP has been

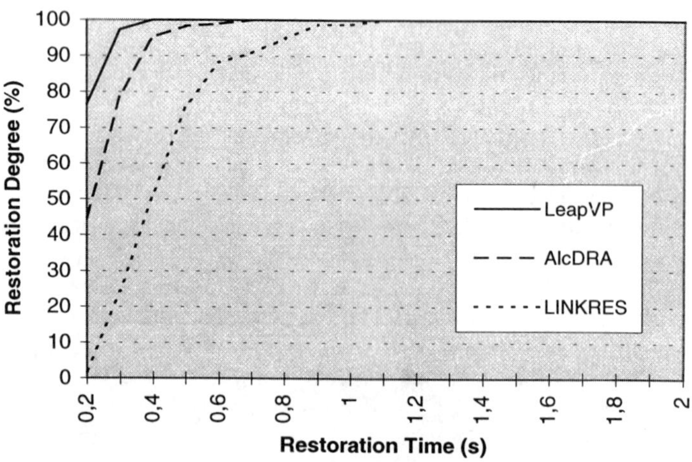

Figure 10: Comparison of LeapVP with Two-Prong algorithms

designed mainly to provide a high restoration speed with still an acceptable level of complexity. The algorithm is specifically intended for ATM networks, as it explicitly makes use of pre-established Virtual Paths for the searching phase. The algorithm accelerates the searching of the routes by taking leaps of two hops at the time by using these Virtual Paths, to speed up the searching phase. The results show that this algorithm obtains better restoration performances compared to algorithms with a Sender-Chooser approach, while it is comparable with respect to Two-Prong restoration algorithms.

Acknowledgement

This work has been supported by the Flemish Government through the IWT project ITA/950214/INTEC.

References

[1] W.D. Grover, "The Self-Healing Network: a fast distributed restoration technique for networks using digital cross-connect machines", Globecom 1987, pp. 1090-1095.

[2] C.E. Chow, J. Bicknell, S. McCaughey, S. Syed, "A Fast Distributed Network Restoration Algorithm", International Phoenix Conference on Computers and Communications, March 23-26, 1993.

[3] C. Han Yang, S. Hasegawa, "FITNESS: Failure Immunization Technology for Network Survivability", Globecom '88, pp. 1549-1554.

[4] C.E. Chow, S. McCaughey, S. Syed, "RREACT: A Distributed Network Restoration Protocol for Rapid Restoration of Active Communication Trunks", 2nd IEEE Network Management and Control Workshop, 1993.

[5] H. Komine, T. Chujo, T. Ogura, K. Miyazaki, T. Soejima, "A Distributed Restoration Algorithm for Multiple-link and Node Failures of Transport networks", IEEE 1990, pp. 459-463.

[6] R. Kawamura, K. Sato, I. Tokizawa, "Self-healing ATM Networks Based on Virtual Path Concept", Networks '92, pp. 129-134.

[7] K. Struyve, P. Demeester, L. Nederlof, L. Van Hauwermeiren, "Design and Evaluation of distributed link and path restoration algorithms for ATM meshed networks", Lecture Notes in Computer Science, Vol. 1044, Proceedings International Zurich Seminar on Digital Communications, ETH Zurich, Switzerland, February 1996.

[8] L. Nederlof, H. Vanderstraeten, P. Vankwikelberge, "A new distributed restoration algorithm to protect ATM meshed networks against link and node failures", Proceedings of International Switching Symposium '95, April 1995.

[9] M. De Prycker, "Asynchronous Transfer Mode, solution for broadband ISDN", 2nd edition, Ellis Horwood Limited, 1993.

[10] MIL3 Inc., USA; info@mil3.com; http://www.mil3.com

[11] J. Tomas, D. Johnson, H. Van Essen, C. Herrmann, G. Ocakoglu, K. Struyve, "Performance Report on Network Restoration Algorithms", RACE project IMMUNE Deliverable D7, September 1995.

[12] C.D.O'Shea, "Requirements and Reference Configurations for Survivability", RACE project IMMUNE Deliverable D2, January 1994.

Core and ATM Networks
D.W. Faulkner and A.L. Harmer (Eds.)
IOS Press 1997 © AKM

An Approach to QoS Monitoring in ATM Networks

Hajo R. Wiltfang
Institute of Telematics, University of Karlsruhe,
Zirkel 2, 76128 Karlsruhe, Germany
Phone: ++49 721 608 6406, Fax: ++49 721 388097
E-mail: wiltfang@telematik.informatik.uni-karlsruhe.de

Abstract. Forthcoming applications, especially in the multimedia area, require communication services with guaranteed quality of service (QoS). In order to guarantee a certain quality of service, supporting mechanisms must be applied inside the communication system. Additionally, due to dynamic changes in the network, it is crucial to monitor continuously the achieved quality of service during data transmission. Based on the monitoring results, supporting mechanisms can be adjusted accordingly. However, QoS monitoring in end systems is not sufficient in order to locate a QoS violation inside an ATM network. A QoS monitor operating at the network interface of end systems is able to measure the achieved end-to-end QoS continuously and to detect QoS violations, but only additional monitoring of each involved ATM switch allows to detect the network component causing the QoS degradation. Therefore, this paper focuses on monitoring of switches inside an ATM network. Firstly, an architecture for management of ATM networks is described. Monitoring agents, currently realized by proxy agents, are proposed for providing a well defined management interface to ATM switches within a network. Moreover, the ATM Network Monitor ANEMON, an entity for the collection of information on topology and connection-related QoS within an ATM network, is presented.

1 Introduction

Service integrated networks such as ATM must be able to provide communication services for a variety of applications with totally different service requirements. Typically, prior to data transfer, applications have to specify their communication requirements in terms of QoS parameters such as throughput, delay, jitter, and reliability. Based on these service requirements, communication resources are reserved and assigned to individual connections in all involved network nodes. The traffic characterization of the application and the resulting QoS guarantees of the network are part of a traffic contract. However, often it is not sufficient to reserve resources only. Additionally, monitoring of QoS parameters must be applied to ensure that the negotiated QoS parameters are really achieved. A QoS monitoring entity [11] has to compute current values of QoS parameters and must be able to detect QoS violations by comparing these values with the negotiated QoS parameters of the traffic contract. Based on the monitoring results, resources can be adapted accordingly. Thus, QoS monitoring plays an integral part in a QoS maintenance feedback loop [3].

QoS monitoring can be applied at different levels of an ATM network where it is able to measure AAL- or ATM-related QoS parameters. Firstly, monitoring at AAL layer in end systems can be used to detect violations of the traffic contract with a network user. At this point, it is possible to monitor both the traffic generated by the user and the QoS achieved by the network [7]. However, AAL level monitoring can only provide an end-to-end view on the QoS of connections since the switches within an ATM network operate on ATM cells in the user plane and thus cannot be integrated into AAL level QoS monitoring. In order to find the reasons for a particular QoS violation or degradation, switches within an ATM network must be monitored at ATM level. Only monitoring inside the ATM network can identify ATM switches in overload situations that are responsible for violations detected at the AAL layer of an end system.

In order to achieve a powerful monitoring at ATM level, all switches within a network must be integrated. To this aim, management and monitoring information can be obtained from switches by communicating with their classical management interface which is designed for traditional network management and realized by management agents. However, management interfaces and provided management information currently differ for switches from individual vendors. Only a well defined interface, however, allows to realize an integrated monitoring for ATM networks since a monitoring entity must be able to access information on all involved switches. Moreover, for QoS monitoring of particular ATM connections, a monitor needs information on the physical network topology as well as on the route of connections within the network in order to identify the switches to be monitored.

In this paper, an approach to QoS monitoring within ATM networks is presented. Information located on ATM switches is accessed by traditional network management. Therefore, Section 2 introduces an architecture for management and monitoring of ATM networks. Based on this architecture, Section 3 proposes a well defined monitoring interface for ATM switches which is currently realized by proxy agents. The *ATM Network Monitor (ANEMON)*, presented in Section 4, uses the interface provided by the monitoring agents in order to realize QoS monitoring within ATM networks. Being located on a dedicated monitoring station, the ANEMON monitor collects information on the topology of the ATM network as well as on QoS of selected connections. Thus, the ANEMON monitor realizes functions in the area of fault and performance management for ATM networks. Finally, Section 5 summarizes some of the central issues of this paper and outlines future work.

2 Management Architecture for ATM Networks

A well known architecture for the area of network management comprises three levels of management entities. This architecture, consisting of element management, management middleware, and management applications, can be applied to the management of ATM networks as well (cf. Figure 1). At the base level, the element management is characterized by the interface, the protocol, and the information provided by each particular ATM switch. Generally, two different approaches can be identified for management interfaces of ATM components. In-band management uses the ATM network for management traffic (e.g., Interim Local Management Interface ILMI [2]), and out-of-band management uses an Ethernet or serial port. At these different interfaces, various management protocols are currently supported. The classical Simple Network Management Protocol (SNMP) is used quite often for in-band [2] as well as for out-of-band management [1]. Moreover, textual-oriented protocols such as Telnet are sometimes provided at out-of-band interfaces. Finally, dedicated solutions such as the General Switch Management Protocol (GSMP) [10] or the Open API [9] were proposed for in-band management. Management information provided at the described interfaces with the various protocols is in general very different, too. For SNMP-based management, a lot of Management Information Bases (MIBs) are currently proposed, and only some of them have already been standardized (e.g., SONET-MIB [4],

ATM-MIB [1], and ILMI-MIB [2]). Moreover, individual ATM components support only parts of well defined MIBs. To summarize, the discussed characteristics of current management interfaces complicate the development of management middleware and management applications for ATM networks since an adaptation to these interfaces has to be integrated.

At the second level of the architecture, management middleware for ATM networks realizes more complex services based on the simple element management. In general, these complex services are related to more than one network element. Therefore, a middleware entity has to access information on each of the involved network elements. The services realized by management middleware entities are provided to management applications. Thus, management middleware does not directly interact with human users. For instance, middleware entities can perform functions related to configuration, accounting, fault, or performance management for ATM networks. In Figure 1, a configuration module and an accounting module are proposed as middleware for ATM network management. Moreover, the ATM Network Monitor ANEMON, presented in Section 4, realizes middleware for performance and fault management of ATM networks.

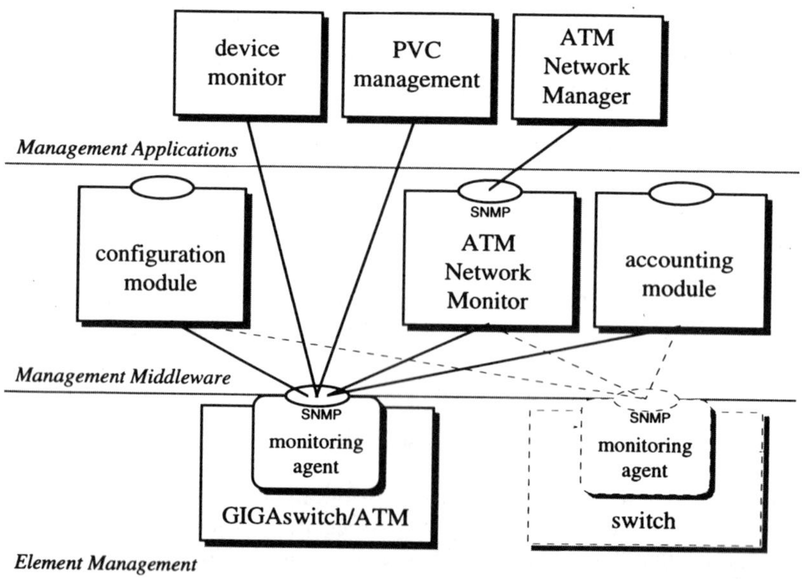

Figure 1: Architecture and proposed Entities for Management of ATM Networks

Management applications are located at the third level of the architecture described. They are characterized by their interaction with human users. Using a graphical or textual interface, management applications directly provide their services to users. In order to obtain the required information, management applications use the services of management middleware as well as the information being provided by the element management. In Figure 1, three management applications are shown as an example. A device monitor is designed to present various monitoring information of a switch graphically. In detail, an first implementation of a device monitor is able to show the configuration of the device and the state of all its ports [12]. Moreover, an application for management of Permanent Virtual Circuits (PVCs) on one particular switch is proposed in Figure 1. Using a graphical interface, this application displays all configured PVCs and is able to set-up or delete PVCs. Finally, the ATM Network Manager is proposed for management of ATM networks. This application provides a graphical representation of the network topology and supports applications located on end systems in monitoring their connections within the ATM network. To this aim, the ATM Network Manager uses the service realized by the ANEMON monitor, and thus the service of management middleware, whereas the device monitor and the PVC management are based on services provided by single network elements.

3 Monitoring Agents

In the presented approach, monitoring agents are used in order to provide the required information for management and monitoring entities in ATM networks (cf. Figure 2). Each monitoring agent realizes a well defined interface for a single switch. At the moment, a proxy agent realization is required for monitoring agents because of the different characteristics of individual management interfaces shown in Section 2. By a proxy agent acting as management gateway, these differences can be hidden and a well defined interface is provided. The proxy agent translates received requests according to the protocol characteristics of the individual switch and forwards them to the switch. Responses of the switch are retranslated and sent to the initiator of the request. For private ATM networks, most of the currently realized management interfaces are based on the SNMP protocol. Moreover, a lot of currently available management applications are using SNMP. Thus, SNMP was chosen as management protocol realizing the well defined interface of the monitoring agent. The information provided at the SNMP interface of the monitoring agent is specified in a separate MIB which has been designed for monitoring purposes.

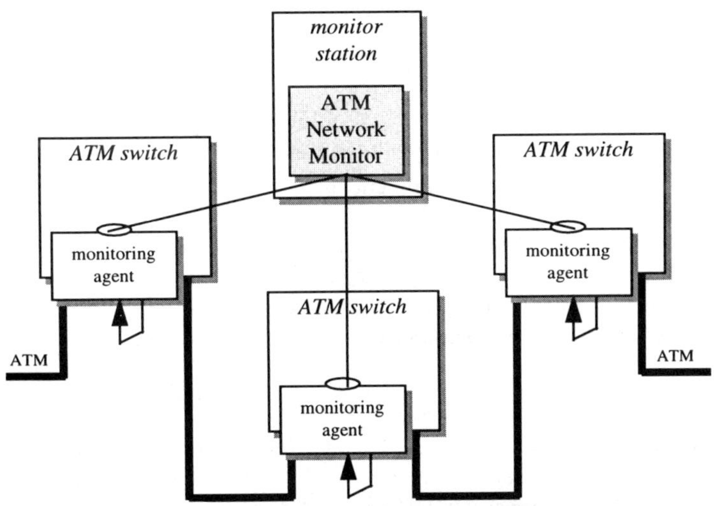

Figure 2: Scenario for the ATM Network Monitor

In the future, the shown monitoring agents may be integrated into ATM switches. This leads to a better performance and more possibilities for information retrieval. For QoS monitoring, especially for time-oriented parameters, additional monitoring information such as time stamps has to be integrated into the regular data flow of connections and must be evaluated by each switch. However, ATM switches operate on the very small ATM cells. The integration of monitoring information in each ATM cell is not useful because of the small cell size and will result in a great protocol overhead. Thus, the only way of transporting monitoring information is the integration of special monitoring cells into the regular data flow. One possible solution may be the usage of OAM cells (Operation, Administration, and Maintenance [6]), but many switches implement a separate data path for OAM cell flows. As a consequence, monitoring results do not reflect the characteristics of individual data connections. A better solution is the integration of monitoring information into some of the regular data cells. For AAL5-based connections, this can be achieved by using a special SSCS (Service Specific Convergence Sublayer), the *QM-SSCS (QoS Monitoring SSCS)* [5], on top of the AAL5 Common Part Convergence Sublayer (CPCS) [8]. As shown in Figure 3, the QM-SSCS integrates a time stamp, a sequence number, and additional SSCS layer padding into the regular data flow of AAL5-based connections. With additional padding at the SSCS layer, all information useful for QoS monitoring is located in fixed positions of the last ATM cell of a segmented AAL5 data unit. The CPCS trailer contains the size of the AAL5 data unit, and the SSCS trailer contains the described sequence number and time stamp. Moreover, the last ATM cell of an AAL5 data unit can easily be identified by the payload type field in the cell header since this is the way AAL5 determines the end of its data units. ATM switches have to be extended in order to detect cells with payload type set to „1“ and to deliver a copy of these cells to the monitoring agent on the switch. Overall, this approach enables to realize a powerful QoS monitoring for connections within an ATM network.

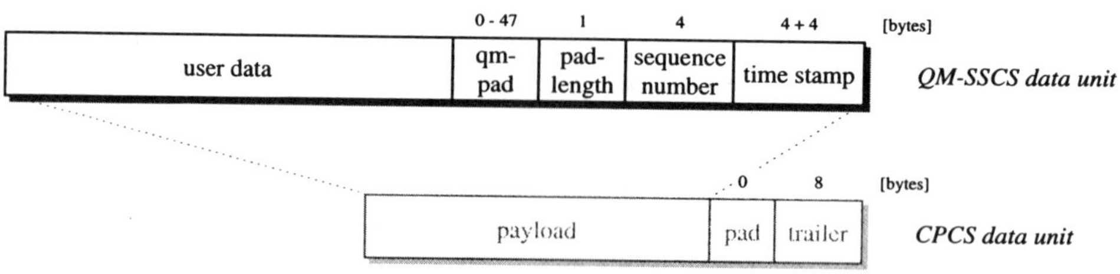

Figure 3: Format of the QM-SSCS data unit for monitoring AAL5-based connections

Currently, the monitoring agent is implemented as proxy agent being located on a workstation. The communication between the proxy agent and the switch has been optimized for the GIGAswitch/ATM, an ATM switch by Digital Equipment Corporation. Two protocols, SNMP as well as Telnet, are used for that communication. The Telnet interface of the monitoring agent is characterized by a very flexible design in order to adapt it to different Telnet interfaces of switches or to changes of the Telnet interface of one particular switch [12].

4 The ATM Network Monitor

The ATM Network Monitor ANEMON represents an entity for monitoring of ATM networks that is designed to provide detailed information on the state of the network and the components included. From an architectural view, this entity can be classified as management middleware for ATM networks since the service of the ANEMON monitor will be used by management applications. Monitoring information is accumulated from all switches within the ATM network and, after successful evaluation, presented to the applications.

Two different types of information are collected by the ANEMON monitor: information on the topology of the network such as configuration and state of switches, and secondly, QoS parameters of selected ATM connections. Consequently, the architecture of the

ANEMON monitor shown in Figure 4 includes two databases, one for topology information and one for QoS-related information.

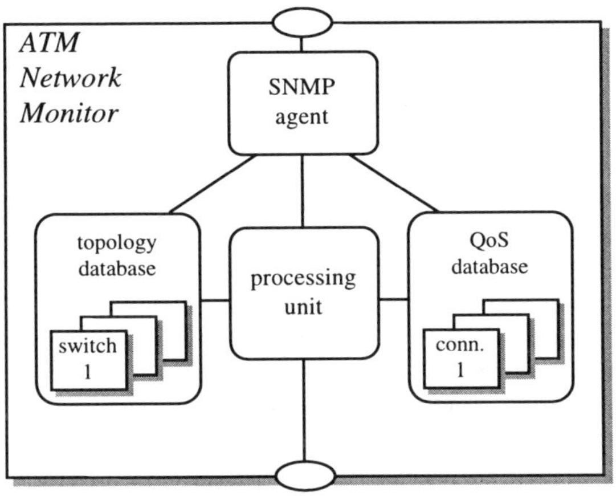

Figure 4: Architecture of the ATM Network Monitor

The topology database of the ANEMON monitor contains information about the monitored switches. By an initial configuration, each switch has to be registered at the ANEMON monitor. After that, further information such as number and state of ATM ports is obtained by communicating with the switch. For each active ATM port, the ANEMON monitor determines the related ATM port at the other end of the physical link, and thus information on network topology is successively collected. Since the type of information contained in the topology database is almost static, monitoring communication required to collect and update this information is very small. The update interval can be set to a value in the range of minutes. The more dynamic information on state and direction of currently established connections is also stored in the topology database in order to precompute routes of connections within the network. This enables a faster reaction if the ANEMON monitor is requested to monitor QoS parameters of one connection. In that case, the monitor only has to check if the precomputed route for this connection is still correct. In general, the described information in the topology database is necessary in order to obtain the information stored in the second database, the so-called QoS database. Moreover, information stored in the topology database can be used to generate a graphical representation of the recent network state and topology.

Information on QoS-related parameters of selected connections is stored in the QoS database. In order to avoid high network load caused by monitoring traffic, information in the QoS database is only collected on demand. Therefore, the ANEMON monitor has to be requested to monitor QoS parameters of a selected connection in a specified interval. Such a request must identify the connection, the QoS parameters to be monitored, and the update interval for monitoring. For instance, transmitted and lost cells of a particular connection can be monitored at each involved switch. Because of the dynamic character of QoS parameters, the polling interval for information contained in the QoS database is very small (e.g. seconds). Since a connection must be specified by its connection end point (Virtual Path Identifier VPI, Virtual Channel Identifier VCI) on one end system, the ANEMON monitor needs information contained in the topology database in order to locate the connection within the network and to identify the involved switches. For instance, QoS monitoring of selected connections can be requested by the ATM Network Manager (cf. Section

2) trying to figure out network problems that were reported by end system QoS monitors [11].

Furthermore, an SNMP agent is located on the ANEMON monitor in order to provide access to the information contained in both databases. By communicating with this agent, monitoring applications can use all information collected by the ANEMON monitor. Additionally, configuration changes in the topology database (e.g. the switches to be monitored) and requests for QoS monitoring of connections are performed by SNMP. Therefore, the information in both the topology and the QoS database is specified in two separate MIBs using the common ASN.1 syntax.

All actions within the ANEMON monitor are controlled by a processing unit. This entity reads the topology database and initiates the collection of the required information by communicating with the monitoring agents of switches. Results of the communication are directed to the topology database as well as to the QoS database. Operations of the processing module can be initiated by making the corresponding settings in the databases using SNMP.

From a historical view, the presented ANEMON monitor can be seen as a further development of a device monitor designed for the GIGAswitch/ATM [12]. At the moment, this device monitor shows the configuration of the device and the state of all its modules (cf. Section 2). By selecting a special module, the user gets more information on it. For instance, type, Ethernet address, and firmware version of a linecard or state, speed, and various counters of each ATM port can be displayed. The ANEMON monitor extends the functionality of this device monitor by integration of other switches and dedicated support for QoS monitoring. However, the device monitor represents a monitoring application whereas the ANEMON monitor can be classified as management middleware.

The implementation of the presented ANEMON monitor is currently done as a first prototype. This prototype is able to collect all useful information about the topology of an ATM network and, up to now, some connection-oriented QoS parameters. At the moment, connection-oriented parameters are restricted by the information being available from ATM switches, especially for the GIGAswitch/ATM.

5 Conclusion

This paper presents an approach to QoS monitoring in ATM-based networks. QoS for ATM connections is described in terms of throughput, delay, jitter, and reliability parameters [7]. Generally, AAL level monitoring on end systems provides an end-to-end view on QoS of ATM connections. Thus, only monitoring inside of each involved ATM switch is able to determine the exact reason for a QoS degradation in an ATM network. Therefore, the presented ATM Network Monitor ANEMON realizes QoS monitoring within ATM networks. This entity was designed to collect information on network topology as well as on QoS parameters of selected connections. In order to access the required information on individual ATM switches, monitoring agents are proposed in this approach. Each monitoring agent realizes a well defined monitoring interface for one ATM switch and translates received requests accordingly to the individual interface characteristics of this switch. Moreover, QoS monitoring performed by the ANEMON monitor is supported by the format of the QoS Monitoring Service Specific Convergence Sublayer (QM-SSCS) data unit defined for AAL5. All useful information, such as a time stamp, a sequence number, and a data length field, is located in fixed positions of the last ATM cell when the AAL5 CPCS data unit is segmented. Moreover, this last cell can easily be identified by the payload type field of the cell header. For AAL5-based connections, this leads to a powerful support of

QoS monitoring within ATM networks as soon as a simple enhancement is integrated into ATM switches.

For the described monitoring agent, a flexible implementation has been developed. This implementation is currently tested for the GIGAswitch/ATM and first results promise a good performance. Furthermore, a first prototype of the ANEMON monitor is almost implemented. In the future, this prototype will be extended in order to measure more QoS parameters for ATM connections. Moreover, the concept of evaluating QM-SSCS layer information within ATM switches will be considered in detail in order to achieve a powerful QoS monitoring within ATM networks. In a next step, the proposed management applications, especially the ATM Network Manager, will be developed.

References

[1] M. Ahmed, K. Tesink: *Definition of Managed Objects for ATM Management Version 8.0 using SMIv2*; Request for Comments 1695; August 1994

[2] ATM-Forum: *Interim Local Management Interface (ILMI) Specification Version 4.0*; ATM-Forum/95-0417R7, April 1996

[3] C. Aurrecoechea, A. Campbell, L. Hauw: *A Survey of Quality of Service Architectures*; Technical Report MPG-95-18; Lancaster University; 1995

[4] T. Brown, K. Tesink: *Definition of Managed Objects for the SONET/SDH Interface Type*; Request for Comments 1595; March 1994

[5] S. Fleck: *Monitoring von Qualitätsparametern in ATM-Netzen (Monitoring of QoS Parameters in ATM Networks*; Diploma Thesis (in German), Institute of Telematics, University of Karlsruhe, Germany, January 1997

[6] ITU-T Recommendation I.610*: B-ISDN Operation and Maintenance - Principles and Functions*; March 1994

[7] J.-I. Jung: *Quality of Service in Telecommunications Part II: Translation of QoS Parameters into ATM Performance Parameters in B-ISDN*; in: IEEE Communications Magazine; vol. 34, no. 8; pp. 112-117; August 1996

[8] J. Klein: *Entwurf und Realisierung einer ATM Programmierschnittstelle (Design and Implementation of an ATM Application Programming Interface*; Diploma Thesis (in German), Institute of Telematics, University of Karlsruhe, Germany, February 1996

[9] A. A. Lazar, F. Marconcini: *Towards an Open API for ATM Switch Control*; Technical Report CU/CTR/TR 441-96-07, Columbia University, April 1996

[10] P. Newman, W. Edwards, R. Hinden, E. Hoffman, F. Ching Liaw, T. Lyon, G. Minshall: *Ipsilon's General Switch Management Protocol Specification Version 1.1*; Request for Comments 1987; August 1996

[11] C. Schmidt, R. Bless: *QoS Monitoring in High Performance Environments*; Proceedings of the Fourth International IFIP Workshop on Quality of Service - IWQoS'96; Paris; 6 - 8 March; 1996

[12] H. R. Wiltfang: *Monitoring the GIGAswitch/ATM*; Proceedings of the 8th IEEE Workshop on Local and Metropolitan Area Networks; Potsdam, Germany; 25 - 28 August; 1996

Author Index

Aarstad, E. 279
Anagnostakis, C.D. 266
Antonopoulos, A. 110
Aranzulla, P. 157
Arijs, P. 94
Autenrieth, A. 303
Bartlett, S.G. 211
Bauschert, T. 312
Bennion, I. 18
Bian, Y. 145
Bisiaux, G. 195
Blondel, M. 161
Boulet, B. 169
Brandt, H. 274
Breuer, D. 33
Brianza, C. 303
Bülow, H. 65
Casoni, M. 139
Caspar, C. 73
Chantrain, D. 295
Charzinski, J. 258
Choquet, Y. 161
Clemente, R. 303
D'Agrosa, M. 134
De Maeyer, D. 186
Demeester, P. 15, 94, 242, 303, 320
Doran, N.J. 18
Ennser, K. 40
Enssle, J. 258
Färber, J. 258
Foisel, H.-M. 73
Fragiacomo, S. 145
Frank, M. 258
Freund, R. 88
Frimout, D. 1
Frings, J. 226
Frisanco, T. 312
Garcia, L.C. 46
Georgatsos, P. 287
Gladisch, A. 73
Govan, D.S. 18
Gruhl, H. 59
Gryseels, M. 303
Hagimoto, K. 118
Hanawa, M. 29, 174
Hanik, N. 178
Hanoteau, J. 195
Harada, Y. 303
Harper, P. 18
Heens, B. 161
Henderson, A.M. 126
Herchenröder, G. 59
Jajszczyk, A. 303
Janssens, G.K. 182
Jaussi, L. 279
Kalbe, G. 303
Kamal, A.E. 182
Kean, P.N. 18
Kinjyou, K. 174
Knight, G. 250
Knox, F.M. 18
Kohn, U. 165
Kremers, E. 165
Kuindersma, P.I. 46
Laming, R.I. 40
Lamquin, M. 195
Lane, P. 110
Lanzoni, D. 134
Le Ligné, M. 169
Leppla, R. 46
Lievens, I. 242, 320
Lorang, M. 279
Lord, A. 81
M. Takahara 174
Macedo, M.M. 153
Mataix, J. 203
Matrakidis, C. 145
Matsuoka, S. 118
Mayeur, A. 195
Mattheus, A. 46, 59

Mellor, J. 157
Meuleman, L. 161, 195
Müller, G. 165
Naldi, M. 218
Nederlof, L. 295
Nunes, M.S. 153
O'Reilly, J. 110, 145
Pelusi, P. 218
Petermann, K. 33
Pettersen, H. 279
Pickavet, M. 94
Point, N. 195
Poppe, F. 94
Popplewell, A. 145
Potts, M. 279
Raffaelli, C. 139
Ranc, D. 287
Ravera, M. 303
Ravet, F. 161
Ray, P. 149
Reid, J.J.E. 46
Renger, T. 279
Rhissa, A.G. 303
Robinson, M. 102
Roszkiewicz, M. 303
Salaün, J. 169
Sartzetakis, S. 287
Scheuing, E.-U. 165
Schulien, C. 46
Shi, R. 250
Siddiqui, A.S. 53
Signorelli, G. 303
Strebel, B. 73, 88
Struyve, K. 303
Sultan, A. 211, 234
T'Joens, Y. 287
Takahara, M. 29
Tang, X.Y. 25, 33
Teichmann, B. 46
Tillerot, F. 169
Tischer, F.-C. 165
Tittel, C. 274
Todorova, P. 274
Tomizawa, M. 118
Uematsu, Y. 118
Vagliani, F.C. 9
Van Hauwermeiren, L. 295
Vanderstraeten, H. 295
Vandoorselaere, K. 303
Venieris, I.S. 266
Vidal, J.R. 203
Vobian, J. 59
Wang, X. 161
Wiltfang, H.R. 328
Wright, I. 102
Wuilmart, L. 195
Yada, T. 303
Yamawaku, J. 29
Yao, B. 29
Zahariadis, Th.B. 266
Zanetta, L. 134
Zervas, M.N. 40